New Insights into Breast Cancer: Current Controversies and Future Perspectives

New Insights into Breast Cancer: Current Controversies and Future Perspectives

Guest Editors

Alejandro Martin Sanchez
Angela Bucaro
Flavia de Lauretis

MDPI

Basel • Beijing • Wuhan • Barcelona • Belgrade • Novi Sad • Cluj • Manchester

Guest Editors
Alejandro Martin Sanchez
Department of Women's and Children's Health Sciences
Fondazione Policlinico Universitario "A. Gemelli" IRCCS
Rome
Italy

Angela Bucaro
Department of Women's and Children's Health Sciences
Fondazione Policlinico Universitario "A. Gemelli" IRCCS
Rome
Italy

Flavia de Lauretis
Department of Women's and Children's Health Sciences
Fondazione Policlinico Universitario "A. Gemelli" IRCCS
Rome
Italy

Editorial Office
MDPI AG
Grosspeteranlage 5
4052 Basel, Switzerland

This is a reprint of the Special Issue, published open access by the journal *Life* (ISSN 2075-1729), freely accessible at: https://www.mdpi.com/journal/life/special_issues/K7ZQ9G3XE2.

For citation purposes, cite each article independently as indicated on the article page online and as indicated below:

Lastname, A.A.; Lastname, B.B. Article Title. *Journal Name* **Year**, *Volume Number*, Page Range.

ISBN 978-3-7258-7412-5 (Hbk)
ISBN 978-3-7258-7413-2 (PDF)
https://doi.org/10.3390/books978-3-7258-7413-2

Contents

About the Editors

Alejandro Martin Sanchez

Alejandro Martin Sanchez is a breast surgeon at Fondazione Policlinico Universitario A Gemelli IRCCS in Rome, Italy, where he is involved in the clinical management and surgical treatment of patients with breast diseases. His professional activity focuses primarily on breast cancer surgery, with particular attention to multidisciplinary approaches to diagnosis and treatment. His research interests include breast oncology, surgical management of breast tumors, and the integration of clinical, pathological, and molecular information to improve patient care. He participates in clinical research projects aiming to advance knowledge in breast cancer biology and treatment, and he collaborates with specialists from different disciplines in order to promote personalized and evidence-based therapeutic strategies. Through his clinical and scientific activity, he contributes to the development of innovative approaches in the management of breast cancer and related conditions.

Angela Bucaro

Angela Bucaro is a breast surgeon at Fondazione Policlinico Universitario A Gemelli IRCCS in Rome, Italy, where she is involved in the diagnosis and surgical management of breast diseases. Her clinical work focuses on the treatment of breast cancer and the comprehensive care of patients within a multidisciplinary clinical setting. Her professional interests include advances in breast surgery, improvements in diagnostic and therapeutic pathways, and the integration of surgical treatment with modern oncologic strategies. She is involved in clinical practice and collaborative research activities aiming to improve outcomes and quality of care for patients with breast cancer. Her work contributes to the continuous development of patient-centered approaches and to the optimization of surgical and clinical management in breast oncology.

Flavia de Lauretis

Flavia de Lauretis is a breast surgeon at Fondazione Policlinico Universitario A Gemelli IRCCS in Rome, Italy, where she works in the field of breast disease diagnosis and surgical treatment. Her clinical activity focuses on the management of breast cancer patients, with particular attention to the multidisciplinary coordination of diagnostic and therapeutic strategies. Her scientific interests include breast oncology, the pathological and clinical characterization of breast tumors, and the development of improved approaches for patient management. She is involved in clinical and research activities aimed at enhancing diagnostic accuracy, optimizing surgical outcomes, and promoting the integration of innovative therapeutic strategies in breast cancer care. Through her work, she contributes to the advancement of knowledge and clinical practice in the field of breast oncology.

Preface

Breast cancer represents a highly complex and heterogeneous disease that continues to challenge both researchers and clinicians. Despite major progress in early detection, molecular profiling, and targeted therapies, several aspects of breast cancer biology, diagnosis, and treatment remain the subject of ongoing debate. Differences in tumor subtypes, variable clinical behavior, and the presence of rare histological variants require increasingly precise diagnostic approaches and individualized therapeutic strategies.

This Reprint entitled "New Insights into Breast Cancer Current Controversies and Future Perspectives" was developed to provide an updated overview of current research addressing key controversial issues and emerging directions in the field of breast cancer.

By bringing together multidisciplinary perspectives, this Reprint aims to highlight recent advances while also identifying areas where further research is needed. The presented studies contribute to improving the understanding of disease mechanisms and support the development of more effective diagnostic and therapeutic approaches.

This Reprint is primarily addressed to oncologists, breast surgeons, radiotherapists, pathologists, and biologists working in the field of targeted therapies, as well as to researchers and healthcare professionals involved in breast cancer research and clinical management.

Alejandro Martin Sanchez, Angela Bucaro, and Flavia de Lauretis
Guest Editors

life

MDPI

Review

Malignant Mesenchymal Tumors of the Breast: Current Challenges and New Perspectives on Primary Sarcomas and Malignant Phyllodes Tumors

Flavia De Lauretis [1], Alejandro Martin Sanchez [1,*], Cristina Accetta [1], Beatrice Carnassale [1], Sabatino D'Archi [1], Alba Di Leone [1], Antonio Franco [1], Federica Gagliardi [1], Stefano Magno [1], Elena Jane Mason [2], Francesca Moschella [1], Lorenzo Scardina [1], Marta Silenzi [1], Angela Bucaro [1], Chiara V. Pirrottina [1], Nicoletta D'Alessandris [3], Antonino Mulè [3], Angela Santoro [3], Fabio Marazzi [4], Valeria Masiello [4], Alessandra Fabi [1], Armando Orlandi [5], Antonella Palazzo [5], Ida Paris [1], Maria Pia Foschini [6], Riccardo Masetti [1] and Gianluca Franceschini [1]

1 Multidisciplinary Breast Center, Dipartimento Scienze della Salute della Donna e del Bambino e di Sanità Pubblica, Fondazione Policlinico Universitario A. Gemelli IRCCS, 00168 Rome, Italy
2 Breast Surgery, Center for Women's and Newborn Health, Isola Tiberina Hospital, Gemelli Isola, 00153 Rome, Italy
3 Unità di Ginecopatologia e Patologia Mammaria, Dipartimento Scienze della Salute della Donna e del Bambino e di Sanità Pubblica, Fondazione Policlinico Universitario A. Gemelli IRCCS, 00168 Rome, Italy
4 Division of Radiotherapy, Dipartimento di Diagnostica per Immagini, Radioterapia Oncologica ed Ematologia, Fondazione Policlinico Universitario A. Gemelli IRCCS, 00168 Rome, Italy
5 Division of Medical Oncology, Fondazione Policlinico Universitario A. Gemelli IRCCS, 00168 Rome, Italy
6 Breast Unit, Bellaria Hospital, AUSL Bologna, Department of Biomedical and Neuromotor Sciences, University of Bologna, 40138 Bologna, Italy
* Correspondence: martin.sanchez@hotmail.it or martin.sanchez@policlinicogemelli.it; Tel.: +39-06-30156328 or +39-33-93259402

Abstract: Mesenchymal tumors of the breast constitute a rare and heterogeneous group of neoplasms, representing only 0.5% to 1% of all breast tumors. Originating from mesenchymal tissues, these tumors include various histological subtypes. They are particularly aggressive, characterized by a high propensity for local recurrence and an overall poor prognosis. The rarity of these cases has impeded the development of comprehensive clinical studies, leading to a lack of standardized diagnostic protocols and treatment guidelines. This review provides a thorough synthesis of current knowledge on breast mesenchymal tumors with a specific focus on malignant variants such as phyllodes tumors and breast sarcomas. It also addresses the diagnostic challenges faced by clinicians, evaluates current therapeutic strategies, and emphasizes the crucial role of surgical treatment. Additionally, it examines the evolving roles of chemotherapy and radiotherapy in enhancing patient outcomes.

Keywords: mesenchymal tumors; breast neoplasms; phyllodes tumors; breast sarcomas; therapeutic approaches; surgical treatment; chemotherapy; radiotherapy; prognosis; local recurrence

1. Introduction

Most breast cancers originate from epithelial tissue, but there is a heterogeneous group of non-epithelial tumors arising from mesenchymal tissues of the breast, which comprise 0.5 to 1% of all breast tumors [1].

According to the World Health Organization (WHO), mesenchymal tumors include 17 lesions of vascular, lipomatous, nerve sheath, myofibroblastic, and myogenic origin. They range from benign to highly malignant and very common to extremely rare [2].

Malignant mesenchymal tumors of the breast have a different natural history, treatment response, and prognosis as compared with epithelial ones. Due to their rarity, most publications consist of case reports or limited retrospective series with inconsistent inclusion criteria and excessively broad observational periods, combining cases from the early 20th century with those treated using the latest diagnostic and therapeutic protocols [3–5] (Table 1). Notably, malignant phyllodes tumors currently fall under breast cancer guidelines [6]. In contrast, breast sarcomas are considered within the broader category of soft tissue sarcomas, following the recommendations for sarcomas arising in other anatomical sites [7,8]. This leads to a heterogeneous interpretation and formulation of current guidelines regarding the clinical management of these rare tumors, particularly on the efficacy of complementary treatments [9].

Table 1. Current evidence on malignant mesenchymal tumors of the breast - focus on sarcomas and phyllodes tumors.

Study (Year)	Global Cohort (Number of Patients)	5y-OS	5y-DFS	5y-LRR	Median Follow-Up (Period of Study)	Key Findings	Study Limitations
Holm et al. [5] (2015)	42	49%	48%	24%	50.4 (1979–2014)	- Tumor size and grade are significant prognostic factors	- Retrospective study - Single institution - Small and disomogeneous cohort - Missing data
Confavreux et al. [3] (2006)	70	49%	28%	-	31.7 (1966–2004)	- Negative surgical margins (R0) significantly improve survival (72% vs. 38%) - Angiosarcomas have the worst DFS (7%)	- Retrospective study - Small and disomogeneous cohort - Lack of uniform treatment protocols
McGowan et al. [4] (2000)	78	57%	47%	26%	114 (1958–1990)	- Negative surgical margins (R0) improve local relapse-free survival (80% vs. 33%) - Radiation dose >48 Gy improves 5-year OS (91% vs. 50%)	- Retrospective study - Single institution, - Small and disomogeneous cohort - Missing or incomplete data

OS: Overall Survival; DFS: Disease Free Survival; LRR: Local Recurrence Rate.

This review provides an overview of current evidence about the most aggressive forms of mammary tumors, specifically malignant phyllodes tumors (MPTs) and primary breast sarcomas (PBSs). It focuses on the proven effectiveness of surgery, as well as the reported results of chemotherapy and radiotherapy.

2. Phyllodes Tumors

Phyllodes tumors (PTs) of the breast are neoplasms composed of both epithelial and stromal elements [4–11].

They are characterized by rapid dimensional growth and often present as fibroadenomas on initial radiological examinations (ultrasound and mammography); cytological examination via fine needle aspiration and histological examination via core biopsy may sometimes be inadequate to reliably distinguish PTs from fibroadenomas [12].

Excisional biopsy is recommended for diagnostic purposes in cases of rapid growth of fibroadenoma-like lesions, which exhibit variable biological behavior. Based on histopathological characteristics, PTs are categorized as benign (60–75%), borderline (15–20%), or malignant (10–20%), which may occasionally transform histologically into sarcomatous lesions [13,14].

National Comprehensive Cancer Network (NCCN) guidelines recommend radical excision of the lesion with ≥1 cm negative margins. Axillary lymph node metastases are rare; sentinel lymph node biopsy and/or axillary lymphadenectomy are, therefore, not indicated unless the lymph nodes are suspected at clinical staging [7].

2.1. Epidemiology and Risk Factors

PTs represent approximately 0.3–0.9% of all breast neoplasms. Among all PTs, malignant ones constitute 10–20% of cases, with an annual incidence of 2.1 cases per million women [15]. No etiological or risk factors have been described for the development of these neoplasms except for Li–Fraumeni syndrome, a rare autosomal dominant condition characterized by the development of multiple tumor types [16].

2.2. Diagnosis

PTs are a complex type of breast lesion that pose challenges for diagnosis. Ultrasound may show hypoechoic, heterogeneous, or complex cystic and solid echo patterns [17]. Features such as lobulated shape, heterogeneous internal echo pattern, and absence of microcalcification are significant sonographic features used to favor PTs over fibroadenomas [11,18].

Specific sonographic features, such as liquefaction and a heterogeneous inner echo, may indicate a higher risk of MPT of the breast [19].

On Magnetic Resonance Imaging (MRI), PTs appear as oval, well-circumscribed, and isointense on T1-weighted images, and heterogeneously hyperintense on T2-weighted images [18]. Certain MRI findings can aid in assessing the risk of malignancy. Features such as non-circumscribed margins, cystic components, irregular cyst walls, peritumoral edema, a low signal intensity on T2-weighted images, and a low apparent diffusion coefficient (ADC) are associated with a higher histologic grade in histopathology [17].

2.3. Pathology Assessment

The World Health Organization categorizes PTs as benign, borderline, or malignant based on factors such as stromal cellularity, stromal overgrowth, nuclear atypia, mitotic activity, tumor margins, and the presence of malignant heterologous elements [20].

Key diagnostic features of benign PTs include a cellular stroma with a subepithelial distribution, a low mitotic count of fewer than five mitoses per 10 high-power fields, and a varied stroma that can include hyalinized or myxoid areas. Borderline tumors, according to the WHO classification, have intermediate characteristics, exhibiting microscopic invasion at the tumor margins along with moderate cellularity and pleomorphism [2,21].

MPTs are characterized by significant nuclear pleomorphism, diffuse stromal cellularity, stromal overgrowth, mitotic activity of more than 10 mitoses per 10 HPFs, and infiltrative borders. Differentiating MPTs from pure or metastatic sarcomas can be quite challenging. There is a notable risk of distant metastasis—reportedly up to 16%—especially to the lungs and bones, while involvement of the axillary lymph nodes is rare.

Histological factors associated with distant metastases include tumor size over 7 cm, infiltrative borders, marked stromal overgrowth, high stromal cellularity, >5 mitoses per 10 HPFs, and necrosis. MPTs are often resistant to chemotherapy, and the prognosis for metastatic cases is poor [19,22].

Immunohistochemical analysis indicates that benign PTs display low positivity for p53, Ki-67, CD117, EGFR, p16, and VEGF, with an increasing positivity in MPTs.

Conversely, Lenhard and colleagues have defined stroma-rich PTs with a bland appearance as typical CD34–positive, while high-grade MPTs usually show loss or a lesser grade of CD34 staining [23].

Mutations in genes such as TP53, RB1, NF1, PIK3CA, and ERBB4 have been found in MPTs, suggesting potential therapeutic targets. Additionally, mutations in genes such as TP53, RB1, NF1, PIK3CA, and ERBB4 have been found in MPTs, suggesting potential therapeutic targets [11,24,25].

In conclusion, accurate histological assessment of PTs, along with immunohistochemical and genetic analyses, help differentiate between benign, borderline, and MPTs. This comprehensive approach guides appropriate clinical intervention and contributes to improved patient outcomes [12,13,23,26,27].

2.4. Surgical Treatment

Surgical excision is the recommended treatment for operable MPTs [11,17,28–31]. As for other breast tumors, the choice of surgery should take into account both oncological and cosmetic outcomes [2,11,17,29]. If an adequate margin and a good cosmetic outcome can be obtained with a wide excision, breast-conserving surgery should be the first choice of treatment. NCCN guidelines recommend at least a 1 cm excision margin if preoperative assessments suggest the presence of a borderline tumor or MPT, while wide margins are not required for excision of benign PTs [7,18,28,31,32].

However, the evidence of a closer margin at final pathology should be discussed in a multidisciplinary context as according to NCCN guidelines, mastectomy is not an absolute indication in these cases [7,11,33–38].

Mastectomy is recommended when the preoperative surgical assessment estimates an inability to obtain adequate margins without causing cosmetic deformities that would be unacceptable to the patient [7,17,19,28,31,39,40]. While tumor size is not associated with LR [12,41], most studies agree that patients with evidence of narrow surgical margins at final pathology have higher LR rates than those with negative margins [11,23,36,37,42–45]. This encompasses a systematic review and meta-analyses of 9234 individual cases, with 18% of these patients having an MPT; among them, a positive surgical margin was significantly linked to an increased risk of local recurrence (OR 6.85; 95% CI 1.58–29.64) [29–32]. However, a meta-analysis of four studies including 162 patients demonstrated no differences in LR between ≥1 cm and <1 cm margins [34,46].

In current NCCN guidelines, there are no clinical scenarios in which axillary staging is recommended [7], so there is no indication to perform axillary surgery [7,17,22,28,47–49]. A palpable lymphadenopathy is noted in 10–15% of patients presenting with an MPT but is usually reactive to tumor necrosis and infected or ulcerated skin lesions, with <1% of lymph nodes being pathologically involved [50].

2.5. Adjuvant Radiotherapy

There is no consensus on whether RT improves overall survival (OS) or disease-free survival (DFS) in patients with an MPT as reported by Chao X et al. in their systematic review and metanalysis [51]. Current guidelines from the NCCN and the European Society for Medical Oncology (ESMO) suggest that radiotherapy may be considered for patients with an MPT that have high-risk features, such as positive surgical margins, tumor size greater than 5 cm, or high-grade pathology [7,52].

A recent update of SEER analysis published in 2024 reported results of a larger cohort of 2261 patients, who were diagnosed and treated in the last two decades. In this larger cohort, at least 20% of patients (20, 12%) underwent radiotherapy. Results showed that in terms of OS and breast cancer-specific survival, RT did not have a significant impact [53].

These findings underscore the role of margin status in the decision to administer RT, suggesting that it may primarily serve as a tool for local control rather than improving long-term survival outcomes. In conclusion, while adjuvant radiotherapy may be beneficial for certain high-risk patients with an MPT—particularly those with positive margins or high-grade histology—the evidence supporting its universal application remains limited. Prospective trials and long-term follow-up studies are needed to refine the guidelines and better delineate which subgroups of patients will benefit most from RT after excision. Until then, treatment decisions should be personalized, carefully considering tumor biology, surgical outcomes, and patient preferences.

2.6. Systemic Treatment

The role of chemotherapy for MPTs is controversial and should only be considered after consultation with an expert center [13].

Many investigators have confirmed the low efficacy of monochemotherapy (cyclophosphamide, ifosfamide, or doxorubicin). Currently, treatments typically involve doxorubicin and ifosfamide, and are similar to those for soft tissue sarcoma (STS) as this combination showed the most activity and greatest impact on survival [49]. Neoadjuvant therapy appears to hold a similar significance, yet some recent case reports indicate that aggressive treatment with alkylating agents should be pursued when feasible [54].

Approximately 20% of patients with MPT develop metastases. Generally, research on systemic therapy is limited to small case reports and series that often highlight positive outcomes. However, in practice, patients appear to receive less benefit from chemotherapy, experiencing relatively short durations of response compared to those with more common sarcoma histologies [55].

Given the rarity of the disease, new therapeutic possibilities are yet to be explored in phase II/III trials.

3. Primary Breast Sarcoma

Breast sarcomas comprise a heterogeneous group of non-epithelial neoplasms. They are rare tumors, accounting for <1% of breast neoplasms and <5% of soft tissue sarcomas. The annual incidence is approximately 4.5 cases per million [56,57].

The average size at diagnosis is larger compared to epithelial tumors, with a mean diameter of about 3 cm at the time of pathological analysis, although lesions up to 40 cm in diameter have been described [58]. The overlying skin and the nipple-areola complex are rarely infiltrated, except for cases of angiosarcoma where thickening, erythema, and/or red-violet discoloration of the skin overlying the lesion may occur [59].

PBSs, like sarcomas in other body sites, are poorly responsive to chemo- and radiotherapeutic treatments, although the role of the latter is still debated; currently, the main treatment remains radical surgical excision [60,61].

3.1. Epidemiology and Risk Factors

PBSs predominantly affect women, with a peak incidence in the fifth to sixth decades of life, although they can occur at any age [60–62]. The most common subtype is angiosarcoma, which accounts for approximately 30–50% of breast sarcomas [62,63].

In some patients, genetic factors (e.g., Li–Fraumeni syndrome or neurofibromatosis type 1) and environmental factors (exposure to radiation, vinyl chloride, alkylating agents, and immunosuppressive agents) may be implicated [58], while chronic lymphedema after mastectomy and axillary lymph node dissection is a reported risk factor for the onset of lymphangiosarcoma (Stewart–Treves syndrome) [64].

3.2. Pathologic Assessment

PBSs are histologically heterogeneous, and different classifications have been proposed [4]. The classification referred to is currently established by the WHO in 2020 [9].

In a Canadian national case series of 991 patients with a PBS—excluding angiosarcoma—the distribution of histological subtypes was as follows: unspecified sarcomas 26%, spindle cell sarcoma 14%, leiomyosarcoma (LMS) 12%, giant cell sarcoma 10%, stromal sarcoma 6%, malignant fibrous histiocytoma 6%, myxoid fibrosarcoma 4%, dermatofibrosarcoma protuberans 3%, fibrosarcoma 3%, undifferentiated sarcoma 3%, liposarcoma (LPS) 3%, pleomorphic liposarcoma 2%, and others 8% [65].

Grading represents an essential prognostic factor in sarcomas, including those of mammary origin: the elements considered for grading are the extent of tissue differentiation, mitotic count, presence/absence of necrosis, cellularity, and pleomorphism; angiosarcoma is an exception for which the literature results are conflicting [66].

Primary angiosarcoma may develop spontaneously but is more usually seen in patients who received radiotherapy to the affected breast [67]. These lesions show poorly formed irregular vascular channels infiltrating the stroma with pleomorphic hyperchromatic endothelial cells exhibiting high mitotic activity [68,69]. They often have hematoma-like areas due to extensive vascular proliferation and infiltrative growth without well-defined borders. Additional features include the presence of epithelioid cells in some variants and frequent hemorrhage [69–71]. Angiosarcomas must be differentiated from benign vascular lesions like hemangiomas, which show well-formed vascular channels lined by regular endothelial cells and lack significant atypia and mitotic activity [69,72]. Tumor cells are positive with a wide range of endothelial markers, such as CD31, CD34, D2-40, and ERG [73]. In post-radiation angiosarcoma, a strong and diffuse nuclear positivity, with a loss or reduced expression of H3K27me3 can be observed [74,75].

The diagnosis of mesenchymal lesions in the breast relies on a combination of histopathological examination, immunohistochemical staining, and molecular studies where necessary [76]. Immunohistochemical markers are critical in differentiating between lesions; for instance, CD34 is useful in distinguishing myofibroblastomas and spindle cell lipomas, while cytokeratins can help identify metaplastic carcinoma in the differential diagnosis of PTs [76,77]. Molecular studies such as fluorescence in situ hybridization FISH or polymerase chain reaction PCR can be employed in challenging cases to detect specific genetic alterations like MDM2 amplification in well-differentiated LPS [78]. Common challenges include histological overlap between benign and malignant lesions, heterogeneity within the same tumor, especially in PTs and LPSs, and ensuring sufficient tissue sampling for accurate diagnosis [76,79]. Understanding these histological features is essential for pathologists and clinicians as accurate histological assessment guides clinical management, ranging from conservative follow-up for benign lesions to aggressive surgical and adjuvant therapy for malignant mesenchymal tumors [80].

3.3. Diagnosis

Although studies concerning imaging findings in breast sarcomas are limited due to their low incidence, it seems that they have some characteristic radiological findings [58,81].

Breast sarcomas on mammography often appear as a noncalcified hyperdense mass, with indistinct or circumscribed margins borders [58,81,82]; these findings may resemble benign fibroadenomas [31,83]. The lack of calcifications and the oval shape of these lesions distinguish them from epithelial tumors; however, these same characteristics are also described in PTs [58]. Furthermore, mammography may appear normal in the presence of a breast sarcomatous lesion [83].

Ultrasound sonography is not specific for radiological diagnosis of breast sarcomas. They usually appear as irregular and hyperechoic lesions with no shadowing [81–83], although hypoechoic lesions with indistinctive borders and a posterior acoustic shadow have been described [58].

Contrast-enhanced MRI is the preferred imaging technique as it helps detect suspicious malignant changes, such as irregularly shaped oval masses that exhibit T2 hyperintensity along with rapid and uneven contrast enhancement and washout [31,58,82,83]. Additionally, MRI thoroughly evaluates the disease's local spread, highlighting the extent of involvement in adjacent skin, fascia, and muscle structures, which is vital for effective surgical planning and subsequent radiotherapy [31].

A core biopsy should be performed with needles 16 G or bigger [58], and immunohistochemical analysis of specific cytokeratins should be performed [31,81].

All patients with a diagnosis of breast sarcoma should be assessed for distant disease. It is advisable to perform computed tomography (CT) of the chest, abdomen, and pelvis, along with a bone scan. Lungs are the most common site of metastases globally, while angiosarcoma has a propensity for secondary spread to the liver and bone [31].

The role of positron emission tomography (PET) scans remains unclear in the assessment of breast sarcomas [31,58].

3.4. Surgical Treatment

It is strongly recommended that breast sarcomas be managed at a specialized sarcoma referral center to improve overall survival (OS) by investigating clinicopathological features and a multidisciplinary approach [31,81].

There is widespread acceptance that surgical resection should be the first modality of treatment for primitive breast sarcoma [58,60,82]. However, the best surgical treatment is still debated, and should be tailored to the individual patient. The goal of surgery is local control, and total mastectomy has been considered the gold standard for many years [31,58,83]. The choice between mastectomy and breast conservative surgery is based on the possibility to obtain negative margins at final histology [31,58,83,84]: in fact, tumor size and excision margin are the two factors that most impact survival and recurrence [3,37,61,85–87]. In a study by Wang et al., the 5-year DFS and OS rates for tumors sized <5 cm, 5–10 cm, and >10 cm, were 75.0% and 88.9%, 70.8% and 75.0%, and 37.5% and 37.5%, respectively [88].

It is also advisable to evaluate the tumor/breast size ratio before surgery to properly perform correct planning of the surgical strategy [60]. While some studies advocate for mastectomy, showing increased OS over conservative surgery, others have demonstrated no significant advantage [3,4,87–92].

For smaller localized breast sarcomas, most authors recommend wide excision with a margin greater than 1 cm [82,83,86]. In the case of angiosarcomas, it is recommended to obtain at least a 3 cm clear margin during surgery since these lesions often have an infiltrative cutaneous disease that extends well beyond the visible tumor [82,83,86]. In this set of patients, mastectomy with delayed reconstruction remains the most prudent surgical approach [58,83].

Because the aesthetic outcome of breast-conserving surgery when approaching large resections is often poor, some authors suggest oncoplastic reconstruction with parenchymal rearrangement or reconstruction with a local flap as a viable option [83,93].

Sentinel lymph node biopsy is not routinely performed in patients with breast sarcomas as nodal dissemination is very rare [58,82–84]. In a study by Fong et al. involving 1772 cases of STS, the rate of nodal metastasis was only 2.6%, while in a study based on SEER and focused on breast sarcomas, 129 out of 333 patients underwent lymphadenectomy, and only 6 patients (4.7%) were found to have nodal metastases [94,95]. Radical dissection of axillary lymph nodes is appropriate in the presence of histologically proven isolated nodal disease or where the purpose of surgery is local control and symptom relief [58,83].

3.5. Adjuvant Radiotherapy

Breast sarcomas are rare malignant tumors for which treatment protocols are not well-established. Adjuvant RT is recommended after a positive margin resection in tumors larger than 5 cm and in any high-grade sarcoma because of the high risk of LR [13,96]. Since PBSs are sometimes associated with p53 mutation and Li–Fraumeni syndrome, testing for these pathogenic variants should occur before administering adjuvant RT [58].

3.6. Systemic Treatment

To the best of our knowledge, there are no randomized controlled trials specifically focused on primary breast sarcomas. As previously noted, the available studies—and those discussed in the following sections—address soft tissue sarcomas more broadly. Current guidelines extend the recommendations from these studies to the treatment of breast sarcomas [7,8].

3.6.1. First-Line Treatments

Surgery remains the cornerstone of treatment for primary breast sarcomas (PBSs), although a subset of patients present with locally advanced or metastatic disease at diagnosis, making resection unfeasible [13,97]. In these cases, neoadjuvant chemotherapy may be considered, especially for high-grade tumors or those larger than 5 cm [98]. The standard perioperative regimen combines anthracyclines and ifosfamide, with doxorubicin as the backbone of first-line therapy [7,99,100]. Histology-driven neoadjuvant approaches have not demonstrated superior outcomes in terms of DFS or OS and are not routinely recommended [100]. Whether doxorubicin should be administered as a monotherapy or in combination remains a matter of debate, with recent studies reporting divergent conclusions [101,102] (Table 2).

The response to chemotherapy remains highly variable and must be tailored to histotype sensitivity in a multidisciplinary setting [55,82]. Among new agents, aldoxorubicin—an albumin-binding prodrug—has shown reduced cardiotoxicity and favorable PFS outcomes in relapsed or refractory STS compared to other treatments including doxorubicin and dacarbazine [103–105] (Table 3). Additional options under investigation include amrubicin, a third-generation anthracycline with potentially reduced cardiac toxicity [106], and gemcitabine/docetaxel, which showed no superiority to doxorubicin and higher toxicity in the GeDDiS trial [107,108] (Table 3).

Weekly paclitaxel has demonstrated sustained activity in angiosarcomas and remains a potential first-line option for this subtype [7,109].

3.6.2. Second-Line Treatments

Second-line options include agents with modest efficacy compared to first-line regimens. Dacarbazine remains one of the oldest available drugs, used as monotherapy or in combination with gemcitabine; the latter is supported by phase II data and NCCN category 2B recommendation [7,110,111]. Eribulin and trabectedin are active mainly in selected subtypes [112,113]. Trabectedin has shown improved PFS over dacarbazine in randomized trials involving patients previously treated with anthracyclines and at least one other systemic regimen [114–116]. Temozolomide, a prodrug of dacarbazine, has demonstrated moderate activity in pretreated sarcomas when administered on a prolonged schedule, though phase III data are lacking [117,118].

Table 2. Randomized Controlled Trials on Sarcomas' First-line treatments.

Study	Patients (n)	Population	Median Follow-Up	Comparative Arms	PFS/OS	Clinical Insights
Judson et al. [101] (EORTC 62012)	455	Advanced, unresectable or metastatic high-grade STS: UPS, myxoid or round cell liposarcoma, pleomorphic and dedifferentiated liposarcoma, pleomorphic rhabdomyosarcoma, synovial sarcoma, myxofibrosarcoma, fibrosarcoma, leiomyosarcoma, angiosarcoma, malignant peripheral nerve sheath tumour, epithelioid sarcoma, unclassified high-grade sarcoma (not otherwise specified)	56 months	Doxorubicin alone (n = 228) vs. Doxorubicin and Ifosfamide (n = 227)	PFS 4.6 vs. 7.4 months (HR = 0.74; $p = 0.003$) OS 12.8 vs. 14.3 months (HR = 0.83; p = 0.076)	Combination improves PFS only. Use in selected, fit patients. Higher toxicity
Pautier et al. [102] (LMS-0)	150	Metastatic or relapsed unresectable leiomyosarcomas (uterine and soft tissue leiomyosarcomas) that had not previously been treated with chemotherapy	37 months	Doxorubicin alone (n = 76) vs. Doxorubicin + Trabectedin (n = 74)	PFS 6.2 vs. 12.2 months (HR = 0.41; $p < 0.0001$)	Strong PFS in LMS. Promising for first-line. Awaiting OS data. *

STS: Soft Tissue Sarcomas; UPS: undifferentiated pleomorphic sarcoma; PFS: Progression Free Survival; OS: Overall Survival; LMS: Leiomyosarcomas. * OS data are not yet mature and further analysis is needed.

Table 3. Randomized Controlled Trials on new cytotoxic drugs for soft tissue sarcomas.

ALDOXORUBICIN—Randomized Controlled Trials						
Study (Year)	**Patients (n)**	**Population**	**Endpoints**	**Median Follow-Up**	**Comparative Arms**	**Key Results**
Chawla et al. [103] 2015	123	Locally advanced, unresectable, and/or metastatic high grade STS (Leiomyosarcoma, Liposarcoma, Fibrosarcoma, Synovial sarcoma, Others *)	PFS, OS	13 months	Aldoxorubicin (n = 83) vs. Doxorubicin (n = 40)	PFS: 8.3 vs. 4.6 months ($p < 0.001$) OS: 15.8 vs. 14.3 months ($p = 0.21$; HR = 0.73)

Table 3. *Cont.*

Study (Year)	Patients (n)	Population	Endpoints	Median Follow-Up	Comparative Arms	Key Results
ALDOXORUBICIN—Randomized Controlled Trials						
Chawla et al. [105] 2017	433	Relapsed/refractory STS (Leiomyosarcoma, Liposarcoma, Synovial sarcoma, L-sarcomas subgroup, Others *)	PFS	-	Aldoxorubicin vs. Investigators' choice (Dacarbazine, Doxorubicin, Pazopanib, Ifosfamide, Gemcitabine/Docetaxel)	Overall PFS: 4.06 vs. 2.9 months ($p = 0.12$; HR = 0.82) L-sarcomas PFS: 5.32 vs. 2.96 months ($p = 0.007$; HR = 0.62)
GEMCITABINE—DOCETAXEL—Randomized Controlled Trials						
Study (Year)	**Patients (n)**	**Population**	**Endpoints**	**Median Follow-Up**	**Comparative Arms**	**Key Results**
Maki et al. [107] 2007	122	Metastatic STS (Leiomyosarcoma, Non-leiomyosarcoma, Liposarcoma, MFH/UPS, Others *)	PFS, OS	-	Gemcitabine-Docetaxel (n = 73) vs. Gemcitabine alone (n = 49)	PFS: 6.2 vs. 3.0 months OS: 17.9 vs. 11.5 months
GeDDis trial [108] 2017	257	Previously untreated advanced unresectable or metastatic STS (Uterine/Non-uterine leiomyosarcoma, Synovial sarcoma, Pleomorphic sarcoma, Others *)	PFS	22 months	Doxorubicin (n = 129) vs. Gemcitabine-Docetaxel (n = 128)	PFS: 23.3 vs. 23.7 weeks ($p = 0.06$; HR = 1.28) No treatment-related deaths (2 deaths due to disease + treatment)

STS: Soft Tissue Sarcomas; MFH: Malignant Fibrous Histiocytoma; UPS: Undifferentiated Pleomorphic Sarcoma; PFS: Progression Free Survival; OS: Overall Survival.
* no otherwise specified.

3.6.3. New Treatment in the Target Therapy Era

To the best of our knowledge, no randomized controlled trials specifically evaluated targeted therapies in primary breast sarcomas. As with other systemic approaches, the available evidence is extrapolated from studies conducted on soft tissue sarcomas, without specifying whether or how many breast sarcomas were included in the analyses.

Among new treatment agents, Tivozanib, another TKI targeting VEGFR and PDGFR, has demonstrated modest PFS in heavily pretreated metastatic STS [119]. Anlotinib, also targeting VEGFR, FGFR, and PDGFR, significantly improved PFS compared to placebo in advanced STS [120]. Regorafenib showed an improvement in PFS for doxorubicin-pretreated nonadipocytic sarcomas [121]. Additionally, several PD-1/L1 inhibitors (e.g., pembrolizumab, nivolumab, camrelizumab, durvalumab, retifanlimab, sintilimab, and toripalimab) are being investigated in combination with chemotherapy, showing promising responses in undifferentiated pleomorphic sarcoma and other subtypes [122–125]. NTRK inhibitors like larotrectinib and entrectinib have also demonstrated efficacy in NTRK fusion-positive sarcomas, advancing personalized medicine [126].

4. Conclusions

Malignant mesenchymal tumors of the breast are rare but extremely aggressive, presenting significant challenges in both diagnosis and treatment. Due to the lack of standardized guidelines, clinical decision-making remains complex and varies widely. Surgical treatment remains the cornerstone of therapy with the primary goal of achieving complete excision to prevent local recurrences and metastases. Surgical options include both breast-conserving surgery and mastectomy, with axillary treatment typically only necessary in cases of biopsy-proven nodal involvement. While the roles of radiotherapy and chemotherapy are still unclear, ongoing research is essential to refine these treatment modalities. The rarity of these tumors complicates the conduction of large-scale studies, making it vital for healthcare professionals to adopt a tailored, multidisciplinary approach, especially in high-volume centers specializing in rare malignancies. These settings provide the best opportunity for improving patient outcomes through expertise and collaboration. Ultimately, advancing our understanding of malignant mesenchymal tumors, enhancing diagnostic precision, and developing comprehensive treatment protocols will be key to improving patient prognoses and managing these challenging malignancies more effectively.

Author Contributions: Conceptualization, F.D.L. and A.M.S.; methodology, F.D.L., F.M. (Fabio Marazzi) and A.M.; software, A.B.; validation, A.M.S., R.M. and G.F.; formal analysis, A.P., L.S., B.C. and S.D.; investigation, A.M.S., A.D.L., A.F. (Antonio Franco), M.P.F., F.G. and E.J.M.; resources, A.O. and I.P.; data curation, A.M.S., A.S. and N.D.; writing—original draft preparation, F.D.L., A.M.S., M.S., V.M., A.O., A.M., A.F. (Alessandra Fabi), and C.V.P.; writing—review and editing, A.M.S. and L.S.; visualization, F.D.L., C.A., M.P.F. and F.M. (Francesca Moschella); supervision, F.D.L., A.M.S., R.M. and G.F.; project administration, F.D.L. and A.M.S.; funding acquisition, L.S, S.M. All authors have read and agreed to the published version of the manuscript.

Funding: This research received no external funding.

Institutional Review Board Statement: Not applicable.

Informed Consent Statement: Not applicable.

Data Availability Statement: Not applicable.

Conflicts of Interest: The authors declare no conflicts of interest.

Abbreviations

The following abbreviations are used in this manuscript:

PTs	Phyllodes Tumors
WHO	World Health Organization
MPT	Malignant Phyllodes Tumors
PBS	Primary Breast Sarcoma
RT	Radiation Therapy
HPF	High-power fields
NCCN	National Comprehensive Cancer Network
MRI	Magnetic Resonance Imaging
ADC	Apparent Diffusion Coefficient
LR	Local Recurrence
OS	Overall Survival
DFS	Disease-free survival
ESMO	European Society for Medical Oncology
SEER	Surveillance, Epidemiology, and End Results
BCSS	Breast Cancer-Specific Survival
STS	Soft Tissue Sarcoma
CT	Computed tomography
PET	Positron Emission Tomography
PFS	Progression-Free Survival
ORR	Objective Response Rate
EORTC	European Organisation for Research and Treatment of Cancer
LPS	Liposarcoma
LMS	Leiomyosarcoma
LVEF	Left Ventricular Ejection Fraction
DCR	Disease Control Rate
MLPS	Metastatic Myxoid-Liposarcoma
PLRs	Platelet–lymphocyte Ratios
BSC	Best Supporting Care
TKI	Tyrosine Kinase Inhibitor
CR	Complete Remission
PR	Partial Response
ICI	Immune Checkpoint Inhibitors
UPS	Undifferentiated Pleomorphic Sarcoma
ASPS	Alveolar Soft Part Sarcoma
O	Olaratumab
G	Gemcitabine
D	Docetaxel

References

1. Tevatia, M.S.; Mishra, P.; Baranwal, A.K.; Nichat, P.B.; Shelly, D.; Awasthi, S.; Sengupta, P. Primary Breast Tumors with Mesenchymal Morphology. *J. Lab. Physicians* **2021**, *13*, 362–367. [CrossRef] [PubMed]
2. Lakhani, S.; Ellis, I.; Schnitt, S.; Tan, P.; Van de Vijver, M. *WHO Classification of Tumors of the Breast*, 4th ed.; International Agency for Research on Cancer: Lyon, France, 2012.
3. Confavreux, C.; Lurkin, A.; Mitton, N.; Blondet, R.; Saba, C.; Ranchère, D.; Sunyach, M.P.; Thiesse, P.; Biron, P.; Blay, J.Y.; et al. Sarcomas and malignant phyllodes tumours of the breast—A retrospective study. *Eur. J. Cancer* **2006**, *42*, 2715–2721. [CrossRef]
4. McGowan, T.S.; Cummings, B.J.; O'Sullivan, B.; Catton, C.N.; Miller, N.; Panzarella, T. An Analysis of 78 Breast Sarcoma Patients Without Distant Metastases at Presentation. *Int. J. Radiat. Oncol. Biol. Phys.* **2000**, *46*, 383–390. [CrossRef] [PubMed]
5. Holm, M.; Aggerholm-Pedersen, N.; Mele, M.; Jørgensen, P.; Baerentzen, S.; Safwat, A. Primary breast sarcoma: A retrospective study over 35 years from a single institution. *Acta Oncol.* **2016**, *55*, 584–590. [CrossRef]
6. Available online: https://www.nccn.org/professionals/physician_gls/pdf/breast.pdf (accessed on 27 January 2025).
7. Available online: https://www.nccn.org/professionals/physician_gls/pdf/sarcoma.pdf (accessed on 27 January 2025).

8. Gronchi, A.; Miah, A.B.; Dei Tos, A.P.; Abecassis, N.; Bajpai, J.; Bauer, S.; Biagini, R.; Bielack, S.; Blay, J.Y.; Bolle, S.; et al. Soft tissue and visceral sarcomas: ESMO-EURACAN-GENTURIS Clinical Practice Guidelines for diagnosis, treatment and follow-up. *Ann. Oncol.* **2021**, *32*, 1348–1365. [CrossRef]
9. AIOM. Linee Guida Sarcomi Dei Tessuti Mollie Gist. 2024. Available online: https://www.aiom.it/linee-guida-aiom-2024-sarcomi-dei-tessuti-molli-e-gist/ (accessed on 11 January 2025).
10. Telli, M.L.; Horst, K.C.; Guardino, A.E.; Dirbas, F.M.; Carlson, R.W. Phyllodes tumors of the breast: Natural history, diagnosis, and treatment. *J. Natl. Compr. Cancer Netw.* **2007**, *5*, 324–330. [CrossRef]
11. Lissidini, G.; Mulè, A.; Santoro, A.; Papa, G.; Nicosia, L.; Cassano, E.; Ashoor, A.A.; Veronesi, P.; Pantanowitz, L.; Hornick, J.L.; et al. Malignant phyllodes tumor of the breast: A systematic review. *Pathologica* **2022**, *114*, 111–120. [CrossRef]
12. Salvadori, B.; Cusumano, F.; Del Bo, R.; Delledonne, V.; Grassi, M.; Rovini, D.; Saccozzi, R.; Andreola, S.; Clemente, C. Surgical treatment of phyllodes tumors of the breast. *Cancer* **1989**, *63*, 2532–2536. [CrossRef]
13. Tse, G.M.K.; Niu, Y.; Shi, H.J. Phyllodes tumor of the breast: An update. *Breast Cancer* **2010**, *17*, 29–34. [CrossRef]
14. Calhoun, K.; Allison, K.H.; Kim, J.N. (PDF) Phyllodes Tumors. In: ResearchGate. Available online: https://www.researchgate.net/publication/262524480_Phyllodes_tumors (accessed on 23 March 2025).
15. Bernstein, L.; Deapen, D.; Ross, R.K. The descriptive epidemiology of malignant cystosarcoma phyllodes tumors of the breast. *Cancer* **1993**, *71*, 3020–3024. [CrossRef]
16. Birch, J.M.; Alston, R.D.; McNally, R.J.; Evans, D.G.; Kelsey, A.M.; Harris, M.; Eden, O.B.; Varley, J.M. Relative frequency and morphology of cancers in carriers of germline TP53 mutations. *Oncogene* **2001**, *20*, 4621–4628. [CrossRef] [PubMed]
17. Papas, Y.; Asmar, A.E.; Ghandour, F.; Hajj, I. Malignant phyllodes tumors of the breast: A comprehensive literature review. *Breast J.* **2020**, *26*, 240–244. [CrossRef]
18. Sars, C.; Sackey, H.; Frisell, J.; Dickman, P.W.; Karlsson, F.; Kindts, I.; Marta, G.N.; Freitas-Junior, R.; Tvedskov, T.F.; Kassem, L.; et al. Current clinical practice in the management of phyllodes tumors of the breast: An international cross-sectional study among surgeons and oncologists. *Breast Cancer Res. Treat.* **2023**, *199*, 293–304. [CrossRef] [PubMed]
19. Logistic regression analysis of ultrasound findings in predicting the malignant and benign phyllodes tumor of breast. *PLoS ONE* **2022**, *17*, e0265952. Available online: https://journals.plos.org/plosone/article?id=10.1371/journal.pone.0265952 (accessed on 11 January 2025).
20. Tan, P.H.; Ellis, I.; Allison, K.; Brogi, E.; Fox, S.B.; Lakhani, S.; Lazar, A.J.; Morris, E.A.; Sahin, A.; Salgado, R.; et al. WHO Classification of Tumours Editorial Board. The 2019 World Health Organization classification of tumours of the breast. *Histopathology* **2020**, *77*, 181–185. [CrossRef]
21. Jagdewsing, D.R.; Murtaza, G.; Jagdewsing, S.A.; Jagdewsing, S.A.; Fahmy, N.S.C.; Silva, F.A.; Koendjbiharie, T.; Djojomoenawi, S.; Kwakye, O.V.; Mahmud, N.M. Evaluation of the Clinicopathological Features Associated with Malignancy of Phyllodes Tumor of the Breast. *Cureus* **2024**, *16*, e76221. [CrossRef]
22. Macdonald, O.K.; Lee, C.M.; Tward, J.D.; Chappel, C.D.; Gaffney, D.K. Malignant phyllodes tumor of the female breast: Association of primary therapy with cause-specific survival from the Surveillance, Epidemiology, and End Results (SEER) program. *Cancer* **2006**, *107*, 2127–2133. [CrossRef]
23. Lenhard, M.S.; Kahlert, S.; Himsl, I.; Ditsch, N.; Untch, M.; Bauerfeind, I. Phyllodes tumour of the breast: Clinical follow-up of 33 cases of this rare disease. *Eur. J. Obstet. Gynecol. Reprod. Biol.* **2008**, *138*, 217–221. [CrossRef]
24. Mohd Ali, N.A.; Nasaruddin, A.F.; Mohamed, S.S.; Wan Rahman, W.F. Ki67 and P53 Expression in Relation to Clinicopathological Features in Phyllodes Tumour of the Breast. *Asian Pac. J. Cancer Prev.* **2020**, *21*, 2653–2659. [CrossRef]
25. Vorotnikov, I.K.; Vysotskaya, I.V.; Denchik, D.A.; Letyagin, V.P.; Davydov, M.M.; Kirsanov, V.Y.; Kim, E.A.; Buseva, V.S. Prognostic Molecular and Biological Characteristics of Phyllodes Tumors of the Breast. *Bull. Exp. Biol. Med.* **2020**, *169*, 806–810. [CrossRef]
26. Moffat, C.J.; Pinder, S.E.; Dixon, A.R.; Elston, C.W.; Blamey, R.W.; Ellis, I.O. Phyllodes tumours of the breast: A clinicopathological review of thirty-two cases. *Histopathology* **1995**, *27*, 205–218. [CrossRef] [PubMed]
27. Barth, R.J.; Wells, W.A.; Mitchell, S.E.; Cole, B.F. A prospective, multi-institutional study of adjuvant radiotherapy after resection of malignant phyllodes tumors. *Ann. Surg. Oncol.* **2009**, *16*, 2288–2294. [CrossRef] [PubMed]
28. Fede ÂBde, S.; Pereira Souza, R.; Doi, M.; De Brot, M.; Aparecida Bueno de Toledo Osorio, C.; Rocha Melo Gondim, G.; Casali-da-Rocha, J.C.; Jbili, R.; Bitencourt, A.G.V.; Alves de Souza, J.; et al. Malignant Phyllodes Tumor of the Breast: A Practice Review. *Clin. Pract.* **2021**, *11*, 205–215. [CrossRef] [PubMed]
29. Spanheimer, P.M.; Murray, M.P.; Zabor, E.C.; Stempel, M.; Morrow, M.; Van Zee, K.J.; Barrio, A.V. Long-Term Outcomes After Surgical Treatment of Malignant/Borderline Phyllodes Tumors of the Breast. *Ann. Surg. Oncol.* **2019**, *26*, 2136–2143. [CrossRef]
30. Bishr, M.K.; Banks, J.; Abdelaziz, M.S.; Badawi, M.; Crane, P.W.; Donigiewicz, U.J.; Elkorety, M.; Girgis, M.; Humphreys, A.; Isherwood, J.; et al. Multidisciplinary Management of Phyllodes Tumours and Breast Sarcoma: A Cross-sectional Survey of Clinical Practice across the UK and Ireland. *Clin. Oncol.* **2024**, *36*, e31–e39. [CrossRef]

31. Esperança-Martins, M.; Melo-Alvim, C.; Dâmaso, S.; Lopes-Brás, R.; Peniche, T.; Nogueira-Costa, G.; Abreu, C.; Luna Pais, H.; de Sousa, R.T.; Torres, S.; et al. Breast Sarcomas, Phyllodes Tumors, and Desmoid Tumors: Turning the Magnifying Glass on Rare and Aggressive Entities. *Cancers* **2023**, *15*, 3933. [CrossRef]
32. Lu, Y.; Chen, Y.; Zhu, L.; Cartwright, P.; Song, E.; Jacobs, L.; Chen, K. Local Recurrence of Benign, Borderline, and Malignant Phyllodes Tumors of the Breast: A Systematic Review and Meta-analysis. *Ann. Surg. Oncol.* **2019**, *26*, 1263–1275. [CrossRef]
33. Liang, M.I.; Ramaswamy, B.; Patterson, C.C.; McKelvey, M.T.; Gordillo, G.; Nuovo, G.J.; Carson, W.E. Giant breast tumors: Surgical management of phyllodes tumors, potential for reconstructive surgery and a review of literature. *World J. Surg. Oncol.* **2008**, *6*, 117. [CrossRef]
34. Breslin, T.M. Risk Factors for Recurrence and Death After Primary Surgical Treatment of Malignant Phyllodes Tumors. *Breast Dis.* **2005**, *16*, 259. Available online: https://pubmed.ncbi.nlm.nih.gov/15525831/ (accessed on 13 January 2025). [CrossRef]
35. Neron, M.; Sajous, C.; Thezenas, S.; Piperno-Neumann, S.; Reyal, F.; Laé, M.; Chakiba, C.; Michot, A.; Penel, N.; Honoré, C.; et al. Surgical Margins and Adjuvant Therapies in Malignant Phyllodes Tumors of the Breast: A Multicenter Retrospective Study. *Ann. Surg. Oncol.* **2020**, *27*, 1818–1827. [CrossRef]
36. Ogunbiyi, S.; Perry, A.; Jakate, K.; Simpson, J.; George, R. Phyllodes tumour of the breast and margins: How much is enough. *Can. J. Surg.* **2019**, *62*, E19–E21. [CrossRef] [PubMed]
37. Kapiris, I.; Nasiri, N.; Healy, V.; Gui, G.P.H. Outcome and Predictive Factors of Local Recurrence and Distant Metastases Following Primary Surgical Treatment of High-Grade Malignant Phyllodes Tumours of the Breast. *Eur. J. Surg. Oncol.* **2001**, *27*, 723–730. Available online: https://pubmed.ncbi.nlm.nih.gov/11735168/ (accessed on 13 January 2025). [CrossRef]
38. Singh, G.; Sharma, R.K. Immediate breast reconstruction for phyllodes tumors. *Breast* **2008**, *17*, 296–301. [CrossRef] [PubMed]
39. Yom, C.K.; Han, W.; Kim, S.W.; Park, S.Y.; Park, I.A.; Noh, D.Y. Reappraisal of conventional risk stratification for local recurrence based on clinical outcomes in 285 resected phyllodes tumors of the breast. *Ann. Surg. Oncol.* **2015**, *22*, 2912–2918. [CrossRef] [PubMed]
40. Bhargav, P.R.K.; Mishra, A.; Agarwal, G.; Agarwal, A.; Verma, A.K.; Mishra, S.K. Phyllodes tumour of the breast: Clinicopathological analysis of recurrent vs. non-recurrent cases. *Asian J. Surg.* **2009**, *32*, 224–228. [CrossRef]
41. Cheng, S.P.; Chang, Y.C.; Liu, T.P.; Lee, J.J.; Tzen, C.Y.; Liu, C.L. Phyllodes tumor of the breast: The challenge persists. *World J. Surg.* **2006**, *30*, 1414–1421. [CrossRef]
42. Lim, R.S.; Cordeiro, E.; Lau, J.; Lim, A.; Roberts, A.; Seely, J. Phyllodes Tumors—The Predictors and Detection of Recurrence. *Can. Assoc. Radiol. J.* **2021**, *72*, 251–257. [CrossRef]
43. Wen, B.; Mousadoust, D.; Warburton, R.; Pao, J.S.; Dingee, C.; Chen, L.; McKevitt, E. Phyllodes tumours of the breast: Outcomes and recurrence after excision. *Am. J. Surg.* **2020**, *219*, 790–794. [CrossRef]
44. Taira, N.; Takabatake, D.; Aogi, K.; Ohsumi, S.; Takashima, S.; Nishimura, R.; Teramoto, N. Phyllodes tumor of the breast: Stromal overgrowth and histological classification are useful prognosis-predictive factors for local recurrence in patients with a positive surgical margin. *Jpn. J. Clin. Oncol.* **2007**, *37*, 730–736. [CrossRef]
45. Shin, E.; Koo, J.S. Prognostic factors of breast phyllodes tumors. *Histol. Histopathol.* **2023**, *38*, 865–878.
46. Thind, A.; Patel, B.; Thind, K.; Isherwood, J.; Phillips, B.; Dhaliwal, K.; Remoundos, D.D. Surgical margins for borderline and malignant phyllodes tumours. *Ann. R. Coll. Surg. Engl.* **2020**, *102*, 165–173. [CrossRef] [PubMed]
47. Chen, W.H.; Cheng, S.P.; Tzen, C.Y.; Yang, T.L.; Jeng, K.S.; Liu, C.L.; Liu, T.P. Surgical treatment of phyllodes tumors of the breast: Retrospective review of 172 cases. *J. Surg. Oncol.* **2005**, *91*, 185–194. [CrossRef]
48. Zhang, Y.; Kleer, C.G. Phyllodes Tumor of the Breast: Histopathologic Features, Differential Diagnosis, and Molecular/Genetic Updates. *Arch. Pathol. Lab. Med.* **2016**, *140*, 665–671. [CrossRef]
49. Mitus, J.W.; Blecharz, P.; Jakubowicz, J.; Reinfuss, M.; Walasek, T.; Wysocki, W. Phyllodes tumors of the breast. The treatment results for 340 patients from a single cancer centre. *Breast* **2019**, *43*, 85–90. [CrossRef]
50. Rowell, M.D.; Perry, R.R.; Hsiu, J.G.; Barranco, S.C. Phyllodes tumors. *Am. J. Surg.* **1993**, *165*, 376–379. [CrossRef]
51. Chao, X.; Chen, K.; Zeng, J.; Bi, Z.; Guo, M.; Chen, Y.; Yao, Y.; Wu, W.; Liang, S.; Nie, Y. Adjuvant radiotherapy and chemotherapy for patients with breast phyllodes tumors: A systematic review and meta-analysis. *BMC Cancer* **2019**, *19*, 372. [CrossRef]
52. Casali, P.G.; Abecassis, N.; Aro, H.T.; Bauer, S.; Biagini, R.; Bielack, S.; Bonvalot, S.; Boukovinas, I.; Bovee, J.V.M.G.; Brodowicz, T.; et al. Soft tissue and visceral sarcomas: ESMO-EURACAN Clinical Practice Guidelines for diagnosis, treatment and follow-up. *Ann. Oncol.* **2018**, *29* (Suppl. 4), iv51–iv67. [CrossRef]
53. Yang, P.; Zhang, G.; Zhang, Y.; Zhao, W.; Tang, J.; Zeng, S.; Lv, X.; Lv, L. Effect of adjuvant radiotherapy on overall survival and breast cancer-specific survival of patients with malignant phyllodes tumor of the breast in different age groups: A retrospective observational study based on SEER. *Radiat. Oncol.* **2024**, *19*, 59. [CrossRef]
54. Moon, S.H.; Jung, J.H.; Lee, J.; Kim, W.W.; Park, H.Y.; Lee, J.W.; Lee, S.J. Complete remission of giant malignant phyllodes tumor with lung metastasis: A case report. *Medicine* **2019**, *98*, e15762. [CrossRef]

55. Li, G.Z.; Raut, C.P.; Hunt, K.K.; Feng, M.; Chugh, R. Breast Sarcomas, Phyllodes Tumors, and Desmoid Tumors: Epidemiology, Diagnosis, Staging, and Histology-Specific Management Considerations. *Am. Soc. Clin. Oncol. Educ. Book* **2021**, *41*, 390–404. [CrossRef]
56. Orlandi, A.; Sanchez, A.M.; Calegari, M.A.; D'Archi, S.; Santoro, A.; Leone, A.D.; Mulè, A.; Arena, V.; Masetti, R.; Barone, C.; et al. Diagnosis and management of breast lymphoma: A single- institution retrospective analysis. *Transl. Cancer Res.* **2018**, *7* (Suppl. 3), S272–S280. Available online: https://tcr.amegroups.org/article/view/17996 (accessed on 29 March 2024). [CrossRef]
57. Lahat, G.; Lev, D.; Gerstenhaber, F.; Madewell, J.; Le-Petross, H.; Pollock, R.E. Sarcomas of the breast. *Expert. Rev. Anticancer. Ther.* **2012**, *12*, 1045–1051. [CrossRef] [PubMed]
58. Nizri, E.; Merimsky, O.; Lahat, G. Optimal management of sarcomas of the breast: An update. *Expert. Rev. Anticancer. Ther.* **2014**, *14*, 705–710. [CrossRef]
59. Zelek, L.; Llombart-Cussac, A.; Terrier, P.; Pivot, X.; Guinebretiere, J.M.; Le Pechoux, C.; Tursz, T.; Rochard, F.; Spielmann, M.; Le Cesne, A. Prognostic factors in primary breast sarcomas: A series of patients with long-term follow-up. *J. Clin. Oncol.* **2003**, *21*, 2583–2588. [CrossRef]
60. Pasta, V.; Monti, M.; Cialini, M.; Vergine, M.; Urciuoli, P.; Iacovelli, A.; Rea, S.; D'Orazi, V. Primitive sarcoma of the breast: New insight on the proper surgical management. *J. Exp. Clin. Cancer Res.* **2015**, *34*, 72. [CrossRef] [PubMed]
61. Shabahang, M.; Franceschi, D.; Sundaram, M.; Castillo, M.H.; Moffat, F.L.; Frank, D.S.; Rosenberg, E.R.; Bullock, K.E.; Livingstone, A.S. Surgical management of primary breast sarcoma. *Am. Surg.* **2002**, *68*, 673–677, discussion 677. [CrossRef] [PubMed]
62. Vorburger, S.A.; Xing, Y.; Hunt, K.K.; Lakin, G.E.; Benjamin, R.S.; Feig, B.W.; Pisters, P.W.T.; Ballo, M.T.; Chen, L.; Trent, J.; et al. Angiosarcoma of the breast. *Cancer* **2005**, *104*, 2682–2688. [CrossRef]
63. Seinen, J.M.; Styring, E.; Verstappen, V.; Vult von Steyern, F.; Rydholm, A.; Suurmeijer, A.J.H.; Hoekstra, H.J. Radiation-associated angiosarcoma after breast cancer: High recurrence rate and poor survival despite surgical treatment with R0 resection. *Ann. Surg. Oncol.* **2012**, *19*, 2700–2706. [CrossRef]
64. Cui, L.; Zhang, J.; Zhang, X.; Chang, H.; Qu, C.; Zhang, J.; Zhong, D. Angiosarcoma (Stewart-Treves Syndrome) in Postmastectomy Patients: Report of 10 cases and Review of Literature. *Int. J. Clin. Exp. Pathol.* **2015**, *8*, 11108. Available online: https://pubmed.ncbi.nlm.nih.gov/26617830/ (accessed on 13 January 2025).
65. Lee, J.S.; Yoon, K.; Onyshchenko, M. Sarcoma of the Breast: Clinical Characteristics and Outcomes of 991 Patients from the National Cancer Database. *Sarcoma* **2021**, *2021*, 8828158. [CrossRef]
66. Deyrup, A.T.; Weiss, S.W. Grading of soft tissue sarcomas: The challenge of providing precise information in an imprecise world. *Histopathology* **2006**, *48*, 42–50. [CrossRef]
67. Cantile, M.; Di Bonito, M.; Cerrone, M.; Pizzolorusso, A.; Apice, G.; Botti, G.; De Chiara, A. Primary breast angiosarcoma in young women from the same geographic region in a short period of time: Only a coincidence or an increased risk? *Breast J.* **2018**, *24*, 91–93. [CrossRef] [PubMed]
68. Chagas, C.; Pinto, R.; Franco, R. Primary Angiosarcoma of the Breast: A Case Report. *Mastology* **2022**, *32*, 100. [CrossRef]
69. Frey, J.D.; Levine, P.G.; Darvishian, F.; Shapiro, R.L. Angiosarcoma of the breast masquerading as hemangioma: Exploring clinical and pathological diagnostic challenges. *Arch. Plast. Surg.* **2015**, *42*, 261–263. [CrossRef] [PubMed]
70. Mulder, L.; Liu, S.; Kopkash, K.; Madrigrano, A. Primary Synchronous Bilateral Angiosarcoma of the Breast. *Am. Surg.* **2017**, *83*, 476–477. [CrossRef]
71. Rao, P.; Lahat, G.; Arnold, C.; Gavino, A.C.; Lahat, S.; Hornick, J.L.; Lev, D.; Lazar, A.J. Angiosarcoma: A Tissue Microarray Study With Diagnostic Implications. *Am. J. Dermatopathol.* **2013**, *35*, 432. [CrossRef]
72. Brodie, C.; Provenzano, E. Vascular Proliferations of the Breast. *Histopathology* **2008**, *52*, 30–44. [CrossRef]
73. Machado, I.; Giner, F.; Lavernia, J.; Cruz, J.; Traves, V.; Requena, C.; Llombart, B.; López-Guerrero, J.A.; Llombart-Bosch, A. Angiosarcomas: Histology, immunohistochemistry and molecular insights with implications for differential diagnosis. *Histol. Histopathol* **2021**, *36*, 3–18.
74. Kuba, M.G.; Xu, B.; D'Angelo, S.P.; Rosenbaum, E.; Plitas, G.; Ross, D.S.; Brogi, E.; Antonescu, C.R. The impact of MYC gene amplification on the clinicopathological features and prognosis of radiation-associated angiosarcomas of the breast. *Histopathology* **2021**, *79*, 836–846. [CrossRef]
75. Mentzel, T.; Kiss, K. Reduced H3K27me3 expression in radiation-associated angiosarcoma of the breast. *Virchows Arch.* **2018**, *472*, 361–368. [CrossRef]
76. Rakha, E.A.; Aleskandarany, M.A.; Lee, A.H.S.; Ellis, I.O. An Approach to the Diagnosis of Spindle Cell Lesions of the Breast. *Histopathology* **2016**, *68*, 33–44. [CrossRef]
77. Shivali, B. Myofibroblastoma Breast with Unusual Morphological Features. Cytohistopathogical Diagnostic Pitfalls and Role of Immunohistochemistry-Review of Literature. *J. Clin. Diagn. Res.* **2013**, *7*, 2323–2325. Available online: http://www.jcdr.net/article_fulltext.asp?issn=0973-709x&year=2013&volume=7&issue=10&page=2323&issn=0973-709x&id=3515 (accessed on 23 March 2025). [CrossRef] [PubMed]
78. Carder, P.J.; Shaaban, A. Mesenchymal lesions of the breast. *Diagn. Histopathol.* **2019**, *25*, 123–131. [CrossRef]

79. Lee, A.H. Use of immunohistochemistry in the diagnosis of problematic breast lesions. *J. Clin. Pathol.* **2013**, *66*, 471–477. Available online: https://jcp.bmj.com/content/66/6/471 (accessed on 23 March 2025). [CrossRef]
80. Collins, L.C.; Tan, P.H.; Tse, G.M. Annual Review Issue: Breast Pathology. *Histopathology* **2016**, *68*, 3–4. [CrossRef] [PubMed]
81. Radu, I.; Scripcariu, V.; Panuța, A.; Rusu, A.; Afrăsânie, V.A.; Cojocaru, E.; Aniței, M.G.; Alexa-Stratulat, T.; Terinte, C.; Șerban, F.; et al. Breast Sarcomas-How Different Are They from Breast Carcinomas? Clinical, Pathological, Imaging and Treatment Insights. *Diagnostics* **2023**, *13*, 1370. [CrossRef]
82. Al-Benna, S.; Poggemann, K.; Steinau, H.U.; Steinstraesser, L. Diagnosis and management of primary breast sarcoma. *Breast Cancer Res. Treat.* **2010**, *122*, 619–626. [CrossRef]
83. Lim, S.Z.; Ong, K.W.; Tan, B.K.T.; Selvarajan, S.; Tan, P.H. Sarcoma of the breast: An update on a rare entity. *J. Clin. Pathol.* **2016**, *69*, 373–381. [CrossRef]
84. Ramelli, E.; Weingertner, N.; Welsch, A.; Reix, N.; Antoni, D.; Amé, S.; Molière, S.; Mathelin, C. Innovative approach to lymphadenectomy in breast sarcoma. *Bull. Cancer* **2022**, *109*, 1017–1028. [CrossRef]
85. Adem, C.; Reynolds, C.; Ingle, J.N.; Nascimento, A. Primary breast sarcoma: Clinicopathologic series from the Mayo Clinic and review of the literature. *Br. J. Cancer* **2004**, *91*, 237–241. Available online: https://pubmed.ncbi.nlm.nih.gov/15187996/ (accessed on 13 January 2025). [CrossRef]
86. Pencavel, T.D.; Hayes, A. Breast sarcoma--A review of diagnosis and management. *Int. J. Surg.* **2009**, *7*, 20–23. [CrossRef]
87. Fields, R.C.; Aft, R.L.; Gillanders, W.E.; Eberlein, T.J.; Margenthaler, J.A. Treatment and outcomes of patients with primary breast sarcoma. *Am. J. Surg.* **2008**, *196*, 559–561. [CrossRef] [PubMed]
88. Wang, F.; Jia, Y.; Tong, Z. Comparison of the clinical and prognostic features of primary breast sarcomas and malignant phyllodes tumor. *Jpn J. Clin. Oncol.* **2015**, *45*, 146–152. [CrossRef] [PubMed]
89. Berg, J.W.; Decrosse, J.J.; Fracchia, A.A.; Farrow, J. Stromal sarcomas of the breast. A unified approach to connective tissue sarcomas other than cystosarcoma phyllodes. *Cancer* **1962**, *15*, 418–524. [CrossRef]
90. Bousquet, G.; Confavreux, C.; Magné, N.; de Lara, C.T.; Poortmans, P.; Senkus, E.; de Lafontan, B.; Bolla, M.; Largillier, R.; Lagneau, E.; et al. Outcome and prognostic factors in breast sarcoma: A multicenter study from the rare cancer network. *Radiother. Oncol.* **2007**, *85*, 355–361. [CrossRef]
91. Ciatto, S.; Bonardi, R.; Cataliotti, L.; Cardona, G. Sarcomas of the breast: A multicenter series of 70 cases. *Neoplasma* **1992**, *39*, 375–379. [PubMed]
92. North, J.H.; McPhee, M.; Arredondo, M.; Edge, S.B. Sarcoma of the breast: Implications of the extent of local therapy. *Am. Surg.* **1998**, *64*, 1059–1061.
93. Crosby, M.A.; Chike-Obi, C.J.; Baumann, D.P.; Sacks, J.M.; Villa, M.T.; Garvey, P.B.; Selber, J.C.; Feig, B.W. Reconstructive outcomes in patients with sarcoma of the breast. *Plast. Reconstr. Surg.* **2010**, *126*, 1805–1814. [CrossRef]
94. Fong, Y.; Coit, D.G.; Woodruff, J.M.; Brennan, M.F. Lymph node metastasis from soft tissue sarcoma in adults. Analysis of data from a prospective database of 1772 sarcoma patients. *Ann. Surg.* **1993**, *217*, 72–77. [CrossRef]
95. Gullett, N.P.; Delman, K.; Folpe, A.L.; Johnstone, P.A.S. National surgical patterns of care: Regional lymphadenectomy of breast sarcomas. *Am. J. Clin. Oncol.* **2007**, *30*, 461–465. [CrossRef]
96. Kuerer, H.M. *Kuerer's Breast Surgical Oncology*; The McGraw-Hill Companies: New York, NY, USA, 2010.
97. Voutsadakis, I.A.; Zaman, K.; Leyvraz, S. Breast sarcomas: Current and future perspectives. *Breast* **2011**, *20*, 199–204. [CrossRef] [PubMed]
98. Gutman, H.; Pollock, R.E.; Ross, M.I.; Benjamin, R.S.; Johnston, D.A.; Janjan, N.A.; Romsdahl, M.M. Sarcoma of the breast: Implications for extent of therapy. The M. D. Anderson experience. *Surgery* **1994**, *116*, 505–509.
99. von Mehren, M.; Randall, R.L.; Benjamin, R.S.; Boles, S.; Bui, M.M.; Ganjoo, K.N.; George, S.; Gonzalez, R.J.; Heslin, M.J.; Kane, J.M.; et al. Soft Tissue Sarcoma, Version 2.2018, NCCN Clinical Practice Guidelines in Oncology. *J. Natl. Compr. Cancer Netw. JNCCN* **2018**, *16*, 536–563. [CrossRef] [PubMed]
100. Gronchi, A.; Palmerini, E.; Quagliuolo, V.; Martin Broto, J.; Lopez Pousa, A.; Grignani, G.; Brunello, A.; Blay, J.Y.; Tendero, O.; Diaz Beveridge, R.; et al. Neoadjuvant Chemotherapy in High-Risk Soft Tissue Sarcomas: Final Results of a Randomized Trial From Italian (ISG), Spanish (GEIS), French (FSG), and Polish (PSG) Sarcoma Groups. *J. Clin. Oncol.* **2020**, *38*, 2178–2186. [CrossRef] [PubMed]
101. Judson, I.; Verweij, J.; Gelderblom, H.; Hartmann, J.T.; Schöffski, P.; Blay, J.Y.; Kerst, J.M.; Sufliarsky, J.; Whelan, J.; Hohenberger, P.; et al. Doxorubicin alone versus intensified doxorubicin plus ifosfamide for first-line treatment of advanced or metastatic soft-tissue sarcoma: A randomised controlled phase 3 trial. *Lancet Oncol.* **2014**, *15*, 415–423. [CrossRef]
102. Pautier, P.; Italiano, A.; Piperno-Neumann, S.; Chevreau, C.; Penel, N.; Firmin, N.; Boudou-Rouquette, P.; Bertucci, F.; Balleyguier, C.; Lebrun-Ly, V.; et al. Doxorubicin alone versus doxorubicin with trabectedin followed by trabectedin alone as first-line therapy for metastatic or unresectable leiomyosarcoma (LMS-04): A randomised, multicentre, open-label phase 3 trial. *Lancet Oncol.* **2022**, *23*, 1044–1054. [CrossRef]

103. Chawla, S.P.; Papai, Z.; Mukhametshina, G.; Sankhala, K.; Vasylyev, L.; Fedenko, A.; Khamly, K.; Ganjoo, K.; Nagarkar, R.; Wieland, S.; et al. First-Line Aldoxorubicin vs Doxorubicin in Metastatic or Locally Advanced Unresectable Soft-Tissue Sarcoma: A Phase 2b Randomized Clinical Trial. *JAMA Oncol.* **2015**, *1*, 1272–1280. [CrossRef]
104. Liu, W.; Jiang, Q.; Zhou, Y. Advances of systemic treatment for adult soft-tissue sarcoma. *Chin. Clin. Oncol.* **2018**, *7*, 42. [CrossRef]
105. Chawla, S.; Ganjoo, K.; Schuetze, S.; Papai, Z.; Tine, B.; Choy, E.; Liebner, D.; Agulnik, M.; Chawla, S.; Wieland, S.; et al. Phase III study of aldoxorubicin vs investigators' choice as treatment for relapsed/refractory soft tissue sarcomas. *J. Clin. Oncol.* **2017**, *35*, 11000. [CrossRef]
106. Gupta, S.; Gouw, L.; Wright, J.; Chawla, S.; Pitt, D.; Wade, M.; Boucher, K.; Sharma, S. Phase II study of amrubicin (SM-5887), a synthetic 9-aminoanthracycline, as first line treatment in patients with metastatic or unresectable soft tissue sarcoma: Durable response in myxoid liposarcoma with TLS-CHOP translocation. *Investig. New Drugs* **2016**, *34*, 243–252. [CrossRef]
107. Maki, R.G.; Wathen, J.K.; Patel, S.R.; Priebat, D.A.; Okuno, S.H.; Samuels, B.; Fanucchi, M.; Harmon, D.C.; Schuetze, S.M.; Reinke, D.; et al. Randomized phase II study of gemcitabine and docetaxel compared with gemcitabine alone in patients with metastatic soft tissue sarcomas: Results of sarcoma alliance for research through collaboration study 002. *J. Clin. Oncol.* **2007**, *25*, 2755–2763. [CrossRef]
108. Seddon, B.; Strauss, S.J.; Whelan, J.; Leahy, M.; Woll, P.J.; Cowie, F.; Rothermundt, C.; Wood, Z.; Benson, C.; Ali, N.; et al. Gemcitabine and docetaxel versus doxorubicin as first-line treatment in previously untreated advanced unresectable or metastatic soft-tissue sarcomas (GeDDiS): A randomised controlled phase 3 trial. *Lancet Oncol.* **2017**, *18*, 1397–1410. [CrossRef] [PubMed]
109. Penel, N.; Bui, B.N.; Bay, J.O.; Cupissol, D.; Ray-Coquard, I.; Piperno-Neumann, S.; Kerbrat, P.; Fournier, C.; Taieb, S.; Jimenez, M.; et al. Phase II trial of weekly paclitaxel for unresectable angiosarcoma: The ANGIOTAX Study. *J. Clin. Oncol.* **2008**, *26*, 5269–5274. [CrossRef]
110. Buesa, J.M.; Mouridsen, H.T.; van Oosterom, A.T.; Verweij, J.; Wagener, T.; Steward, W.; Poveda, A.; Vestlev, P.M.; Thomas, D.; Sylvester, R. High-dose DTIC in advanced soft-tissue sarcomas in the adult. A phase II study of the E.O.R.T.C. Soft Tissue and Bone Sarcoma Group. *Ann. Oncol.* **1991**, *2*, 307–309. [CrossRef] [PubMed]
111. García-Del-Muro, X.; López-Pousa, A.; Maurel, J.; Martín, J.; Martínez-Trufero, J.; Casado, A.; Gómez-España, A.; Fra, J.; Cruz, J.; Poveda, A.; et al. Randomized phase II study comparing gemcitabine plus dacarbazine versus dacarbazine alone in patients with previously treated soft tissue sarcoma: A Spanish Group for Research on Sarcomas study. *J. Clin. Oncol.* **2011**, *29*, 2528–2533. [CrossRef] [PubMed]
112. Schöffski, P.; Ray-Coquard, I.L.; Cioffi, A.; Bui, N.B.; Bauer, S.; Hartmann, J.T.; Krarup-Hansen, A.; Grünwald, V.; Sciot, R.; Dumez, H.; et al. Activity of eribulin mesylate in patients with soft-tissue sarcoma: A phase 2 study in four independent histological subtypes. *Lancet Oncol.* **2011**, *12*, 1045–1052. [CrossRef]
113. Iwai, T.; Hoshi, M.; Oebisu, N.; Takada, N.; Ban, Y.; Yao, H.; Nakamura, H. Optimal Prognostic Factors for Metastatic and Inoperable Sarcomas Treated With Pazopanib, Eribulin, and Trabectedin. *In Vivo* **2023**, *37*, 2634–2641. [CrossRef]
114. Delaloge, S.; Yovine, A.; Taamma, A.; Riofrio, M.; Brain, E.; Raymond, E.; Cottu, P.; Goldwasser, F.; Jimeno, J.; Misset, J.L.; et al. Ecteinascidin-743: A marine-derived compound in advanced, pretreated sarcoma patients—Preliminary evidence of activity. *J. Clin. Oncol.* **2001**, *19*, 1248–1255. [CrossRef]
115. Demetri, G.D.; von Mehren, M.; Jones, R.L.; Hensley, M.L.; Schuetze, S.M.; Staddon, A.; Milhem, M.; Elias, A.; Ganjoo, K.; Tawbi, H.; et al. Efficacy and Safety of Trabectedin or Dacarbazine for Metastatic Liposarcoma or Leiomyosarcoma After Failure of Conventional Chemotherapy: Results of a Phase III Randomized Multicenter Clinical Trial. *J. Clin. Oncol.* **2016**, *34*, 786–793. [CrossRef]
116. Le Cesne, A.; Blay, J.Y.; Cupissol, D.; Italiano, A.; Delcambre, C.; Penel, N.; Isambert, N.; Chevreau, C.; Bompas, E.; Bertucci, F.; et al. A randomized phase III trial comparing trabectedin to best supportive care in patients with pre-treated soft tissue sarcoma: T-SAR, a French Sarcoma Group trial. *Ann. Oncol.* **2021**, *32*, 1034–1044. [CrossRef]
117. Garcia del Muro, X.; Lopez-Pousa, A.; Martin, J.; Buesa, J.M.; Martinez-Trufero, J.; Casado, A.; Poveda, A.; Cruz, J.; Bover, I.; Maurel, J.; et al. A phase II trial of temozolomide as a 6-week, continuous, oral schedule in patients with advanced soft tissue sarcoma: A study by the Spanish Group for Research on Sarcomas. *Cancer* **2005**, *104*, 1706–1712. [CrossRef]
118. Talbot, S.M.; Keohan, M.L.; Hesdorffer, M.; Orrico, R.; Bagiella, E.; Troxel, A.B.; Taub, R.N. A phase II trial of temozolomide in patients with unresectable or metastatic soft tissue sarcoma. *Cancer* **2003**, *98*, 1942–1946. [CrossRef] [PubMed]
119. Agulnik, M.; Costa, R.L.B.; Milhem, M.; Rademaker, A.W.; Prunder, B.C.; Daniels, D.; Rhodes, B.T.; Humphreys, C.; Abbinanti, S.; Nye, L.; et al. A phase II study of tivozanib in patients with metastatic and nonresectable soft-tissue sarcomas. *Ann. Oncol.* **2017**, *28*, 121–127. [CrossRef]
120. Chi, Y.; Yao, Y.; Wang, S.; Huang, G.; Cai, Q.; Shang, G.; Wang, G.; Qu, G.; Wu, Q.; Jiang, Y.; et al. Anlotinib for metastasis soft tissue sarcoma: A randomized, double-blind, placebo-controlled and multi-centered clinical trial. *J. Clin. Oncol.* **2018**, *36* (Suppl. 15), 11503. [CrossRef]

121. Berry, V.; Basson, L.; Bogart, E.; Mir, O.; Blay, J.Y.; Italiano, A.; Bertucci, F.; Chevreau, C.; Clisant-Delaine, S.; Liegl-Antzager, B.; et al. REGOSARC: Regorafenib versus placebo in doxorubicin-refractory soft-tissue sarcoma—A quality-adjusted time without symptoms of progression or toxicity analysis. *Cancer* **2017**, *123*, 2294–2302. [CrossRef]
122. Tian, Z.; Yao, W. PD-1/L1 inhibitor plus chemotherapy in the treatment of sarcomas. *Front. Immunol.* **2022**, *13*, 898255. [CrossRef] [PubMed]
123. Tawbi, H.A.; Burgess, M.; Bolejack, V.; Van Tine, B.A.; Schuetze, S.M.; Hu, J.; D'Angelo, S.; Attia, S.; Riedel, R.F.; Priebat, D.A.; et al. Pembrolizumab in advanced soft-tissue sarcoma and bone sarcoma (SARC028): A multicentre, two-cohort, single-arm, open-label, phase 2 trial. *Lancet Oncol.* **2017**, *18*, 1493–1501. Available online: https://pubmed.ncbi.nlm.nih.gov/28988646/ (accessed on 13 January 2025). [CrossRef]
124. Burgess, M.A.; Bolejack, V.; Schuetze, S.; Van Tine, B.A.; Attia, S.; Riedel, R.F.; Hu, J.S.; Davis, L.E.; Okuno, S.H.; Priebat, D.A.; et al. Clinical activity of pembrolizumab (P) in undifferentiated pleomorphic sarcoma (UPS) and dedifferentiated/pleomorphic liposarcoma (LPS): Final results of SARC028 expansion cohorts. *J. Clin. Oncol.* **2019**, *37* (Suppl. 15), 11015. [CrossRef]
125. D'Angelo, S.P.; Mahoney, M.R.; Van Tine, B.A.; Atkins, J.; Milhem, M.M.; Jahagirdar, B.N.; Antonescu, C.R.; Horvath, E.; Tap, W.D.; Schwartz, G.K.; et al. Nivolumab with or without ipilimumab treatment for metastatic sarcoma (Alliance A091401): Two open-label, non-comparative, randomised, phase 2 trials. *Lancet Oncol.* **2018**, *19*, 416–426. [CrossRef]
126. Solomon, J.P.; Benayed, R.; Hechtman, J.F.; Ladanyi, M. Identifying patients with NTRK fusion cancer. *Ann. Oncol.* **2019**, *30* (Suppl. 8), viii16–viii22. [CrossRef]

life

MDPI

Review

Protective Role of Lipid-Lowering Drugs in Breast Cancer: Effects on Cancer Incidence and Cardiotoxicity

Alfredo Mauriello [1,†], Adriana Correra [2,†], Anna Chiara Maratea [3], Celeste Fonderico [1], Arianna Amata [4], Valeria Cetoretta [5], Vincenzo Russo [6] and Antonello D'Andrea [7,*]

1 S.C. Cardiology, Institute National Cancer, IRCCS, Foundation "G. Pascale", Via M. Semmola 52, 80131 Naples, Italy; alfredo.mauriello93@libero.it (A.M.); celeste.fonderico@istitutotumori.na.it (C.F.)
2 Cardiology Department, Ospedali Riuniti University Hospital, Viale Pinto 1, 71122 Foggia, Italy; adrianacorrera@gmail.com
3 Department of Cardiovascular Disease, ASL Napoli 1 Centro, Via Comunale del Principe, 13a, 80145 Napoli, Italy; annachiara.maratea@gmail.com
4 Department of Cardiovascular Medicine, Fondazione Policlinico Universitario A. Gemelli, IRCCS, L.go A. Gemelli 8, 00168 Rome, Italy; arianna.amata2@gmail.com
5 Cardiology and Arrhythmology Clinic, University Hospital "Ospedali Riuniti", Marche Polytechnic University, Via Conca 71, 60126 Ancona, Italy; vcetoretta@gmail.com
6 Cardiology Unit, Department of Medical and Translational Sciences, Monaldi Hospital, University of Campania "Luigi Vanvitelli", Via L. Bianchi snc, 80131 Naples, Italy; vincenzo.russo@unicampania.it
7 Cardiology and Intensive Care Unit, Department of Cardiology, Umberto I Hospital, Via A. de Nicola 1, 84014 Nocera Inferiore, Italy
* Correspondence: antonellodandrea@libero.it
† These authors contributed equally to this work.

Abstract

The bidirectional link between cancer and cardiovascular disease presents a growing clinical challenge, often exacerbated by the cardiotoxic effects of many anti-cancer drugs. Simultaneously, the incidence of cardiovascular pathologies in cancer patients is on the rise. Recently, lipid-lowering drugs, particularly statins, have emerged not only as first-line cardiovascular drugs but also as potential agents with pleiotropic effects that could influence cancer progression and mitigate cardiotoxic side effects. We will analyze data from epidemiological studies, clinical trials, and laboratory research. This comprehensive review aims to examine and synthesize the current evidence regarding the use of lipid-lowering drugs in breast cancer patients, focusing on two main areas: their potential protective role in reducing the incidence of breast cancer and their impact on the prevention and management of treatment-induced cardiotoxicity.

Keywords: statins; chemotherapy; cardiotoxicity; cancer prevention; cardio-oncology; lipid-lowering drugs; PCSK9

1. Introduction

The growing number of cancer survivors has led to increased attention to the long-term sequelae of oncological treatment, particularly cardiovascular complications, which represent a significant source of morbidity and mortality [1]. In parallel, there has been a rising interest in the repurposing of lipid-lowering drugs, traditionally used for the prevention of atherosclerotic cardiovascular diseases (ASCVD) [2,3], for their pleiotropic effects that could influence both cancer progression and chemotherapy-related cardiotoxicity (CTRCD) [4]. Beyond their cholesterol-lowering effect, statins exert pleiotropic properties, including anti-inflammatory and antiproliferative effects [5,6]. These effects can attenuate the inflammatory response, inhibit tumor cell proliferation, and reduce cancer progression and micrometastasis [7,8].

The use of lipid-lowering drugs in oncological patients is focused on two main areas: the potential for chemoprevention or improvement of cancer prognosis, and protection against CTRCD. While preclinical investigations into the application of lipid-lowering drugs in the field of cardio-oncology are emerging [9], a critical need remains for dedicated clinical trials, especially those focusing on novel agents such as anti- proprotein convertase subtilisin/kexin type 9 (PCSK9) drugs. Cardiotoxicity related to drugs like anthracyclines and anti-HER2 agents is a significant clinical challenge that can lead to cardiac dysfunction and heart failure [10,11].

This narrative review examines the actual scientific evidence on the use of lipid-lowering drugs, particularly statins, ezetimibe, and PCSK9 inhibitors, in breast cancer patients. The review focuses on the dual action of these agents: the potential reduction of breast cancer incidence, recurrence, and mortality and the cardioprotective effects against toxicity induced by therapies such as anthracyclines, trastuzumab, and radiotherapy. Figure 1 illustrates the direct and indirect effects of lipid-lowering drugs on reducing breast cancer mortality and cardiotoxicity.

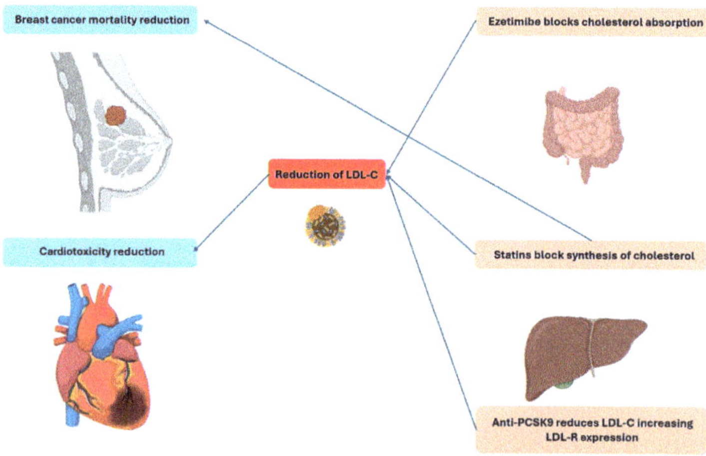

Figure 1. Relationship between lipid-lowering drugs and breast cancer mortality and cardiotoxicity. LDL-C: low-density-lipoprotein circulant; LDL-R: low-density-lipoprotein receptor; PCSK9: proprotein convertase subtilisin/kexin type 9.

2. Material and Methods

This article is structured as a narrative, clinically oriented review regarding the state of the art about the relationship between the use of lipid-lowering drugs and breast cancer. Its primary aim is to establish the pathophysiological link between use of lipid-lowering drugs and breast cancer and to explore potential preventive and therapeutic strategies in breast cancer patients with a rational synthesis of evidence, avoiding the production of a formal systematic review or meta-analysis.

To guarantee authoritativeness and transparency in the selection of sources, we used a search strategy that guided, but did not solely determine, the narrative synthesis. A comprehensive search was conducted in PubMed/MEDLINE, EMBASE and Web of Sciences for articles published from January 2015 to October 2025, employing juxtaposition of the following keywords and Boolean operators: "breast cancer" OR "cancer" AND ("lipid-lowering drugs" OR "statin" OR "ezetimibe" OR "PCSK9" OR "fibrates" OR "lipid"). The search was limited to English-language documents with visible abstracts.

Notably, we included prospective and retrospective clinical studies, randomized controlled trials, meta-analyses, and registry reports addressing the relationship between the use of lipid-lowering drugs and the risk of breast cancer and the incidence of risk of cardiotoxicity in breast cancer patients.

We excluded case reports, editorials, commentaries, comprehensive reviews, experimental pre-clinical studies without clear clinical implications, and conference abstracts without clinical complete data.

This review is not systematic, so no aggregate quantitative synthesis was performed. However, we assessed the quality of the included evidence by considering key domains relevant to clinical trials: patient selection (representativeness of study populations, inclusion/exclusion criteria); outcome ascertainment and follow-up (definition and measurement of cardiotoxicity or clinical endpoints); and control for confounding (adjustment for baseline cardiovascular risk factors, treatment exposure, and competing risks). For observational studies, we performed a qualitative appraisal based on the Joanna Briggs Institute (JBI) Critical Appraisal Checklists [12] and, when applicable, by the Cochrane Risk-of-Bias framework for randomized trials. The use of these aspects was used to define the relative weight assigned to different types of evidence in the discussion.

The search strategy we used was reproducible to ensure transparency. However, due to the heterogeneity of the study designs, which spanned everything from clinical trials and registries to models of organization and policy analyses, we could not follow the complete PRISMA guidelines. Specifically, we omitted steps like screening for duplicates, formally assessing the risk of bias, and performing a meta-analysis, because our main objective was the practical implementation and management of health systems. We determined that a PRISMA-based systematic review would be inappropriate for this work. To emphasize this methodological difference, we state explicitly that we performed a structured narrative synthesis based on the search results, rather than a systematic review. Throughout the text, we have made an effort to indicate the strength of the evidence, noting if conclusions originated from meta-analyses, extensive multicenter studies, or smaller observational series, so the reader can assess the credibility of the findings.

3. Molecular Mechanisms in Cancer Incidence Reduction

Most of the lipid-lowering drugs studied for anticancer activity, such as statins, act primarily on the mevalonate (MVA) pathway [13].

3.1. Statins

Statins are candidate drugs for repurposing in cancer treatment. Their primary mechanism is the inhibition of 3-hydroxy-3-methylglutaryl-CoA (HMG-CoA) reductase, the rate-limiting enzyme in the mevalonate pathway. The anticancer molecular mechanisms include:

3.1.1. Inhibition of Isoprenoids and Oncogenic Pathways

The MVA pathway is essential for the synthesis of cholesterol and isoprenoids, such as farnesyl diphosphate (FPP) and geranylgeranyl pyrophosphate (GGPP). Statins deplete intracell reserves of isoprenoids, which are important regulators of tumor cell proliferation and metastasis [14]. The inhibition of geranylgeranylation plays an important role in antitumoral activity, as it is involved in activating oncogenic kinase signaling [7]. In human epidermal growth factor receptor (HER)-2 positive breast cancer cells, the combination of statins with anti-HER2 agents induces apoptosis by inhibiting AKT and extracellular signal-regulated kinases (ERK) phosphorylation. This effect is mediated by the inhibition of geranylgeranylation, which restores the dephosphorylation of AKT and ERK [7]. The geranylgeranylated protein Rac1 indirectly activates the AKT and ERK pathways. Statins block the membrane localization of the geranylgeranylated Rac1 protein [7]. Figure 2 represents this mechanism.

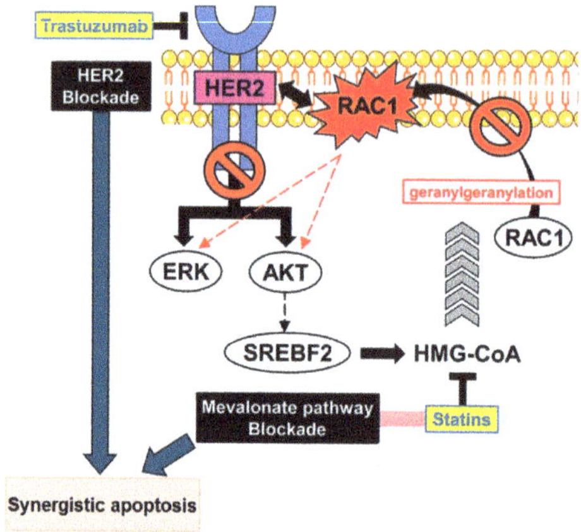

Figure 2. Illustration of synergistic action of trastuzumab and statins in induce apoptosis. In human epidermal growth factor receptor (HER)-2 positive breast cancer cells, the combination of statins with anti-HER2 agents induces apoptosis by inhibiting protein kinase B (AKT) and extracellular signal-regulated kinases phosphorylation. This effect is mediated by the inhibition of geranylgeranylation, which restores the dephosphorylation of AKT and ERK. The geranylgeranylated protein Ras-related C3 botulinum toxin substrate 1 (Rac1) indirectly activates the AKT and ERK pathways. AKT: protein kinase B; EKR: extracellular signal-regulated kinases; HER2: human epidermal growth factor receptor 2; HMG-CoA: β-Hydroxy β-methylglutaryl-CoA; Rac1: Ras-related C3 botulinum toxin substrate 1; SREBP-2: sterol regulatory element-binding protein 2.

3.1.2. Induction of Apoptosis and Ferroptosis

Combined treatment with statins and anti-HER2 agents significantly increases Caspase 3/7 activity, indicating the induction of apoptosis. In Triple-Negative Breast Cancer (TNBC) cells, simvastatin (SIM) can induce ferroptosis. This occurs because SIM inhibits HMG-CoA reductase expression, downregulating the MVA pathway and glutathione peroxidase 4 (GPX4), an enzyme that prevents cellular oxidative damage [15].

3.1.3. Immune Modulation and Anti-Inflammation

Inhibition of the HMG-CoA reductase target is associated with a reduced risk of breast cancer. This suggests that part of statins' action may occur through pathways not directly related to lipid lowering, such as modulating immune function and reducing inflammation. Statins can enhance the anti-tumor immune response by reducing cluster of differentiation (CD)8+ T cell exhaustion and suppressing the expression of programmed death-ligand 1 (PD-L1) [16].

3.2. Ezetimibe

Ezetimibe inhibits the absorption of cholesterol through the Niemann-Pick C1-Like 1 (NPC1L1) receptor [17]. It has shown anticancer mechanisms that include anti-angiogenesis effect by inhibiting neovascular growth, cell proliferation, and promoting apoptosis. Therefore, ezetimibe seems to have anti-inflammatory effects through promoting anti-cancer immunity by enhancing the memory function and infiltration of CD8+ T lymphocytes and reducing mTORC2 signaling in these lymphocytes [18].

4. Molecular Mechanisms in Cardiotoxicity Reduction

Lipid-lowering drugs, particularly statins and PCSK9 inhibitors, can be used to mitigate chemotherapy-induced cardiotoxicity [19].

4.1. Statins

Statins exert a cardioprotective effect through their pleiotropic properties:

4.1.1. Anti-Inflammatory and Antioxidant Effects

Statins determine anti-inflammatory and antioxidant benefits that improve maladaptive cardiac remodeling and myocardial fibrosis [20]. Some evidence suggests that rosuvastatin has antioxidant and anti-inflammatory effects that may help prevent deoxyribonucleic acid (DNA) damage and provide a cardioprotective effect [21]. Mechanisms include the ability of statins to reduce the gene expression of myeloperoxidase (MPO), a marker of inflammation and cardiotoxicity. Rosuvastatin, in randomized controlled trial, has been shown to reduce the levels of interleukin (IL)-6 and high-sensitivity cardiac troponin I (hs-cTnI) in 50 newly diagnosed HER2-positive breast cancer patients and chemotherapy-treated patients [22,23].

4.1.2. Inhibition of Fibrosis and Cardiomyocyte Protection

Preclinical studies suggest that statins preserve systolic function and calcium management and reduce oxidative stress and myocardial cell death during anthracycline exposure [24–26]. In a subgroup including 171 participants aged 52 years (53% men) in the Statins to Prevent the Cardiotoxicity From Anthracyclines (STOP-CA) trial [26], atorvastatin significantly reduced the likelihood of an increase in myocardial extracellular volume (ECV) in patients with regimen of anthracycline-based chemotherapy ($p < 0.001$) [27]. Statins inhibit Rho GTPases, reducing the production of reactive oxygen species (ROS) and limiting topoisomerase II-mediated DNA damage. These damages are caused by anthracycline-induced cardiotoxicity.

4.1.3. Protection Against Immune Checkpoint Inhibitor (ICI) Cardiotoxicity

High-dose atorvastatin has been shown to mitigate functional deficits by inhibiting coronary microvascular endothelial dysfunction, a phenomenon that precedes cardiotoxicity induced by pembrolizumab [28]. Therefore, atorvastatin prevented endothelial toxicity, decreasing endothelial activation markers like intercellular adhesion molecule 1 (ICAM-1) and vascular adhesion molecule 1 (VCAM-1) induced by pembrolizumab in in vitro models [28].

4.2. PCSK9 Inhibitors

PCSK9 inhibitors, including inclisiran, a silencing ribonucleic acid [29], and alirocumab [30] and evolocumab [31], two monoclonal antibodies, are known for their powerful lipid-lowering capabilities and demonstrate significant pleiotropic benefits [32].

4.2.1. Modulation of Inflammation and Oxidative Stress

Inclisiran attenuates doxorubicin (DOXO)- and trastuzumab-induced cardiotoxicity by reducing NLRP3 inflammasome activation and MyD88 signaling [9]. These pathways are essential mediators of cardiac inflammation. PCSK9i have been shown to significantly reduce intracellular ROS and lipid peroxidation products, malondialdehyde (MDA) and 4-hydroxy-2-nonenal (4-HNA) induced by combined chemotherapy [9]. Therefore,

PCSK9 inhibition suppresses the upregulation of pro-inflammatory cytokines such as IL-1a, IL-1b, IL-6, TNF-α, IL-12, and IL-17A, and increases the anti-inflammatory cytokine IL-10 [9].

4.2.2. Structural Protection and Apoptosis Prevention

Inclisiran exerts cardioprotective effects by improving cell viability and reducing apoptotic activation, decreasing in caspase-3 activity [9]. Furthermore, PCSK9is preserve mitochondrial integrity restoring mitofusin-2 levels and calcium homeostasis [9]. PCSK9 inhibition also protects against cytoskeleton disruption, preserving F-actin expression induced by sequential DOXO-trastuzumab therapy [9].

4.2.3. PCSK9 and Nuclear Transport

DOXO-induced cardiotoxicity upregulates PCSK9 expression in myocardial tissue [33]. The suggested mechanism is that DOXO-induced PCSK9 accumulates in the cardiomyocyte nucleus, where it interacts with the nuclear transport protein importin subunit beta-1 (KPNB1). This interaction is believed to influence the expression of apoptotic genes, such as Bax/Bcl-2 and cleaved Caspase-3/Pro-Caspase-3, thereby promoting cardiomyocyte apoptosis [33]. PCSK9 inhibitors downregulate the expression of these apoptotic proteins and the nuclear accumulation of PCSK9 [33].

5. Protective Effects of Lipid-Lowering Drugs in Reducing Breast Cancer Incidence and Progression and Protection Against Cardiotoxicity

5.1. Statins

5.1.1. Protective Effects on Cancer Incidence and Progression

Preclinical Studies

Statins have been shown to suppress growth and induce apoptosis in tumor cells. Lovastatin inhibits cancer growth and lung metastasis in a murine breast carcinoma model, through a mitochondrial-mediated, p53-independent apoptotic mechanism [34].

Simvastatin exhibited greater cytotoxicity against MDA-MB-231 cells, a type of triple-negative breast cancer model, than MCF-7 cells, a type of normal breast cancer, due to higher HMG-CoA-reductase expression. Simvastatin induces ferroptosis in TNBC cells by inhibiting HMG-CoA reductase, downregulating the MVA pathway, and deactivating GPX4 [35]. While simvastatin demonstrated effects on MDA-MB-231 and MCF-7 breast cancer cells, there is a clear gap between these in vitro findings and clinical applicability, given that the concentrations used were at least one order of magnitude higher than human therapeutic doses.

Statins block the membrane expression of the geranylgeranylated protein Rac1 in HER2-positive BC cells. Blocking HMG-CoA-reductase with statins enhances the efficacy of trastuzumab by modulating the phosphorylation status of PAK [7].

Simvastatin appears to radiosensitize stem cells and differentiated breast cancer cells. Additionally, simvastatin appears to reduce local inflammation caused by post-mastectomy radiation therapy [36].

In vitro and in vivo studies with simvastatin have demonstrated the induction of pyroptosis through activation of the ROS/caspase-1/GSDMD pathway in colon cancer cells, suggesting that statins may act as potential therapeutic agents for this malignancy. Simvastatin has also been shown to suppress cell proliferation in vitro [37].

Clinical Studies

The association between statin use and cancer outcomes has been the subject of numerous clinical studies and meta-analyses [38–42].

Scott et al. [38], in their meta-analysis including forty-one studies, concluded that there was a significant protective association between statin use and both breast cancer death (21 studies, hazard ratio (HR) = 0.81, 95% confidence interval (CI): 0.75–0.87; $p < 0.001$) and breast cancer recurrence (20 studies, HR = 0.81, 95% CI: 0.74–0.89, $p < 0.001$). In a cohort of elderly cancer survivors, the use of lipophilic statins was significantly associated with a lower risk of all-cause and cancer-related mortality (HR = 0.75, 95% CI: 0.62–0.91, $p = 0.003$).

In the Breast International Group (BIG) 1–98 trial [43], including 6193 postmenopausal women with hormone receptor–positive early breast cancer, the use of cholesterol-lowering drugs, including statins, was associated with improved outcomes in postmenopausal patients with early-stage hormone receptor-positive breast cancer (HR 0.79 for disease-free survival). The benefit appears to be more pronounced in ER+ patients [41].

On the other hand, a post hoc multivariate analysis in the Adjuvant Pertuzumab and Trastuzumab in Early HER2-Positive Breast Cancer (APHINITY) Trial [40], including 4804 patients with a median follow-up of 6.2 years, showed that statin use was not associated with invasive disease-free survival (IDFS) (HR, 1.11; 95% CI, 0.80–1.52), distant relapse-free interval (DRFI) (HR, 1.21; 95% CI, 0.81–1.81), or overall survival (OS) (HR, 1.16; 95% CI, 0.78–1.73).

5.1.2. Protective Effects Against Cardiotoxicity

Preclinical Studies and Mechanisms

Statins offer anti-inflammatory and antioxidant benefits that are important in mitigating chemotherapy-induced cardiotoxicity, particularly that caused by anthracyclines. Statins inhibit Rho guanosine triphosphate (GTP)-ases, reducing the production of ROS and limiting topoisomerase II-mediated DNA damage, both contributing factors to anthracycline-induced cardiotoxicity [44]. Therefore, atorvastatin protects cardiomyocytes from DOXO toxicity by modulating survivin expression through the inhibition of Forkhead box protein O1 (FOXO1) [45].

In murine models, rosuvastatin has shown potential for modulating oxidative stress and alleviating DOXO-induced cardiotoxicity. It has also been shown to target pro-inflammatory cytokines and oxidative stress to mitigate trastuzumab cardiotoxicity in rats [46].

Atorvastatin demonstrated a prophylactic role against pembrolizumab-induced cardiotoxicity in murine models. Coronary microvascular endothelial dysfunction precedes pembrolizumab-induced cardiotoxicity in vivo, and atorvastatin mitigates functional deficits by inhibiting this dysfunction. Atorvastatin prevented immune-related cytotoxicity in human endothelial cells [27].

Clinical Studies

Statin prophylaxis for CTRCD associated with anthracyclines has received attention, although with mixed results in randomized controlled trials.

Despite initially discordant results from some RCTs, statins have received a Class IIa recommendation in the 2022 European Society of Cardiology (ESC) guidelines for patients at high or very high risk of developing cancer therapy-related cardiovascular toxicity [10].

The STOP-CA [47], a multicenter, randomized, double-blind trial compared atorvastatin (40 mg/day) with placebo in 300 lymphoma patients treated with anthracycline-based chemotherapy. The study showed a significant reduction in the primary endpoint (absolute decline of $\geq$10% in left ventricular ejection fraction (LVEF) to a final value of <55% at 12 months) in the atorvastatin group compared to the placebo group (9% vs. 22%; $p = 0.002$).

A secondary study of the STOP-CA [27], using cardiac magnetic resonance (CMR), provided a mechanistic insight. Atorvastatin significantly reduced the odds of a clinically relevant increase in myocardial extracellular volume (ECV), a marker for expanded myocardial interstitial fibrosis, after anthracycline-based chemotherapy. The odds of a $\geq$3% increase in ECV were reduced by 80% in the atorvastatin group compared to the placebo group (OR: 0.20; $p = 0.002$). ECV expansion was associated with a clinically relevant decrease in LVEF.

Nabati et al. [48] in their single-blind RCT, including 89 women with newly diagnosed breast cancer, showed that rosuvastatin (20 mg) given 24 h before the first cycle of DOXO and continued for 6 months in HER2-positive breast cancer patients also treated with trastuzumab reduced the decline in LVEF, levels of Hs-cTnI and IL-6 at 3 and 6 months, and MPO at 6 months (intergroup $p = 0.012$).

On the other hand, clinical studies such as PREVENT [49], including 279 patients treated with anthracyclines predominantly for breast cancer (85%) and a smaller cohort of lymphoma patients (15%), and Statins for the primary prevention of heart failure in patients with cancer receiving anthracycline-based chemotherapy (SPARE-HF) [50], including 112 patients, also using atorvastatin (40 mg/day) in patients with solid tumors (mainly breast cancer, 65%) and hematologic malignancies, found no statistically significant difference in CTRCD (LVEF decline $\geq$ 10%) compared to placebo. The discrepancy in results may be due to differences in populations, chemotherapy regimens, and endpoints.

Statin therapy, particularly rosuvastatin, in a retrospective study [9] including 1471 patients, demonstrated a reduction in the risk of major adverse cardiovascular events (MACE) in breast cancer patients undergoing adjuvant radiotherapy (adjusted HR, 0.34 [95% CI, 0.25–0.44]). Hydrophilic statins, particularly rosuvastatin and pravastatin, demonstrated the most significant dose-dependent risk reduction. Indeed, increasing daily doses were associated with a lower risk of MACE over 5 years compared to statin non-users (12.24% versus 31.70%).

5.2. Ezetimibe

5.2.1. Protective Effects on Breast Cancer Incidence and Progression

Preclinical Studies and Mechanisms

Ezetimibe has shown potential for treating and preventing various types of cancer. Ezetimibe inhibits angiogenesis, apoptosis, proliferation, inflammation, stem cell inhibition and enhances the immune response. For example, ezetimibe reduces hepatic cholesterol levels and prevents steatosis-induced liver fibrosis and tumorigenesis. In a hypercholesterolemia murine model, it reduced breast tumor growth accelerated by hypercholesterolemia itself. Ezetimibe promotes anti-tumor immunity and attenuates prostate tumor growth and metastasis by suppressing Akt phosphorylation [51].

Clinical Studies

The clinical history of ezetimibe in oncology is initially linked to a controversy [52–55].

The Simvastatin Ezetimibe in Aortic Stenosis (SEAS) trial [52], a randomized, double-blind trial involving 1873 patients with mild-to-moderate, asymptomatic aortic stenosis, during a follow up of 4 years, had as primary outcome a composite of MACE, including death from cardiovascular causes, aortic-valve replacement, nonfatal myocardial infarction, hospitalization for unstable angina pectoris, heart failure, coronary-artery bypass grafting, percutaneous coronary intervention, and nonhemorrhagic stroke. The study suggested a possible increase in cancer incidence in the simvastatin–ezetimibe group (105 vs. 70, $p = 0.01$) and mortality (HR, 1.00; 95% CI, 0.84 to 1.18; $p = 0.97$).

Subsequent large-scale studies, such as the Study of Heart and Renal Protection (SHARP) [53] including 4650 patients, and IMPROVE-IT [54], including 18,144 patients, did not provide credible evidence of an increased cancer risk related to ezetimibe, used alone or in combination with simvastatin (respectively, $p = 0.89$ and $p = 0.68$). The prospective analysis of the IMPROVE-IT trial showed no effect on the incidence of malignancy [55].

Data regarding breast cancer are not available.

5.3. PCSK9 Inhibitors

Preclinical Studies

Despite the lack of published clinical studies on PCSK9 inhibition during chemotherapy in humans, the preclinical data are robust and indicate a mechanism of action independent of lowering lipids.

DOXO treatment increases the expression of the PCSK9 protein in myocardial tissue and induces its nuclear aggregation in H9C2 cardiomyocytes, suggesting its role in DOXO-induced cardiotoxicity [33].

Evolocumab improved cardiac function, myocardial damage, and fibrosis in DOXO-treated mice, showing a protective effect against DOXO-induced cardiotoxicity. The mechanism involves the reduction of PCSK9 expression and cardiomyocyte apoptosis, reducing in the Bax/Bcl-2 and Cleaved Caspase-3/Pro Caspase-3 ratios [56].

Inclisiran attenuates cardiotoxicity induced by sequential exposure to DOXO and trastuzumab in induced pluripotent stem cell-derived cardiomyocytes (hiPSC-CMs) [9]. Inclisiran significantly improved cell viability and reduced apoptotic activity. It significantly reduced cardiac damage and heart failure biomarkers, including heart-type fatty acid binding protein (H-FABP), troponin T and brain natriuretic peptide (BNP) [9]. Furthermore, inclisiran significantly reduces the intracellular expression of NLRP3 and MyD88, key components in inflammatory pathways involved in cardiotoxicity. This suggests a lipid-independent cardioprotective effect [9].

There are no available data regarding other lipid-lowering drugs.

6. Drug Interactions and Contraindications of Lipid-Lowering Drugs in Cancer Patients

The main limitations to statin use in cancer patients are the significant risk of drug–drug interactions (DDIs) with oncological therapies and the potential for hepatic dysfunction, either pre-existing or treatment-induced [57].

The primary concern stems from the fact that many statins and several chemotherapy agents, targeted therapies, and hormone therapies are metabolized by the same liver enzymes, primarily the cytochrome P450 (CYP450) system [58].

Many oncological drugs are CYP inhibitors. When taken concurrently, they can slow down the metabolism of the statin, leading to dangerously elevated plasma concentrations of the statin. This significantly increases the risk of statin-related side effects, particularly myopathy and, less commonly, hepatotoxicity. Conversely, a statin might affect the metabolism of an anticancer drug, potentially reducing its effectiveness or increasing its toxicity. To mitigate this risk, clinicians often prefer statins with minimal CYP interaction, such as pravastatin or rosuvastatin. Close therapeutic drug monitoring and dose adjustments are essential when concurrent use is unavoidable [59].

Statins carry a low risk of causing dose-dependent and reversible elevations in liver enzymes [60]. This risk becomes a major limitation in cancer patients due to several factors. Most chemotherapy and many newer targeted/immuno-therapies are themselves hepatotoxic. Adding a statin can potentially lead to cumulative liver injury, making it difficult to determine the source of transaminase elevation and potentially necessitating the interruption of essential cancer treatment. Cancer patients may already have compromised liver function due to liver metastases or underlying chronic conditions. Statins are generally contraindicated or require extreme caution in cases of acute liver disease, persistent and significant elevations of transaminases, or decompensated cirrhosis [61].

7. Future Perspectives

Despite the accumulation of preclinical and observational evidence supporting the repurposing of lipid-lowering drugs, particularly statins, generalized clinical implementation requires more robust data from randomized trials across heterogeneous populations.

The MAmmary cancer STatins in ER positive breast cancer (MASTER) trial [62], a phase III, randomized, placebo-controlled trial, is currently ongoing to assess whether adding atorvastatin (80 mg/day) to standard therapy improves IDFS in patients with ER-positive breast cancer. The results of this study will be essential in establishing whether statins can be integrated into oncology guidelines as a standard adjuvant therapy.

It is vital to understand the lipid-lowering independent anticancer mechanisms better. For example, in triple negative breast cancer, simvastatin induces ferroptosis, while in colon cancer, it induces pyroptosis. Future research should identify which patient subgroups benefit the most from the pleiotropic effects of statins (e.g., ER+ patients). The potential of statins to modulate inflammation and immune function (as suggested by MR studies on BC and ICI/Pembrolizumab models) requires further investigation.

Despite highly encouraging preclinical findings demonstrating that inclisiran and evolocumab can attenuate CTRCD (DOXO/trastuzumab) through NLRP3/MyD88 inhibition and reduced oxidative stress, independent of lipids, there's a lack of clinical studies on PCSK9 inhibition in chemotherapy patients. Prospective RCTs are urgently needed to evaluate the long-term impact of these agents on cardiac outcomes and their oncological safety.

The optimal dosage and timing for initiating cardioprotective therapy (before, during, or after chemotherapy or radiotherapy) are not yet fully defined. For atorvastatin, only the high dose has been studied for cardioprotection. Future studies should also explore the potential for synergy with other cardioprotective agents (such as beta-blockers or SGLT2i).

The conflicting evidence regarding metastatic colorectal cancer treated with regorafenib and the potential protective role in ICI toxicity suggests the need to study drug–drug interactions with targeted therapies and immunotherapy specifically.

Developing medications with reduced propensity for drug interactions and fewer adverse effects on metabolic clearance systems is a significant challenge, but one that is essential for optimizing care in cancer patients undergoing highly metabolically demanding chemotherapy.

8. Conclusions

Statins have shown robust preclinical anticancer effects through the modulation of the mevalonate pathway, and consistent clinical benefits in reducing cancer-related mortality and recurrence, especially for ER-positive breast cancer (ER+). Furthermore, statins have received a Class IIa recommendation in the 2022 ESC guidelines for the prevention of CTRCD in high-risk patients, supported by clinical studies demonstrating a significant reduction in cardiac dysfunction and myocardial fibrosis. PCSK9 inhibitors, inclisiran and evolocumab, although lacking definitive clinical data in cardio-oncology, have shown promising preclinical cardioprotective effects, independent of lowering lipids, through the modulation of oxidative stress and inflammatory pathways, such as NLRP3/MyD88.

Despite the promising data, clinical results are sometimes discordant due to the heterogeneity of study designs and breast cancer populations. It is fundamental that future studies, such as the MASTER trial, provide definitive randomized evidence to establish the optimal integration of these drugs into oncological and cardio-oncological clinical practice.

Author Contributions: Conceptualization, A.M. and A.C.; data curation, V.C.; writing—original draft preparation, A.M., A.C., C.F., A.C.M., V.C., A.D., V.R., A.A.; writing—review and editing, A.M., A.C., C.F., A.C.M., V.C., A.D., V.R., A.A.; supervision, A.D. All authors have read and agreed to the published version of the manuscript.

Funding: This research received no external funding.

Institutional Review Board Statement: Not applicable.

Informed Consent Statement: Not applicable.

Data Availability Statement: No new data were created or analyzed in this study. Data sharing is not applicable to this article.

Conflicts of Interest: The authors declare no conflicts of interest.

References

1. Siegel, R.L.; Kratzer, T.B.; Giaquinto, A.N.; Sung, H.; Jemal, A. Cancer statistics, 2025. *CA Cancer J. Clin.* **2025**, *75*, 10–45. [CrossRef] [PubMed]
2. Mach, F.; Baigent, C.; Catapano, A.L.; Koskinas, K.C.; Casula, M.; Badimon, L.; Chapman, M.; De Backer, G.; Delgado, V.; Ference, B.; et al. 2019 ESC/EAS Guidelines for the management of dyslipidaemias: Lipid modification to reduce cardiovascular risk. *Eur. Heart J.* **2020**, *41*, 111–188. [CrossRef] [PubMed]
3. Mauriello, A.; Maratea, A.C.; Fonderico, C.; Quagliariello, V.; Maurea, F.; Maurea, N. Factor XI and Cancer: Physiopathological Linkage and Clinical Perspectives. *J. Clin. Med.* **2025**, *14*, 6341. [CrossRef]
4. Mach, F.; Koskinas, K.C.; Roeters van Lennep, J.E.; Tokgözoğlu, L.; Badimon, L.; Baigent, C.; Benn, M.; Binder, C.J.; Catapano, A.; De Backer, G.; et al. 2025 Focused Update of the 2019 ESC/EAS Guidelines for the management of dyslipidaemias. *Eur. Heart. J.* **2025**, *46*, 4359–4378. [CrossRef]
5. Mauriello, A.; Correra, A.; Quagliariello, V.; Iovine, M.; Di Micco, P.; Imbalzano, E.; Giallauria, F.; Giordano, A.; Russo, V.; D'Andrea, A.; et al. Atrial Fibrillation and Cancer: Pathophysiological Mechanism and Clinical Implications. *J. Clin. Med.* **2025**, *14*, 5600. [CrossRef]
6. Russo, V.; Napolitano, N.; Ascrizzi, A.; Leonardi, S.; Pisacane, F.; Di Micco, P.; Imbalzano, E.; Sasso, F.C.; D'Andrea, A.; Caturano, A.; et al. The Lipid-Lowering Efficacy of a Nutraceutical Combination Including Leucoselect Phytosome, Red Yeast Rice, Policosanol and Folic Acid in Dyslipidaemia Patients: Real-World Insights. *Pharmaceuticals* **2024**, *17*, 447. [CrossRef]
7. Kato, C.; Iizuka-Ohashi, M.; Honda, M.; Konishi, E.; Yokota, I.; Boku, S.; Mizuta, N.; Morita, M.; Sakaguchi, K.; Taguchi, T.; et al. Additional statin treatment enhances the efficacy of HER2 blockade and improves prognosis in Rac1-high/HER2-positive breast cancer. *Biochim. Biophys. Acta (BBA)-Mol. Basis Dis.* **2024**, *1870*, 167458. [CrossRef]
8. Mauriello, A.; Correra, A.; Maratea, A.C.; Fonderico, C.; Amata, A.; Quagliariello, V.; Russo, V.; D'Andrea, A.; Maurea, N. The Crucial Role of Tinzaparin in Managing Venous Thromboembolism in the Cancer Population. *J. Clin. Med.* **2025**, *14*, 6695. [CrossRef]
9. Lyon, A.R.; López-Fernández, T.; Couch, L.S.; Asteggiano, R.; Aznar, M.C.; Bergler-Klein, J.; Borian, G.; Cardinale, D.; Cordoba, R.; Cosyns, B.; et al. 2022 ESC Guidelines on cardio-oncology developed in collaboration with the European Hematology Association (EHA), the European Society for Therapeutic Radiology and Oncology (ESTRO) and the International Cardio-Oncology Society (IC-OS). *Eur. Heart J.* **2022**, *43*, 4229–4361. [CrossRef]
10. Mauriello, A.; Ascrizzi, A.; Roma, A.S.; Molinari, R.; Caturano, A.; Imbalzano, E.; D'Andrea, A.; Russo, V. Effects of Heart Failure Therapies on Atrial Fibrillation: Biological and Clinical Perspectives. *Antioxidants* **2024**, *13*, 806. [CrossRef] [PubMed]
11. Roqué, M.; Martínez-García, L.; Solà, I.; Alonso-Coello, P.; Bonfill, X.; Zamora, J. Toolkit of methodological resources to conduct systematic reviews. *F1000Res* **2020**, *9*, 82. [CrossRef]
12. Iizuka-Ohashi, M.; Watanabe, M.; Sukeno, M.; Morita, M.; Hoang, N.T.H.; Kuchimaru, T.; Kizaka-Kondoh, S.; Sowa, Y.; Sakaguchi, K.; Taguchi, T.; et al. Blockage of the mevalonate pathway overcomes the apoptotic resistance to MEK inhibitors with suppressing the activation of Akt in cancer cells. *Oncotarget* **2018**, *9*, 19597–19612. [CrossRef]
13. Ashida, S.; Kawada, C.; Inoue, K. Stromal regulation of prostate cancer cell growth by mevalonate pathway enzymes HMGCS1 and HMGCR. *Oncol. Lett.* **2017**, *14*, 6533–6542. [CrossRef]

14. Liu, T.; Liu, W.; Zhang, M.; Yu, W.; Gao, F.; Li, C.; Wang, S.; Jun, F.; Zhang, X. Ferrous-Supply-Regeneration Nanoengineering for Cancer-Cell-Specific Ferroptosis in Combination with Imaging-Guided Photodynamic Therapy. *ACS Nano* **2018**, *12*, 12181–12192. [CrossRef] [PubMed]
15. Wang, T.; Yao, Y.; Gao, X.; Luan, H.; Wang, X.; Liu, L.; Sun, C. Genetic association of lipids and lipid-lowering drug target genes with breast cancer. *Discov. Oncol.* **2025**, *16*, 331. [CrossRef]
16. Leosdottir, M.; Schubert, J.; Brandts, J.; Gustafsson, S.; Cars, T.; Sundström, J.; Jernberg, T.; Ray, K.K.; Hagström, E. Early Ezetimibe Initiation After Myocardial Infarction Protects Against Later Cardiovascular Outcomes in the SWEDEHEART Registry. *J. Am. Coll. Cardiol.* **2025**, *85*, 1550–1564. [CrossRef] [PubMed]
17. Gu, J.; Zhu, N.; Li, H.-F.; Zhang, C.-J.; Gong, Y.-Z.; Liao, D.-F.; Qin, L. Ezetimibe and Cancer: Is There a Connection? *Front. Pharmacol.* **2022**, *13*, 831657. [CrossRef]
18. Kettana, K.M.; El-Haggar, S.M.; Alm El-Din, M.A.; El-Afify, D.R. Possible protective effect of rosuvastatin in chemotherapy-induced cardiotoxicity in HER2 positive breast cancer patients: A randomized controlled trial. *Med. Oncol.* **2024**, *41*, 196. [CrossRef]
19. Henriksen, P.A. Anthracycline cardiotoxicity: An update on mechanisms, monitoring and prevention. *Heart* **2018**, *104*, 971–977. [CrossRef] [PubMed]
20. Schupp, N.; Schmid, U.; Heidland, A.; Stopper, H. Rosuvastatin protects against oxidative stress and DNA damage in vitro via upregulation of glutathione synthesis. *Atherosclerosis* **2008**, *199*, 278–287. [CrossRef]
21. Katsurada, K.; Ichida, M.; Sakuragi, M.; Takehara, M.; Hozumi, Y.; Kario, K. High-sensitivity troponin T as a marker to predict cardiotoxicity in breast cancer patients with adjuvant trastuzumab therapy. *Springerplus* **2014**, *3*, 620. [CrossRef]
22. Alyasiry, E.; Janabi, A.; Hadi, N. Dipyridamole ameliorates doxorubicin-induced cardiotoxicity. *J. Med. Life* **2022**, *15*, 1184–1190. [CrossRef]
23. Kim, Y.-H.; Park, S.-M.; Kim, M.; Kim, S.H.; Lim, S.-Y.; Ahn, J.-C.; Song, W.; Shim, W. Cardioprotective effects of rosuvastatin and carvedilol on delayed cardiotoxicity of doxorubicin in rats. *Toxicol. Mech. Methods* **2012**, *22*, 488–498. [CrossRef]
24. Riad, A.; Bien, S.; Westermann, D.; Becher, P.M.; Loya, K.; Landmesser, U.; Schultheiss, H.; Tschöpe, C. Pretreatment with Statin Attenuates the Cardiotoxicity of Doxorubicin in Mice. *Cancer Res.* **2009**, *69*, 695–699. [CrossRef]
25. SVVSR; Trivedi, P.P.; Kushwaha, S.; Vikram, A.; Jena, G.B. Protective role of atorvastatin against doxorubicin-induced cardiotoxicity and testicular toxicity in mice. *J. Physiol. Biochem.* **2013**, *69*, 513–525. [CrossRef] [PubMed]
26. Juhasz, V.; Quinaglia, T.; Drobni, Z.D.; Heemelaar, J.C.; Neuberg, D.S.; Han, Y.; Ky, B.; Kwong, R.Y.; Januzzi, J.L.; Asnani, A.; et al. Atorvastatin and Myocardial Extracellular Volume Expansion During Anthracycline-Based Chemotherapy. *JACC CardioOncol* **2025**, *7*, 125–137. [CrossRef] [PubMed]
27. Efentakis, P.; Choustoulaki, A.; Kwiatkowski, G.; Varela, A.; Kostopoulos, I.V.; Tsekenis, G.; Ntanasis-Stathopoulos, I.; Georgoulis, A.; Vorgias, C.E.; Gakiopoulou, H.; et al. Early microvascular coronary endothelial dysfunction precedes pembrolizumab-induced cardiotoxicity. Preventive role of high dose of atorvastatin. *Basic Res. Cardiol.* **2024**, *120*, 263–286. [CrossRef]
28. Ray, K.K.; Kallend, D.; Leiter, L.A.; Raal, F.J.; Koenig, W.; Jaros, M.J.; Schwartz, G.; Landmesser, U.; Garcia Conde, L.; Scott Wright, R.; et al. Effect of inclisiran on lipids in primary prevention: The ORION-11 trial. *Eur. Heart J.* **2022**, *43*, 5047–5057. [CrossRef]
29. Goodman, S.G.; Steg, P.G.; Poulouin, Y.; Bhatt, D.L.; Bittner, V.A.; Diaz, R.; Garon, G.; Harrington, R.A.; Wouter Jukema, J.; Manvelian, G.; et al. Long-Term Efficacy, Safety, and Tolerability of Alirocumab in 8242 Patients Eligible for 3 to 5 Years of Placebo-Controlled Observation in the ODYSSEY OUTCOMES Trial. *J. Am. Heart Assoc.* **2023**, *12*, e029216. [CrossRef]
30. Sabatine, M.S.; Giugliano, R.P.; Keech, A.C.; Honarpour, N.; Wiviott, S.D.; Murphy, S.A.; Kuder, J.; Wang, H.; Liu, T.; Wasserman, S.M.; et al. Evolocumab and Clinical Outcomes in Patients with Cardiovascular Disease. *N. Engl. J. Med.* **2017**, *376*, 1713–1722. [CrossRef] [PubMed]
31. Mauriello, A.; Correra, A.; Maratea, A.C.; Caturano, A.; Liccardo, B.; Perrone, M.A.; Giordano, A.; Nigro, G.; D'Andrea, A.; Russo, V. Serum Lipids, Inflammation, and the Risk of Atrial Fibrillation: Pathophysiological Links and Clinical Evidence. *J. Clin. Med.* **2025**, *14*, 1652. [CrossRef]
32. Quagliariello, V.; Berretta, M.; Bisceglia, I.; Iovine, M.; Barbato, M.; Arianna, R.; Canale, M.L.; Paccone, A.; Inno, A.; Scherillo, M.; et al. PCSK9 Inhibitor Inclisiran Attenuates Cardiotoxicity Induced by Sequential Anthracycline and Trastuzumab Exposure via NLRP3 and MyD88 Pathway Inhibition. *Int. J. Mol. Sci.* **2025**, *26*, 6617. [CrossRef]
33. Shi, S.; Qin, Z.; Liu, C.; Zhao, Y.; Bai, X.; Sun, C.; Li, X.; Cong, W.; Yuan, X.; Sun, L.; et al. The function of PCSK9 in doxorubicin-induced cardiotoxicity and its underlying mechanism. *Sci. Rep.* **2025**, *15*, 22067. [CrossRef] [PubMed]
34. Shibata, M.-A. Lovastatin inhibits tumor growth and lung metastasis in mouse mammary carcinoma model: A p53-independent mitochondrial-mediated apoptotic mechanism. *Carcinogenesis* **2004**, *25*, 1887–1898. [CrossRef] [PubMed]
35. Tripathi, V.; Jaiswal, P.; Verma, R.; Sahu, K.; Majumder, S.K.; Chakraborty, S.; Jha, H.; Parmar, H.S. Therapeutic influence of simvastatin on MCF-7 and MDA-MB-231 breast cancer cells via mitochondrial depletion and improvement in chemosensitivity of cytotoxic drugs. *Adv. Cancer Biol.-Metastasis* **2023**, *9*, 100110. [CrossRef]

36. Lacerda, L.; Reddy, J.P.; Liu, D.; Larson, R.; Li, L.; Masuda, H.; Brewer, T.; Debeb, B.G.; Xu, W.; Hortobágyi, G.N.; et al. Simvastatin Radiosensitizes Differentiated and Stem-Like Breast Cancer Cell Lines and Is Associated with Improved Local Control in Inflammatory Breast Cancer Patients Treated with Postmastectomy Radiation. *Stem Cells Transl. Med.* **2014**, *3*, 849–856. [CrossRef] [PubMed]
37. Xie, W.; Peng, M.; Liu, Y.; Zhang, B.; Yi, L.; Long, Y. Simvastatin induces pyroptosis via ROS/caspase-1/GSDMD pathway in colon cancer. *Cell Commun. Signal* **2023**, *21*, 329. [CrossRef]
38. Scott, O.W.; Tin, S.T.; Cavadino, A.; Elwood, J.M. Statin use and breast cancer-specific mortality and recurrence: A systematic review and meta-analysis including the role of immortal time bias and tumour characteristics. *Br. J. Cancer* **2025**, *133*, 539–554. [CrossRef]
39. von Minckwitz, G.; Procter, M.; de Azambuja, E.; Zardavas, D.; Benyunes, M.; Viale, G.; Suter, T.; Arahmani, A.; Rouchet, N.; Clark, E.; et al. Adjuvant Pertuzumab and Trastuzumab in Early HER2-Positive Breast Cancer. *N. Engl. J. Med.* **2017**, *377*, 122–131. [CrossRef]
40. Maurer, C.; Agostinetto, E.; Ameye, L.; Lambertini, M.; Martel, S.; Ponde, N.; Brandão, M.; Poggio, F.; Ferreira, A.; Schiff, R.; et al. Association of statin use on survival outcomes of patients with early-stage HER2-positive breast cancer in the APHINITY trial. *Breast Cancer Res. Treat.* **2025**, *212*, 57–69. [CrossRef]
41. Chirgwin, J.H.; Giobbie-Hurder, A.; Coates, A.S.; Price, K.N.; Ejlertsen, B.; Debled, M.; Gelber, R.D.; Goldhirsch, A.; Smith, I.; Rabaglio, M.; et al. Treatment Adherence and Its Impact on Disease-Free Survival in the Breast International Group 1-98 Trial of Tamoxifen and Letrozole, Alone and in Sequence. *J. Clin. Oncol.* **2016**, *34*, 2452–2459. [CrossRef]
42. Yu, J.-M.; Chang, C.-L.; Lin, K.-C.; Chen, W.-M.; Shia, B.-C.; Wu, S.-Y. Statin Use During Concurrent Chemoradiotherapy for Advanced Nasopharyngeal Cancer. *J. Natl. Compr. Cancer Netw.* **2024**, *22*, e247046. [CrossRef]
43. Breast International Group (BIG) 1-98 Collaborative Group; Thürlimann, B.; Keshaviah, A.; Coates, A.S.; Mouridsen, H.; Mauriac, L.; Forbes, J.F.; Paridaens, R.; Castiglione-Gertsch, M.; Gelber, R.D.; et al. A Comparison of Letrozole and Tamoxifen in Postmenopausal Women with Early Breast Cancer. *N. Engl. J. Med.* **2005**, *353*, 2747–2757. [CrossRef]
44. Meléndez, G.C.; Riojas, A.M. Unveiling the Mechanisms of Anthracycline Cardiotoxicity. *JACC CardioOncol* **2025**, *7*, 138–140. [CrossRef]
45. Oh, J.; Lee, B.S.; Lim, G.; Lim, H.; Lee, C.J.; Park, S.; Lee, S.; Chung, J.H.; Kang, S.M. Atorvastatin protects cardiomyocyte from doxorubicin toxicity by modulating survivin expression through FOXO1 inhibition. *J. Mol. Cell Cardiol.* **2020**, *138*, 244–255. [CrossRef]
46. Rajangam, J.; Krishnan, N.; Palei, N.N.; Bhatt, S.; Das, M.K.; Das, S.; Mathusoothanan, K. Ameliorative Potential of Rosuvastatin on Doxorubicin-induced Cardiotoxicity by Modulating Oxidative Damage in Rats. *Turk. J. Pharm. Sci.* **2022**, *19*, 28–34. [CrossRef] [PubMed]
47. Neilan, T.G.; Quinaglia, T.; Onoue, T.; Mahmood, S.S.; Drobni, Z.D.; Gilman, H.K.; Smith, A.; Heemelaar, J.C.; Brahmbhatt, P.; Jor Ho, S.; et al. Atorvastatin for Anthracycline-Associated Cardiac Dysfunction. *JAMA* **2023**, *330*, 528–536. [CrossRef] [PubMed]
48. Nabati, M.; Janbabai, G.; Esmailian, J.; Yazdani, J. Effect of Rosuvastatin in Preventing Chemotherapy-Induced Cardiotoxicity in Women with Breast Cancer: A Randomized, Single-Blind, Placebo-Controlled Trial. *J. Cardiovasc. Pharmacol. Ther.* **2019**, *24*, 233–241. [CrossRef] [PubMed]
49. Hundley, W.G.; D'Agostino, R.; Crotts, T.; Craver, K.; Hackney, M.H.; Jordan, J.H.; Ky, B.; Wagner, L.I.; Herrington, D.M.; Yeboah, J.; et al. Statins and Left Ventricular Ejection Fraction Following Doxorubicin Treatment. *NEJM Evid.* **2022**, *1*, evidoa2200097. [CrossRef]
50. Thavendiranathan, P.; Houbois, C.; Marwick, T.H.; Kei, T.; Saha, S.; Runeckles, K.; Huang, F.; Shalmon, T.; Thorpe, K.E.; Pezo, R.; et al. Statins to prevent early cardiac dysfunction in cancer patients at increased cardiotoxicity risk receiving anthracyclines. *Eur. Heart J. Cardiovasc. Pharmacother.* **2023**, *9*, 515–525. [CrossRef]
51. Vieira, R.F.; Sanchez, S.R.; Arumugam, M.; Mower, P.D.; Curtin, M.C.; Jackson, A.E.; Wright, J.; Bowles, A.; Ducker, G.S.; Hilgendorf, K.I.; et al. Hyperlipidemia drives tumor growth in a mouse model of obesity-accelerated breast cancer growth. *Cancer Metab.* **2025**, *13*, 39. [CrossRef] [PubMed]
52. Rossebø, A.B.; Pedersen, T.R.; Boman, K.; Brudi, P.; Chambers, J.B.; Egstrup, K.; Gerdts, E.; Gohlke-Bärwolf, C.; Holme, I.; Kesäniemi, Y.A.; et al. Intensive Lipid Lowering with Simvastatin and Ezetimibe in Aortic Stenosis. *N. Engl. J. Med.* **2008**, *359*, 1343–1356. [CrossRef]
53. Baigent, C.; Landray, M.J.; Reith, C.; Emberson, J.; Wheeler, D.C.; Tomson, C.; Wanner, C.; Krane, V.; Cass, A.; Craig, J.; et al. The effects of lowering LDL cholesterol with simvastatin plus ezetimibe in patients with chronic kidney disease (Study of Heart and Renal Protection): A randomised placebo-controlled trial. *Lancet* **2011**, *377*, 2181–2192. [CrossRef]
54. Cannon, C.P.; Blazing, M.A.; Giugliano, R.P.; McCagg, A.; White, J.A.; Theroux, P.; Darius, H.; Lewis, B.S.; Oude Ophuis, T.; Jukema, J.W.; et al. Ezetimibe Added to Statin Therapy after Acute Coronary Syndromes. *N. Engl. J. Med.* **2015**, *372*, 2387–2397. [CrossRef]

55. Pradhan, A.; Bhandari, M.; Sethi, R. Ezetimibe and Improving Cardiovascular Outcomes: Current Evidence and Perspectives. *Cardiol. Res. Pract.* **2020**, *2020*, 9815016. [CrossRef]
56. Ahmed, E.I.; Moneam Shamardl, H.A.; Elsayed, A.M.; Sadik, S.A. Evolocumab ameliorates myocardial fibrosis and improves metabolic syndrome–induced cardiac dysfunction in rats via inhibiting PCSK9/NLRP3 inflammasome and Caspase-1/IL-1β pathways. *Eur. J. Pharmacol.* **2025**, *998*, 177499. [CrossRef] [PubMed]
57. Radhakrishnan, A.; Kuppusamy, G.; Ponnusankar, S.; Shanmukhan, N.K. Pharmacogenomic phase transition from personalized medicine to patient-centric customized delivery. *Pharmacogenomics J.* **2020**, *20*, 1–18. [CrossRef] [PubMed]
58. Asiimwe, I.G.; Pirmohamed, M. Drug–Drug–Gene Interactions in Cardiovascular Medicine. *Pharmgenomics Pers. Med.* **2022**, *15*, 879–911. [CrossRef]
59. Chen, Z.; Wu, P.; Wang, J.; Chen, P.; Fang, Z.; Luo, F. The association of statin therapy and cancer: A meta-analysis. *Lipids Health Dis.* **2023**, *22*, 192. [CrossRef]
60. Vell, M.S.; Loomba, R.; Krishnan, A.; Wangensteen, K.J.; Trebicka, J.; Creasy, K.T.; Trautwein, C.; Scorletti, E.; Seeling, K.S.; Hehl, L.; et al. Association of Statin Use with Risk of Liver Disease, Hepatocellular Carcinoma, and Liver-Related Mortality. *JAMA Netw. Open* **2023**, *6*, e2320222. [CrossRef]
61. Zhou, L.; Wu, B.; Bian, Y.; Lu, Y.; Zou, Y.; Lin, S.; Li, Q.; Liu, C. Hepatotoxicity associated with statins: A retrospective pharmacovigilance study based on the FAERS database. *PLoS ONE* **2025**, *20*, e0327500. [CrossRef] [PubMed]
62. Borgquist, S.; Jensen, M.-B.; Bendorff, C.; Christiansen, P.; Offersen, B.; Kodahl, A.; Ewertz, M.; Bonde Jensen, A.; Ahern, T.P.; Cronin-Fenton, D.; et al. Statin Therapy in Early Breast Cancer: The MASTER Trial; A Randomized Phase III, Placebo-Controlled Comparison of Standard (Neo)Adjuvant Therapy Plus Atorvastatin versus Standard (Neo)Adjuvant Therapy Plus Placebo. *Clin. Epidemiol.* **2025**, *17*, 409–419. [CrossRef] [PubMed]

life MDPI

Review

Breast-Conserving Surgery in *BRCA* Mutation Carriers: A Paradigm Shift Toward Individualized, Multidisciplinary Care

Calogero Cipolla [1,2], Giuseppa Scandurra [3,4], Daniela Sambataro [3,5], Chiara Mesi [6], Martina Greco [6], Eleonora D'Agati [6], Vittorio Gebbia [3,7,*], Luca Giacomelli [8] and Maria Rosaria Valerio [6,9]

1 Breast Unit, AOUP Paolo Giaccone, 90127 Palermo, Italy; calogero.cipolla@unipa.it
2 Department of Biomedicina, Neuroscienze e Diagnostica Avanzata, University of Palermo, 90127 Palermo, Italy
3 Medical Oncology, Department of Medicine and Surgery, Kore University of Enna, 94019 Enna, Italy; giuseppa.scandurra@unikore.it (G.S.); daniela.sambataro@unikore.it (D.S.)
4 Medical Oncology Unit, Ospedale Cannizzaro, 94126 Catania, Italy
5 Medical Oncology Unit, Ospedale Umberto I, 94019 Enna, Italy
6 UOC Medical Oncology, AOUP Paolo Giaccone Palermo, 90127 Palermo, Italy; chiaramesi95@gmail.com (C.M.); doc.martinagreco@gmail.com (M.G.); dagatieleonora@gmail.com (E.D.); mariarosaria.valerio@unipa.it (M.R.V.)
7 Medical Oncology Unit, CDC Torina, 90100 Palermo, Italy
8 Polistudium SRL, 20121 Milan, Italy; luca.giacomelli@polistudium.it
9 Department of Precision Medicine in Critial Area (MEDPREC), University of Palermo, 90127 Palermo, Italy
* Correspondence: vittorio.gebbia@unikore.it

Abstract

Breast cancer associated with *BRCA1* and *BRCA2* mutations presents unique therapeutic challenges, traditionally favoring mastectomy due to concerns over recurrence and new primaries. However, evolving evidence and advances in multimodal therapy have reshaped this paradigm, positioning breast-conserving surgery (BCS) as a viable option for selected carriers. This narrative review synthesizes current data from meta-analyses, retrospective cohorts, and pivotal studies, including a multicenter analysis which affirmed oncologic equivalence between BCS and mastectomy when combined with radiotherapy and systemic therapy. While meta-analyses confirm higher local events following BCS, survival remains comparable, indicating that recurrence reflects genetic predisposition rather than surgical inadequacy. Optimized systemic treatments, including chemotherapy, endocrine therapy, risk-reducing salpingo-oophorectomy, and PARP inhibitors, further mitigate recurrence risk. Meanwhile, patient-centered outcomes favor BCS: studies consistently link it to improved body image, psychosocial well-being, and quality of life, especially for younger *BRCA* carriers. Fertility-preserving options remain viable, with evidence supporting the safety of pregnancy, breastfeeding, and assisted reproductive technologies in *BRCA*-mutated survivors. These findings support individualized surgical planning for *BRCA* carriers within multidisciplinary care, balancing oncologic safety, systemic strategies, and psychosocial priorities. BCS should be considered a standard option for well-selected patients in hereditary breast cancer management.

Keywords: *BRCA* mutation; breast-conserving surgery; hereditary breast cancer; radiotherapy; systemic therapy; quality of life; psychosocial outcomes

1. Introduction

Breast cancer is the most common malignancy among women worldwide, with 5–10% of cases attributable to hereditary predisposition, primarily due to pathogenic variants in the *BRCA1* and *BRCA2* genes [1,2]. These genes are crucial for homologous recombination repair, and their dysfunction results in genomic instability and substantially elevated lifetime risks of breast cancer, contralateral breast cancer, and ovarian cancer [3]. *BRCA*-associated breast cancers often exhibit distinct biological characteristics, including higher rates of triple-negative phenotype and early onset, necessitating tailored therapeutic approaches [4].

Historically, the detection of a *BRCA* mutation often prompted bilateral mastectomy as a therapeutic or prophylactic strategy, driven by concerns over local recurrence and new primaries in the conserved breast [3,5]. While risk-reducing mastectomy effectively lowers incidence, its significant physical and psychosocial impact has led to growing interest in breast-conserving surgery (BCS) for selected patients [3]. Current guidelines from the American Society of Clinical Oncology, the American Society for Radiation Oncology, and the Society of Surgical Oncology endorse BCS as an acceptable option for *BRCA* carriers meeting conventional criteria, if radiotherapy (RT) and systemic therapies are appropriately integrated [6].

Emerging evidence supports the oncologic feasibility of BCS in *BRCA* mutation carriers, with studies demonstrating comparable survival outcomes with mastectomy when adjuvant treatments are optimized [7,8]. Nevertheless, concerns persist regarding higher ipsilateral and contralateral breast cancer risks and the potential impact of RT in this genetically predisposed population [4,9]. Despite these increased risks, recent large cohort studies highlight that most *BRCA* carriers treated with BCS remain event-free over long-term follow-up and avoid subsequent mastectomy [9].

However, the suitability and safety of BCS in patients with *BRCA1* or *BRCA2* pathogenic variants remain relatively uncertain, reflecting heterogeneity in study designs, follow-up durations, and evolving treatment paradigms [10]. Furthermore, the expanding availability of genetic testing and the introduction of targeted therapies, such as poly(ADP-ribose) polymerase (PARP) inhibitors, underscore the need to re-examine surgical decision-making within a modern multidisciplinary context, considering fertility preservation [6,9,11].

Given these factors, there is a pressing need to synthesize current evidence on BCS for *BRCA* mutation carriers, critically appraise its oncologic and quality-of-life (QoL) implications, and clarify areas where knowledge gaps persist. This review aims to provide a comprehensive evaluation of the literature to inform clinical decision-making and guide personalized management strategies for this high-risk population.

2. Search Strategy

For this narrative review, relevant studies were identified through targeted searches of the PubMed, Scopus, and Web of Science databases. A Mac Microsoft Word, Version 16.102.2 (25102623) licenced to V.G. was employed to write the paper. We used a combination of free-text keywords and Medical Subject Headings (MeSH) to capture a comprehensive set of relevant articles. The core search terms included "*BRCA* mutations", "breast-conserving surgery", "breast-conserving therapy", "mastectomy", "local recurrence", and "survival outcomes". Boolean operators (AND, OR) were used to refine the search strategy. We included articles published up to August 2025. Only English-language publications were considered. This exclusion was based on concerns about potential translation inaccuracies and interpretation variability, which may introduce bias in a narrative synthesis. References cited in the retrieved articles were screened manually to identify additional relevant sources.

3. Evidence from Meta-Analyses

Systematic reviews and meta-analyses provide the highest-level evidence currently available on the comparative safety of BCS versus mastectomy in *BRCA* mutation carriers. Given the absence of randomized controlled trials in this setting, these pooled analyses are critical for informing surgical decision-making, offering insights into long-term local recurrence risk, contralateral breast cancer incidence, and survival outcomes (Table 1) [7,8,12,13].

Table 1. Summary of key meta-analyses comparing BCS and mastectomy in *BRCA* mutation carriers.

Study (Year)	Design	*n* (*BRCA* Carriers)	Follow-Up (Median)	LRR/IBTR Risk (BCS vs. Mastectomy)	OS Difference	Contralateral BC Risk
Nara 2022 [12]	Meta-analysis	701	≥10 years	RR 1.59 (95% CI, 1.25–2.02)	Qualitative: No deterioration in OS	Not assessed
Davey 2021 [13]	Meta-analysis	3807	Up to 15 years	HR 4.54 (95% CI, 2.77–7.42)	HR 1.10 (95% CI, 0.72–1.69)	HR 1.51 (95% CI, 0.44–5.11)
Co 2020 [7]	Systematic review	2157	5–15 years	LRR 8.2%, 15.5%, 23% vs. 3.4%, 4.9%, 6.4%	5-, 10-, 15-year OS comparable	Not reported
Wang 2022 [8]	Meta-analysis	1254	Variable	HR 3.84 (95% CI, 2.38–6.20)	HR 1.02 (95% CI, 0.77–1.36)	Not significant

BC, breast cancer; BCS, breast-conserving surgery; HR, hazard ratio; IBTR, ipsilateral breast tumor recurrence; LRR, locoregional recurrence; OS, overall survival; RR, risk ratio.

3.1. Local Recurrence Risk

Several meta-analyses have consistently reported that BCS in *BRCA* mutation carriers is associated with an increased risk of local recurrence compared with mastectomy. Nara et al. synthesized data from 13 studies including 701 *BRCA* carriers and 4788 controls, demonstrating a significantly higher rate of ipsilateral breast tumor recurrence (IBTR) in carriers undergoing BCS (risk ratio [RR] 1.59; 95% CI, 1.25–2.02; $p < 0.001$), with the effect more pronounced at extended follow-up (≥7 years: RR 1.50; ≥10 years: RR 1.60) [12]. Davey et al., in the largest meta-analysis to date ($n = 3807$ carriers across 23 studies), reported a pooled hazard ratio (HR) of 4.54 (95% CI, 2.77–7.42) for locoregional recurrence (LRR) after BCS compared with mastectomy, with cumulative LRR rates of 14.7%, 15.5%, and 27.5% at 5, 10 and 15 years for BCS versus 4.8%, 4.7%, and 6.2% for mastectomy [13]. Similarly, Co et al. documented pooled ipsilateral recurrence rates of 8.2% at 5 years, 15.5% at 10 years, and 23% at 15 years for BCS, compared with 3.4%, 4.9% and 6.4% for mastectomy [7].

3.2. Contralateral Breast Cancer and Survival Outcomes

Despite increased local recurrence risk, survival outcomes appear unaffected. Wang et al. pooled four studies comprising 1254 *BRCA*-mutated patients [8]. They confirmed no significant differences between BCS and mastectomy in disease-free survival (DFS), metastasis-free survival, breast cancer-specific survival, or overall survival (OS), with HRs consistently near unity (e.g., OS: HR 1.02; 95% CI, 0.77–1.36) (Wang 2022 [8]). Davey et al. similarly reported equivalent 15-year OS between BCS (22.2% mortality) and mastectomy (23.9% mortality; HR 1.10; 95% CI, 0.72–1.69) [13]. Importantly, Nara et al. performed only a qualitative synthesis of OS data across three cohort studies, reporting no evidence of deterioration in survival among *BRCA* carriers treated with BCS [12]. Co et al. corroborated these findings, demonstrating overlapping 15-year OS rates (83.6% for BCS vs. 83.2% for mastectomy) and equivalent contralateral breast cancer risk (HR 1.51; 95% CI, 0.44–5.11) [7].

All meta-analyses emphasize the absence of randomized trials directly comparing BCS and mastectomy in *BRCA* carriers, reflecting ethical and logistical barriers to such studies. Consequently, available evidence derives from retrospective designs subject to selection bias, variability in systemic therapy use, and evolving surgical practices [7,13]. Nonetheless, the consistent observation of survival equivalence across pooled studies supports BCS as an oncologically safe alternative for selected *BRCA* carriers, if patients receive careful counseling on recurrence risk and undergo long-term surveillance.

Taken together, these meta-analyses establish the foundation for current understanding of BCS outcomes in *BRCA* carriers and highlight the need to examine complementary evidence from large retrospective cohort studies, which provide granular insights into real-world practice patterns, treatment era effects, and long-term follow-up.

4. Retrospective and Historical Cohort Studies in *BRCA* Mutation Carriers

In addition to meta-analyses, retrospective and historical cohort studies provide essential complementary evidence on the oncologic safety of BCS in *BRCA* mutation carriers. Spanning over four decades, these studies capture the evolution of clinical practice from the pre-genetic testing era to the modern framework of precision oncology. They collectively illustrate how improvements in early detection, systemic therapy, and genetic counseling have shaped patterns of recurrence and survival, offering a nuanced understanding of surgical decision-making for this high-risk group (Table 2) [9,10,14–19].

Table 2. Summary of retrospective and historical cohort studies evaluating BCS versus mastectomy in *BRCA* mutation carriers.

Study (Year)	Design	*n* (*BRCA* Carriers)	Follow-Up (Median)	Key Findings
Pierce 2010 [14]	Multi-institutional cohort	655 (302 BCS; 353 mastectomy)	8–15 years	15-year IBTR: 23.5% (BCS) vs. 5.5% (mastectomy); no OS difference (10-year OS: 92.1% vs. 91.8%); most IBTRs were new primaries.
Nilsson 2014 [15]	Single-institution cohort	162 (45 BCS; 117 mastectomy)	12.9 years	15-year IBTR: 32% (BCS) vs. 9% (mastectomy); HR 2.9 (95% CI, 1.1–7.8); OS equivalent between groups.
van den Broek 2019 [16]	Population-based cohort	191 *BRCA1* carriers (mixed control cohort)	10 years	IBTR after BCS: 7.3% (*BRCA1*) vs. 7.9% (non-carriers); OS: HR 0.80 (95% CI, 0.42–1.51); comparable outcomes with non-carriers.
Gentile 2022 [17]	Single-institution cohort	124 (69 BCS; 55 mastectomy)	~10 years	No difference in DFS, distant DFS, or OS; BSO significantly improved distant DFS and OS ($p = 0.033$ and $p = 0.040$).
Shubeck 2022 [18]	Single-institution cohort	395 (149 BCS; 246 mastectomy)	7.9 years	10-year LRR: 7.3% (BCS) vs. 6.3% (mastectomy); OS: 88.8% vs. 87.5%; contralateral BC risk higher after unilateral surgery.
Wanis 2024 [9]	Single-institution cohort	172 (all BCS)	11.8 years	10-year IBTR: 12.2%; contralateral BC: 21.3%; OS: 88.5%; majority remained bilateral mastectomy-free.
Mahiou 2024 [19]	Registry-based cohort	104 (69 BCS; 35 mastectomy)	~10 years	No OS difference (HR 1.49; $p = 0.25$); no difference in RFS (HR 1.40; $p = 0.22$).
Lee 2025 [10]	Multicenter propensity-matched cohort	575 (159 matched pairs)	8.3 years	No difference in locoregional RFS (HR 0.96), distant RFS (HR 0.62), RFS (HR 0.63), or OS (HR 0.82); contralateral BC: 14.2%; outcomes consistent across *BRCA1/2* subgroups.

BC, breast cancer; BCS, breast-conserving surgery; BSO, bilateral salpingo-oophorectomy; DFS, disease-free survival; HR, hazard ratio; IBTR, ipsilateral breast tumor recurrence; LRR, locoregional recurrence; OS, overall survival; RFS, recurrence-free survival.

4.1. Historical Long-Term Data: Survival Equivalence Despite Higher In-Breast Events

Early studies established that although BCS was associated with higher IBTR, it did not compromise survival. In a multi-institutional cohort of 655 carriers with stage I–III breast cancer, Pierce et al. reported a 15-year IBTR risk of 23.5% following BCS compared with 5.5% after mastectomy ($p < 0.0001$), while 10-year OS was virtually identical (92.1% for BCS vs. 91.8% for mastectomy) [14]. Detailed pathological reviews showed that most IBTRs were new primary tumors rather than true recurrences, reflecting the inherent predisposition of *BRCA* carriers to develop new malignancies rather than a failure of local therapy.

Nilsson et al. similarly analyzed 162 Swedish carriers over a mean follow-up of nearly 13 years and reported a 15-year cumulative IBTR rate of 32% for BCS compared with 9% for mastectomy (HR 2.9; 95% CI, 1.1–7.8), with OS equivalence (HR 1.2; 95% CI, 0.5–2.9) [15]. These early findings, arising before widespread use of adjuvant chemotherapy or endocrine therapy in carriers, set the precedent that survival equivalence can be maintained despite increased local events.

4.2. *Transitional Evidence: The Mitigating Impact of Systemic Therapy and Prophylactic Strategies*

Later cohorts demonstrated that adjuvant systemic therapies and prophylactic interventions narrowed the recurrence gap between surgical modalities. Van den Broek et al. assessed 191 *BRCA1* carriers diagnosed between 1970 and 2003 and found comparable 10-year IBTR rates for BCS (7.3%) and non-carriers (7.9%), with no OS difference (HR 0.80; 95% CI, 0.42–1.51) [16]. Gentile et al. reinforced these findings in a series of 124 carriers treated between 2008 and 2018, reporting no significant differences in DFS or OS between BCS and mastectomy, while noting improved distant DFS and OS among women who underwent bilateral salpingo-oophorectomy (BSO) [17]. These results underscore how risk-reducing surgery and optimized adjuvant treatments have reshaped outcomes in carriers over the past two decades.

4.3. *Contemporary Cohorts: Outcomes in the Era of Precision Oncology*

Recent studies, conducted within the context of routine germline testing, intensified surveillance, and modern systemic therapy, have provided the strongest reassurance of BCS safety in *BRCA* carriers. Shubeck et al. analyzed 395 carriers treated between 2006 and 2015, finding no differences in 10-year LRR (7.3% after BCS vs. 6.3% after mastectomy) or OS (88.8% vs. 87.5%) [18]. Wanis et al., in a single-institution series of 172 BCS-treated carriers with a median follow-up of 11.8 years, reported 10-year ipsilateral and contralateral breast event rates of 12.2% and 21.3%, respectively. Most patients avoided subsequent bilateral mastectomy, and the OS rate was 88.5% [9]. Mahiou et al. corroborated these findings in a French registry study ($n = 104$), observing no significant differences in OS (HR 1.49; 95% CI, 0.76–2.93) or recurrence-free survival (RFS; HR 1.40; 95% CI, 0.81–2.40) between BCS and mastectomy [19].

4.4. *The Lee et al. Multicenter Study: High-Quality Evidence in the Genomic Era*

Among contemporary analyses, the multicenter study by Lee et al. represents the most comprehensive evaluation of surgical outcomes in *BRCA* carriers to date [10]. This study enrolled 575 carriers diagnosed between 2008 and 2015 across multiple high-volume centers. It used rigorous propensity score matching to control for confounding factors, including age, tumor size, nodal status, hormone receptor subtype, and systemic therapy. Following matching, 159 patients per group were compared, with a median follow-up of 8.3 years.

Lee et al. reported no significant differences in locoregional RFS (HR 0.96; 95% CI, 0.36–2.59), distant RFS (HR 0.62; 95% CI, 0.28–1.38), RFS (HR 0.63; 95% CI, 0.33–1.22), or OS (HR 0.82; 95% CI, 0.34–1.98) between BCS and mastectomy [10]. Importantly, subgroup analyses stratified by *BRCA* mutation type (*BRCA1* vs. *BRCA2*) and tumor biology (triple-negative vs. hormone receptor-positive) confirmed these findings, supporting the applicability of BCS across biologically distinct subgroups within the *BRCA*-mutated population. Notably, while 14.2% of patients developed contralateral breast cancer during follow-up, this was independent of initial surgical approach and consistent with known contralateral risk profiles in this population.

A key strength of this study lies in the integration of systemic therapies: over 85% of patients received chemotherapy, and 80% of eligible patients received endocrine therapy, reflecting modern treatment standards. The inclusion of robust RT further mirrors current practice, addressing earlier concerns about radio resistance in *BRCA* carriers. Together, these factors underscore the relevance of this study in informing contemporary decision-making.

These data collectively illustrate an evolving trajectory: while early studies confirmed survival equivalence despite higher in-breast events, more recent cohorts show that with modern systemic therapies, optimized RT, and rigorous surveillance, recurrence risks associated with BCS are reduced and survival remains unaffected. The study by Lee et al. offers compelling evidence that, in the genomic era, BCS is an oncologically safe alternative to mastectomy for appropriately selected *BRCA* carriers, provided that vigilant follow-up and complementary risk-reducing strategies are implemented [10]. Persistent contralateral breast cancer risk remains a defining feature of this population, underscoring the importance of ongoing surveillance and shared decision-making when selecting local treatment.

5. Impact of RT and Locoregional Treatment Modalities

RT remains a cornerstone of breast-conserving therapy and plays a crucial role in optimizing local control for *BRCA* mutation carriers. Historically, concerns regarding increased radiosensitivity in *BRCA*-associated breast cancers raised questions about the safety and efficacy of RT. However, RT has been shown to be safe and effective in this population, with no evidence of increased toxicity or adverse survival impact.

More recent data have reinforced and refined this understanding, particularly in the context of evolving surgical techniques such as skin-sparing and nipple-sparing mastectomy (SSM/NSM). Bernstein-Molho et al. analyzed 484 breast tumors in 464 *BRCA* carriers treated between 2006 and 2022. Post-mastectomy RT significantly reduced LRR rates compared with mastectomy alone (1.25% vs. 13%, $p = 0.003$), despite the post-mastectomy RT group presenting with more advanced disease [20]. By contrast, mastectomy without RT, often performed in early-stage patients or those identified via genetic testing after lumpectomy, was associated with elevated early LRR rates, particularly within the first 2 years post-surgery. These results support the role of RT in eradicating residual microscopic disease in high-risk carriers, even following mastectomy with reconstructive techniques that may leave behind residual breast tissue.

Further insights come from Moshe et al., who investigated 255 patients (127 *BRCA* and 128 non-*BRCA*) undergoing SSM/NSM with or without RT [21]. They reported that patients not receiving RT had significantly higher LRR rates (15.3%) compared with those treated with RT (0%), independent of *BRCA* status. Multivariable analysis confirmed that RT was an independent protective factor against LRR (HR 0.67; 95% CI, 0.31–0.91). Interestingly, the study highlighted the early timing of recurrences in non-irradiated patients, suggesting residual breast tissue within skin flaps or dermal lymphatics as a plausible source of failure. These data underscore the critical importance of meticulous surgical planning and the integration of RT in cases where residual microscopic disease is likely.

Collectively, these findings emphasize that RT is integral to modern locoregional management in *BRCA* carriers. In the genomic era, its use extends beyond traditional indications to address the interplay between advanced surgical approaches (e.g., SSM/NSM), reconstruction techniques, and intrinsic genetic risk. Importantly, RT provides durable local control without increasing contralateral breast cancer risk or compromising survival, reinforcing its role within a multimodal treatment paradigm tailored to hereditary breast cancer biology.

6. Integration with Systemic and Risk-Reducing Strategies

Adjuvant systemic therapies and prophylactic interventions are key determinants of oncologic outcomes for *BRCA* mutation carriers undergoing BCS, complementing local therapy in reducing recurrence risk and improving survival. Modern management paradigms integrate chemotherapy, endocrine therapy, and risk-reducing BSO, alongside the emerging role of PARP inhibitors, to address the intrinsic biology of *BRCA*-associated breast cancer.

The prognostic value of risk-reducing BSO has been consistently demonstrated. Gentile et al. retrospectively analyzed 124 *BRCA* mutation carriers treated between 2008 and 2018, finding no difference in DFS or OS between BCS and mastectomy, but significantly improved distant DFS (p = 0.033) and OS (p = 0.040) among women who underwent BSO [17]. These results underscore the dual impact of systemic endocrine modulation and ovarian ablation in reducing recurrence risk, particularly in hormone receptor-positive carriers.

Chemotherapy and endocrine therapy remain central to modern BCS outcomes. In the contemporary series by Shubeck et al., involving 395 *BRCA* carriers treated between 2006 and 2015, receipt of adjuvant systemic therapy was high (74% chemotherapy, 54% endocrine therapy), and no differences were observed in LRR (7.3% for BCS vs. 6.3% for mastectomy) or OS (10-year OS: 88.8% vs. 87.5%) [18]. These findings suggest that optimized systemic therapy mitigates local recurrence risk and supports oncologic equivalence between surgical approaches in *BRCA* carriers.

The single-institution cohort by Wanis et al. extended these findings to long-term follow-up. Among 172 carriers treated with BCS over four decades, the 10-year OS was 88.5% (95% CI, 83.1–94.2%), with ipsilateral and contralateral breast cancer event risks of 12.2% and 21.3%, respectively [9]. Importantly, adjuvant RT (HR 0.09; 95% CI, 0.02–0.54) and BSO (HR 0.29; 95% CI, 0.09–0.90) were independently associated with lower ipsilateral recurrence risk. Endocrine therapy showed a trend toward reduced contralateral breast events (HR 0.61; 95% CI, 0.31–1.20), while chemotherapy was associated with lower distant recurrence risk (HR 0.48; 95% CI, 0.22–0.96). These results reinforce the synergistic interplay between systemic and local interventions in hereditary breast cancer management.

Emerging targeted therapies further reshape this landscape. PARP inhibitors, particularly olaparib, have demonstrated substantial benefits in *BRCA*-associated early breast cancer, with the OlympiA trial showing significant improvements in invasive DFS and distant RFS in high-risk carriers receiving adjuvant olaparib [9]. Although direct data on PARP inhibitors in the context of BCS are limited, their integration into standard adjuvant regimens is expected to further reduce recurrence risk, particularly in patients with residual disease or triple-negative tumors.

Collectively, these findings illustrate that the favorable outcomes of BCS in *BRCA* carriers are closely tied to modern systemic therapy utilization and prophylactic strategies (Table 3) [9,17,18]. The combination of optimized adjuvant treatment, BSO, and, increasingly, PARP inhibition provides a comprehensive multimodal framework that supports BCS as an oncologically sound option for appropriately selected carriers.

Table 3. Summary of studies evaluating systemic therapy and risk-reducing interventions in *BRCA* mutation carriers undergoing BCS.

Study (Year)	Design	N (*BRCA* Carriers)	Key Systemic/Risk-Reducing Interventions	Key Outcomes
Gentile 2022 [17]	Single-institution cohort	124 (69 BCS; 55 mastectomy)	BSO (58.1%); adjuvant chemotherapy and endocrine therapy per subtype	No DFS (p = 0.39) or OS (p = 0.265) difference between BCS and mastectomy; BSO improved distant DFS (p = 0.033) and OS (p = 0.040).
Shubeck 2022 [18]	Single-institution cohort	395 (149 BCS; 246 mastectomy)	Chemotherapy (74%), endocrine therapy (54%), RT (65% BCS)	10-year LRR: 7.3% (BCS) vs. 6.3% (mastectomy); 10-year OS: 88.8% vs. 87.5%; systemic therapy supported equivalence.
Wanis 2024 [9]	Single-institution cohort	172 (all BCS)	Adjuvant RT (100%), BSO (54%), endocrine therapy, chemotherapy; PARP inhibitor context (OlympiA)	10-year OS: 88.5% (95% CI, 83.1–94.2); ipsilateral recurrence: 12.2%; contralateral events: 21.3%; RT (HR 0.09; 95% CI, 0.02–0.54) and BSO (HR 0.29; 95% CI, 0.09–0.90) reduced ipsilateral recurrence; endocrine therapy (HR 0.61; 95% CI, 0.31–1.20) lowered contralateral risk; chemotherapy (HR 0.48; 95% CI, 0.22–0.96) reduced distant recurrence.

BCS, breast-conserving surgery; BSO, bilateral salpingo-oophorectomy; DFS, disease-free survival; HR, hazard ratio; LRR, locoregional recurrence; OS, overall survival; PARP, poly(ADP-ribose) polymerase; RT, radiotherapy.

7. Pregnancy, Breastfeeding, and Assisted Reproductive Technologies in *BRCA*-Mutated Breast Cancer Survivors

Reproductive considerations represent a central component in the holistic management of young women with *BRCA*-mutated breast cancer. In recent years, a series of large-scale, collaborative studies have reshaped the evidence base surrounding the safety and feasibility of pregnancy, breastfeeding, and assisted reproductive technologies (ART) in this population. Collectively, these data reinforce the concept of personalized survivorship care and support the integration of fertility-preserving strategies and family planning in *BRCA*-mutated breast cancer management.

A recent large international cohort study assessed the safety of breastfeeding among 178 *BRCA1/2* carriers who delivered a child after breast cancer and for whom lactation data were available. In this cohort, 61.8% breastfed their child, with a median duration of 5 months. After a median follow-up of 7 years post-delivery, no significant differences were observed in DFS (adjusted HR 0.83; 95% CI, 0.49–1.41) or OS (adjusted HR 1.32; 95% CI, 0.31–5.66) between women who breastfed and those who did not. Furthermore, the cumulative incidence of locoregional or contralateral recurrences was numerically lower in the breastfeeding group (29% vs. 37%), though not statistically significant [22].

Assisted reproduction represents another critical concern, particularly for *BRCA* mutation carriers who may experience impaired ovarian reserve due to both genetic predisposition and oncologic treatment. A multicenter study by Magaton et al. examined outcomes in 543 *BRCA1/2* carriers who became pregnant after breast cancer, of whom 107 conceived through ART. Among these, 42.1% used oocytes or embryos cryopreserved at diagnosis, 30.8% underwent in vitro fertilization or ovulation induction after treatment, and 19.6% used donor oocytes. At a median follow-up of 5.2 years, the use of ART was not associated with worse DFS (adjusted HR 0.72; 95% CI, 0.39–1.34), providing reassuring evidence for the oncologic safety of ART in this setting [23].

Pregnancy-related outcomes were also comparable between the ART and non-ART groups. The live birth rate was 83.0% among ART users and 79.8% among those who conceived spontaneously. Most pregnancies in both groups were delivered at term (85.5% vs. 91.5%, respectively), and complication rates were low. Notably, a slightly higher incidence of miscarriage was observed in the ART group (11.3% vs. 8.8%), which may reflect older maternal age at conception in this subgroup [23].

In parallel, Perachino et al. explored reproductive outcomes following a diagnosis of pregnancy-associated breast cancer (PrBC) in *BRCA* carriers—a scenario historically associated with limited data. Among 282 patients diagnosed with PrBC, 68 experienced a subsequent pregnancy, with a 10-year cumulative incidence of 36.6%. The majority of pregnancies were full-term (75.9%) and without complications (74.1%). Importantly, no differences in DFS (adjusted HR 1.06; 95% CI, 0.49–2.31) or OS (adjusted HR 0.63; 95% CI, 0.19–2.05) were noted between women with or without subsequent pregnancy, suggesting that a history of PrBC does not preclude safe future childbearing in this high-risk group [24].

Finally, these findings build upon a growing body of work, including earlier observational data by Blondeaux et al., which have consistently shown that *BRCA* mutation status should not, per se, limit reproductive planning. For example, factors such as nulliparity at diagnosis, hormone receptor status, and time to conception were explored in these cohorts and support a nuanced approach to counseling [25].

Altogether, these studies converge to reinforce that pregnancy, breastfeeding, and ART can be safely considered in selected *BRCA*-mutated breast cancer survivors. These findings lend further support to the paradigm shift toward personalized treatment strategies and away from blanket recommendations for risk-reducing mastectomy or reproductive deferral. In this context, breast-conserving strategies gain additional relevance—not only for their oncologic equivalence but also for preserving body image and function in women who may wish to breastfeed or consider future pregnancy. Importantly, these data provide clinicians with much-needed reassurance to support shared decision-making in a setting where reproductive autonomy and oncologic safety must be balanced.

8. Psychosocial and Patient-Centered Outcomes

Specific evidence addressing psychosocial and QoL outcomes in *BRCA* mutation carriers undergoing BCS remains limited. Consequently, insights must largely be extrapolated from recent prospective and cross-sectional studies conducted in broader breast cancer populations, where breast-conserving approaches consistently demonstrate favorable patient-reported outcomes compared with mastectomy and reconstruction. These findings are particularly relevant given the oncologic equivalence of BCS and mastectomy in *BRCA* carriers and the critical importance of patient-centered decision-making in this context.

Recent studies have underscored the QoL advantages of BCS over more extensive surgical approaches. In a large propensity score–matched study of over 2600 patients, Kim et al. reported that women undergoing BCS scored significantly higher on BREAST-Q domains—including satisfaction with breasts, psychosocial and sexual well-being—throughout 3 years of follow-up compared with those undergoing postmastectomy reconstruction ($p < 0.05$ for all) [26]. Similarly, Liu et al., in a prospective cohort of 471 patients, observed that BCS was associated with the highest absolute postoperative scores across multiple health-related QoL dimensions, including breast satisfaction and psychosocial and physical well-being ($p < 0.01$), while mastectomy patients exhibited greater declines in pain and functional domains [27]. Importantly, younger age, which is common among *BRCA* mutation carriers, was independently associated with higher levels of postoperative anxiety, depression, and pain, suggesting that hereditary breast cancer patients may face amplified psychosocial challenges regardless of surgical choice [27].

Complementary cross-sectional data reinforce these trends. Tomita et al. reported that patients treated with BCS achieved significantly better scores for breast satisfaction (mean 58.3) and sexual well-being (mean 45.8) than those undergoing mastectomy (41.8 and 29.9, respectively). However, psychosocial well-being was highest in reconstruction cohorts, reflecting the complex interplay between body image, surgical choice, and psychological adaptation [28]. Furthermore, van Zeelst et al. demonstrated that integration of oncoplastic refinements, such as perforator flap reconstruction, can enhance esthetic outcomes within BCS, with median BREAST-Q scores of 87/100 for psychosocial well-being and 82/100 for breast satisfaction at 17 months, alongside low complication rates and high outpatient feasibility [29].

Although not *BRCA*-specific, these recent studies suggest that BCS may confer measurable QoL advantages relative to mastectomy, particularly in terms of breast satisfaction, psychosocial adjustment, and functional well-being. For *BRCA* carriers, these benefits must be weighed against their distinct cancer risk profiles and the need for long-term surveillance. Nonetheless, they underscore the importance of incorporating patient-reported outcomes into shared decision-making. Given that younger age and intensive adjuvant therapy are linked to higher distress [27], structured psychosocial support and counseling may be particularly crucial for hereditary breast cancer patients selecting BCS.

9. BCS in *BRCA* Mutation Carriers: A Paradigm Shift Toward Individualized Care

The surgical management of breast cancer in *BRCA* mutation carriers is undergoing a transformative shift. Historically, mastectomy was considered the default approach due to concerns about elevated recurrence risks and the propensity for new primary tumors in this high-risk group. However, recent evidence demonstrates that with the integration of modern RT, systemic therapy, and vigilant surveillance, BCS can be considered an oncologically safe option for appropriately selected carriers [1,10,13].

This evolution reflects broader advances in multidisciplinary care. Improved systemic treatments, including chemotherapy, endocrine therapy, and more recently PARP inhibitors, have significantly contributed to reducing recurrence risk and improving long-term outcomes in *BRCA* carriers [1,9]. Furthermore, risk-reducing strategies such as BSO remain integral in comprehensive management, particularly in hormone receptor-positive disease [17]. These developments collectively redefine the balance between local and systemic control in hereditary breast cancer.

Beyond oncologic safety, psychosocial and QoL outcomes are becoming central to surgical decision-making. Studies in broader breast cancer populations consistently show that BCS is associated with greater satisfaction with breast appearance, improved body image, and more favorable psychosocial well-being compared with mastectomy and reconstruction [2]. Although such data in *BRCA* carriers are limited, these insights are relevant and likely applicable, given similar surgical contexts. For many women, retaining their breasts may mitigate the psychological burden associated with hereditary cancer, reducing feelings of loss and improving sexual and emotional adjustment. These potential benefits highlight the importance of integrating psychosocial considerations into surgical counseling and tailoring choices not solely to risk reduction, but also to patient-reported priorities.

Emerging data also affirm the safety of pregnancy and breastfeeding in *BRCA* carriers after treatment. Recent international studies show no increased risk of recurrence or mortality following either outcome, and breastfeeding was not associated with adverse oncologic events [22,24]. These findings are especially relevant for women who pursue BCS, which may allow for future lactation.

Similarly, ART appears safe in this population. A multicenter study reported no significant differences in DFS between women conceiving through ART and those with spontaneous pregnancies [23], further supporting reproductive autonomy in survivorship care.

Together, these findings endorse a more personalized surgical approach that considers not only oncologic risk but also reproductive goals and long-term QoL. For *BRCA* carriers, particularly younger patients, BCS can offer oncologic equivalence with potential functional and psychosocial advantages.

Nonetheless, challenges remain. The persistent risk of contralateral breast cancer necessitates lifelong enhanced surveillance and underscores the importance of comprehensive genetic counseling and psychosocial support [9]. Furthermore, differences in tumor biology and systemic therapy response between *BRCA1* and *BRCA2* carriers require individualized strategies [10].

Importantly, not all *BRCA* mutation carriers face the same risk profile or respond to treatment similarly. Breast cancer patients carrying a *BRCA* mutation who may safely avoid mastectomy can consider breast-conserving surgery followed by radiation therapy. This option may be particularly suitable for *BRCA2* carriers, as several studies suggest that BCS with radiation may yield comparable or even favorable outcomes relative to mastectomy. In contrast, *BRCA1* carriers are known to have a higher risk of contralateral breast cancer, which may increase the likelihood of recommending prophylactic mastectomy.

Other key selection criteria include tumor characteristics—such as size (T stage), nodal involvement (N status), grade, and hormone receptor profile. Luminal tumors, typically associated with hormone receptor positivity, often have better prognoses and may be more amenable to breast conservation, while triple-negative cancers, more common in *BRCA1* carriers, may favor a more aggressive surgical approach. Furthermore, individual psychological tolerance for risk plays a critical role. Some patients may prioritize maximum risk reduction and prefer mastectomy, while others may find the trade-off between oncologic safety and body preservation acceptable in choosing BCS. Looking forward, BCS for *BRCA* carriers represents a paradigm shift from uniform recommendations toward personalized, patient-centered care. While a growing body of retrospective evidence suggests that breast conservation may be considered as part of multimodal treatment planning for appropriately selected individuals, this must be balanced against emerging data indicating a potential survival advantage with risk-reducing mastectomy. Notably, Blondeaux et al. (2025) [25] reported improved DRFS and OS with contralateral risk-reducing mastectomy in young *BRCA* carriers, highlighting the importance of individualized decision-making based on both oncologic and patient-centered factors.

Ultimately, surgical choice in *BRCA* mutation carriers must balance oncologic safety, systemic therapy integration, and the profound psychological implications of breast loss or preservation. By embedding BCS within comprehensive multidisciplinary care, including genetic counseling, risk-reduction planning, and psychosocial support, we can better align treatment with both medical evidence and individual patient values. In this context, incorporating formal psychological assessment and structured support into preoperative counseling is essential to help patients navigate the complex emotional and decisional aspects of choosing between BCS and mastectomy.

10. Conclusions

BCS is increasingly recognized as a valid and oncologically acceptable option for selected *BRCA* mutation carriers, particularly when delivered within a comprehensive, multidisciplinary framework. Accumulating evidence, though largely retrospective, consistently supports equivalent long-term survival outcomes between BCS and mastectomy, especially when radiotherapy and systemic therapies are appropriately integrated.

Importantly, BCS may offer meaningful advantages in terms of body image, quality of life, and reproductive planning—factors that are particularly relevant for younger patients navigating the complex implications of hereditary breast cancer. However, we recognize that most existing QoL data are derived from studies involving non-*BRCA* populations. While these findings provide useful context, they should be interpreted with caution when applied to *BRCA* carriers, given their distinct psychosocial and clinical challenges. This underscores the need for future *BRCA*-specific research to inform patient-centered care more accurately. At the same time, we acknowledge the limitations inherent in the current bulk of evidence, including the retrospective nature of most studies and heterogeneity in design and follow-up. The recent international cohort study by Blondeaux et al. (2025) [25], which demonstrated improved distant recurrence-free and overall survival with risk-reducing mastectomy in young *BRCA* carriers, provides important prospective data that must be considered in surgical counseling.

Taken together, these findings reinforce the need for nuanced, patient-centered decision-making that balances oncologic safety with psychosocial well-being and personal values. As evidence continues to evolve, BCS should be presented as a standard treatment option—not a compromise—for appropriately selected *BRCA* mutation carriers.

Funding: This research received no external funding.

Data Availability Statement: No new data were created or analyzed in this study.

Acknowledgments: Editorial assistance was provided by Valentina Attanasio and Aashni Shah.

Conflicts of Interest: Author Luca Giacomelli was employed by the company Polistudium SRL. The remaining authors declare that the research was conducted in the absence of any commercial or financial relationships that could be construed as a potential conflict of interest.

References

1. Edaily, S.; Abdel-Razeq, H. Management Strategies of Breast Cancer Patients with *BRCA1* and *BRCA2* Pathogenic Germline Variants. *Onco Targets Ther.* **2022**, *15*, 815–826. [CrossRef] [PubMed]
2. Magnoni, F.; Sacchini, V.; Veronesi, P.; Bianchi, B.; Bottazzoli, E.; Tagliaferri, V.; Mazzotta, E.; Castelnovo, G.; Deguidi, G.; Rossi, E.M.C.; et al. Surgical Management of Inherited Breast Cancer: Role of Breast-Conserving Surgery. *Cancers* **2022**, *14*, 3245. [CrossRef]
3. Cipolla, C.; Scandurra, G.; Sambataro, D.; Mesi, C.; Greco, M.; D'Agati, E.; Gebbia, V.; Valerio, M.R. Risk-reducing mastectomy in healthy women with *BRCA* mutation: A narrative review. *Expert. Rev. Anticancer. Ther.* **2025**, *25*, 797–807. [CrossRef]
4. Vallard, A.; Magné, N.; Guy, J.B.; Espenel, S.; Rancoule, C.; Diao, P.; Deutsch, E.; Rivera, S.; Chargari, C. Is breast-conserving therapy adequate in BRCA 1/2 mutation carriers? The radiation oncologist's point of view. *Br. J. Radiol.* **2019**, *92*, 20170657. [CrossRef]
5. Terribile, D.A.; Mason, E.J.; Murando, F.; DILeone, A.; Sanchez, A.M.; Scardina, L.; Magno, S.; Franco, A.; D'Archi, S.; Natale, M.; et al. Surgical management of BRCA pathogenic variant carriers with breast cancer: A recent literature review and current state of the art. *Minerva Surg.* **2021**, *76*, 564–574. [CrossRef]
6. Tung, N.M.; Boughey, J.C.; Pierce, L.J.; Robson, M.E.; Bedrosian, I.; Dietz, J.R.; Dragun, A.; Gelpi, J.B.; Hofstatter, E.W.; Isaacs, C.J.; et al. Management of Hereditary Breast Cancer: American Society of Clinical Oncology, American Society for Radiation Oncology, and Society of Surgical Oncology Guideline. *J. Clin. Oncol.* **2020**, *38*, 2080–2106. [CrossRef]
7. Co, M.; Liu, T.; Leung, J.; Li, C.H.; Tse, T.; Wong, M.; Kwong, A. Breast Conserving Surgery for BRCA Mutation Carriers-A Systematic Review. *Clin. Breast Cancer* **2020**, *20*, e244–e250. [CrossRef] [PubMed]
8. Wang, C.; Lin, Y.; Zhu, H.; Zhou, Y.; Mao, F.; Huang, X.; Zhou, X.; Cao, X.; Sun, Q. Breast-conserving therapy for breast cancer with BRCA mutations: A meta-analysis. *Breast Cancer* **2022**, *29*, 314–323. [CrossRef]
9. Wanis, K.N.; Kuerer, H.M.; Sun, S.X.; Hunt, K.K.; Glencer, A.C.; Teshome, M.; Lucci, A.; Weiser, R.; Johnson, H.; Smith, B.D.; et al. Clinical Outcomes for BRCA Pathogenic Variant Carriers With Breast Cancer Undergoing Breast Conservation. *JAMA Netw. Open* **2024**, *7*, e2418486. [CrossRef] [PubMed]

10. Lee, J.; Ryu, J.M.; Kim, H.K.; Park, H.S.; Kang, B.; Ahn, S.G.; Chung, M.S.; Shin, S.H.; Go, J.; Kim, S.; et al. Long-Term Oncologic Outcome of Breast-Conserving Treatment in Patients With Breast Cancer With BRCA Variants. *JAMA Netw. Open* **2025**, *8*, e259840. [CrossRef]
11. Magaton, I.M.; Arecco, L.; Mariamidze, E.; Jankovic, K.; Stana, M.; Buzzatti, G.; Trevisan, L.; Scavone, G.; Ottonello, S.; Fregatti, P.; et al. Fertility and Pregnancy-Related Issues in Young *BRCA* Carriers With Breast Cancer. *Breast Cancer* **2024**, *18*, 11782234241261429. [CrossRef]
12. Nara, M.; Ishihara, S.; Kitano, A.; Tamura, N.; Aruga, T.; Kobayashi, D.; Nakamura, S.; Yamauchi, H. Does breast-conserving surgery with radiotherapy in *BRCA*-mutation carriers significantly increase ipsilateral breast tumor recurrence? A systematic review and meta-analysis. *Breast Cancer* **2022**, *29*, 394–401. [CrossRef]
13. Davey, M.G.; Davey, C.M.; Ryan, É.J.; Lowery, A.J.; Kerin, M.J. Combined breast conservation therapy versus mastectomy for *BRCA* mutation carriers—A systematic review and meta-analysis. *Breast* **2021**, *56*, 26–34. [CrossRef]
14. Pierce, L.J.; Phillips, K.A.; Griffith, K.A.; Buys, S.; Gaffney, D.K.; Moran, M.S.; Haffty, B.G.; Ben-David, M.; Kaufman, B.; Garber, J.E.; et al. Local therapy in BRCA1 and BRCA2 mutation carriers with operable breast cancer: Comparison of breast conservation and mastectomy. *Breast Cancer Res. Treat.* **2010**, *121*, 389–398. [CrossRef] [PubMed]
15. Nilsson, M.P.; Hartman, L.; Kristoffersson, U.; Johannsson, O.T.; Borg, A.; Henriksson, K.; Lanke, E.; Olsson, H.; Loman, N. High risk of in-breast tumor recurrence after *BRCA1*/2-associated breast cancer. *Breast Cancer Res. Treat.* **2014**, *147*, 571–578. [CrossRef] [PubMed]
16. van den Broek, A.J.; Schmidt, M.K.; van't Veer, L.J.; Oldenburg, H.S.A.; Rutgers, E.J.; Russell, N.S.; Smit, V.T.H.B.M.; Voogd, A.C.; Koppert, L.B.; Siesling, S.; et al. Prognostic Impact of Breast-Conserving Therapy Versus Mastectomy of *BRCA1*/2 Mutation Carriers Compared with Noncarriers in a Consecutive Series of Young Breast Cancer Patients. *Ann. Surg.* **2019**, *270*, 364–372. [CrossRef]
17. Gentile, D.; Losurdo, A.; Sagona, A.; Zuradelli, M.; Gatzemeier, W.; Barbieri, E.; Testori, A.; Errico, V.; Bianchi, P.; Biondi, E.; et al. Surgical management of *BRCA*-mutation carriers: A single institution experience. *Eur. J. Surg. Oncol.* **2022**, *48*, 1706–1712. [CrossRef]
18. Shubeck, S.; Sevilimedu, V.; Berger, E.; Robson, M.; Heerdt, A.S.; Pilewskie, M.L. Comparison of Outcomes Between *BRCA* Pathogenic Variant Carriers Undergoing Breast-Conserving Surgery Versus Mastectomy. *Ann. Surg. Oncol.* **2022**, *29*, 4706–4713. [CrossRef] [PubMed]
19. Mahiou, K.; Jankowski, C.; Vincent, L.; Costaz, H.; Padeano, M.M.; Mamguem, A.; Dabakuyo, S.; Coutant, C. Impact of breast surgical procedure on survival in BRCA mutated patients with invasive breast cancer: Mastectomy versus conservative treatment. *J. Gynecol. Obstet. Hum. Reprod.* **2024**, *53*, 102760. [CrossRef]
20. Bernstein-Molho, R.; Haisraely, O.; Galper, S.; Abu-Shhada, N.; Nili Gal-Yam, E.; Menes, T.S.; Poortmans, P.; Kaidar-Person, O. The impact of locoregional treatment on ipsilateral breast cancer recurrence and outcomes in carriers of *BRCA1*/2 pathogenic variants. *Breast Cancer Res. Treat.* **2025**, *211*, 431–439. [CrossRef]
21. Moshe, N.; Haisraely, O.; Globus, O.; Faermann, R.; Abu-Shehada, N.; Anaby, D.; Gal Yam, E.; Balint Lahat, N.; Galper, S.; Menes, T.; et al. Breast cancer outcomes after skin- and nipple-sparing mastectomy in BRCA pathogenic mutation carriers versus non-BRCA carriers. *Radiother. Oncol.* **2025**, *205*, 110710. [CrossRef]
22. Blondeaux, E.; Delucchi, V.; Mariamidze, E.; Bernstein-Molho, R.; Frank, S.; Ferrari, A.; Linn, S.; Jeong Kim, H.; Agostinetto, E.; Paluch-Shimon, S.; et al. Breastfeeding after breast cancer in young *BRCA* carriers. *J. Natl. Cancer Inst.* **2025**, djaf177. [CrossRef] [PubMed]
23. Magaton, I.M.; Blondeaux, E.; Hamy, A.S.; Linn, S.; Bernstein-Molho, R.; Peccatori, F.A.; Ferrari, A.; Carrasco, E.; Paluch-Shimon, S.; Agostinetto, E.; et al. Assisted reproductive technology in young BRCA carriers with a pregnancy after breast cancer: An international cohort study. *Eur. J. Cancer* **2025**, *222*, 115434. [CrossRef]
24. Perachino, M.; Blondeaux, E.; Delucchi, V.; Bernstein Molho, R.; Frank, S.; Paluch-Shimon, S.; Linn, S.; Di Meglio, A.; Peccatori, F.; Moore, H.C.F.; et al. Safety of having a subsequent pregnancy after prior diagnosis of breast cancer during pregnancy in young BRCA carriers. *ESMO Open* **2025**, *10*, 105513. [CrossRef]
25. Blondeaux, E.; Sonnenblick, A.; Agostinetto, E.; Bas, R.; Kim, H.J.; Franzoi, M.A.; Bernstein-Molho, R.; Linn, S.; Kwong, A.; Pogoda, K.; et al. Association between risk-reducing surgeries and survival in young BRCA carriers with breast cancer: An international cohort study. *Lancet Oncol.* **2025**, *26*, 759–770. [CrossRef]
26. Kim, M.; Tadros, A.B.; Boe, L.A.; Vingan, P.; Allen, R.J., Jr.; Mehrara, B.J.; Morrow, M.; Nelson, J.A. Breast-Conserving Therapy Versus Postmastectomy Breast Reconstruction: Propensity Score-Matched Analysis. *Ann. Surg. Oncol.* **2024**, *31*, 8030–8039. [CrossRef]
27. Liu, C.; Beresford, A.; Saleeb, M.; Liu, G.; Crump, T.; Warburton, R.; Pao, J.S.; Dingee, C.K.; Bazzarelli, A.; Sutherland, J.M.; et al. Preoperative and Postoperative Change in Patient-Reported Health-Related Quality of Life Outcomes in Breast Cancer Surgery Patients Across Surgical Modalities: A Prospective Study. *Cancers* **2025**, *17*, 1409. [CrossRef]

28. Tomita, S.; Yoshitake, T.; Matsunaga, N.; de Kerckhove, M.; Fujii, M.; Terao, Y. Patient-reported outcomes and quality of life after breast-conserving surgery, mastectomy, and breast reconstruction assessed using the BREAST-Q questionnaire. *Breast Cancer Res. Treat.* **2024**, *207*, 641–648. [CrossRef] [PubMed]
29. van Zeelst, L.J.; Straten, R.; van Eekeren, R.R.J.P.; van Uden, D.J.P.; de Wilt, J.H.W.; Strobbe, L.J.A. Chest wall perforator flap reconstruction in breast conserving surgery: Quality of life and limited complications in outpatient treatment. *World J. Surg. Oncol.* **2025**, *23*, 17. [CrossRef] [PubMed]

life

MDPI

Review

Translating Metabolic Interventions into Breast Cancer Therapy: A Comprehensive Review

Luxi Chen [1], Stephen L. Shiao [2] and Yuan Yuan [1,*]

1 Department of Medicine, Samuel Oschin Comprehensive Cancer Institute, Cedars-Sinai Medical Center, Los Angeles, CA 90048, USA; luxi.chen@cshs.org

2 Department of Radiation Oncology, Cedars-Sinai Medical Center, Los Angeles, CA 90048, USA; stephen.shiao@cshs.org

* Correspondence: yuan.yuan@cshs.org

Abstract

Breast cancer remains a leading cause of morbidity and mortality in women worldwide. Despite significant advances in targeted therapies, therapeutic resistance, metabolic toxicities, and disease recurrence continue to limit long-term efficacy. Metabolic syndrome is a major epidemiologic risk factor for the development of breast cancer, with metabolic dysregulation strongly linked to tumor progression, recurrence, and mortality. Crosstalk between insulin and insulin-like growth factor (IGF) signaling and oncogenic pathways such as PI3K/AKT/mTOR provides a mechanistic basis for these associations, highlighting the interplay between metabolism and tumor biology. Given this context, anti-diabetic and anti-obesity agents are being investigated as novel therapeutic strategies in breast cancer. Beyond their established metabolic benefits, these agents can directly modulate tumor cell growth, immune responses, and signaling pathways central to breast cancer pathogenesis. In this review, we summarize the current knowledge on the intersection of metabolic dysregulation and breast cancer as well as critically evaluate preclinical and clinical evidence supporting the use of metabolic therapies in this space.

Keywords: breast cancer; metabolic syndrome; signaling pathways; tumor microenvironment; cancer therapies

1. Introduction

Breast cancer is the most frequently diagnosed malignancy and remains the leading cause of cancer-related mortality among women worldwide, with an estimated 2.3 million new cases and 665,684 deaths reported in 2022 [1]. Established breast cancer risk factors include age, genetic predisposition, hormonal and reproductive history, obesity, and lifestyle factors such as diet and physical inactivity [2]. Notably, a recent meta-analysis reported that higher dietary intake of essential fatty acids may be associated with an increased risk of breast cancer, highlighting the complex contribution of metabolic and nutritional factors to disease development [3]. While there has been remarkable progress in the treatment of breast cancer in the last decade, resistance to therapy and recurrence remain significant challenges.

The phosphatidylinositol 3-kinase (PI3K)/AKT/mammalian target of rapamycin (mTOR) pathway is a crucial signal transduction system that is aberrantly activated in breast cancer, mediating cell proliferation, survival, metabolism, and metastasis [4,5]. Targeting this signaling pathway has resulted in major breakthroughs in the treatment landscape of metastatic hormone receptor (HR)-positive breast cancer. As such, many Phase III clinical trials have led to the FDA approval of PI3K/AKT/mTOR inhibitors including alpelisib, capivasertib, inavolisib, and everolimus [6–9]. Utilization of these inhibitors has resulted in significantly prolonged progression-free survival and/or overall survival in breast cancer patients. However, given the important role of the PI3K/AKT/mTOR pathway in normal cell metabolism, major on-target, off-tumor side effects, particularly hyperglycemia and hyperlipidemia, have impaired wide adaptation of these inhibitors in clinical practice [10].

Obesity and type 2 diabetes mellitus (T2DM) are independent risk factors for the development of various cancers, including breast, colorectal, pancreatic, esophageal, gastric, and endometrial cancers [11,12]. Increased body weight and obesity augment the risk of triple negative breast cancer (TNBC) in premenopausal women by 67% and premenopausal luminal B breast cancer by 73%, in contrast to women of normal weight [13,14]. In addition, obese women with breast cancer have an approximately 30% increased risk of cancer recurrence or death compared to those with normal weight [15]. Both obesity and T2DM are associated with a higher risk of postmenopausal breast cancer [16]. Several studies have shown an increased breast cancer specific mortality in patients with metabolic syndrome, likely due to crosstalk between the insulin-like growth factor (IGF) family and signaling pathways that drive tumorigenesis, including the PI3K/AKT/mTOR pathway [17,18].

Given the above, anti-diabetic and anti-obesity medications such as metformin and glucagon-like peptide-1 receptor agonists (GLP-1RAs) have recently emerged as potential therapeutic candidates in the treatment of breast cancer [19]. By simultaneously modulating metabolic pathways and directly influencing tumor and immune cell signaling, these agents hold the potential to complement existing targeted therapies, reduce toxicities, and improve long-term outcomes in breast cancer patients. This review will explore the mechanistic underpinnings, preclinical evidence, and emerging clinical data supporting the role of metabolic interventions in redefining the landscape of breast cancer treatment.

2. Methodology

We conducted a targeted search of PubMed, ClinicalTrials.gov, and Google Scholar using combinations of relevant keywords ("metabolic syndrome," "breast cancer," "insulin signaling," "tumor signaling pathways," "metformin," "GLP-1 receptor agonists," "SGLT2 inhibitors"). Both preclinical and clinical studies were considered, with emphasis on high-impact epidemiological analyses, seminal mechanistic investigations, and Phase II–III clinical trials with defined endpoints. ClinicalTrials.gov was systematically reviewed to capture ongoing and completed studies of metabolic interventions in breast cancer, and reference lists of key publications were screened to identify additional relevant reports. This strategy enabled a rigorous synthesis of current evidence across epidemiologic, mechanistic, and translational domains.

3. Metabolic Syndrome and Breast Cancer

3.1. Epidemiological Evidence

Multiple studies have demonstrated a significant association between the development of breast cancer and metabolic syndrome, a condition characterized by central obesity, dyslipidemia, hypertension, and impaired glucose tolerance [20,21]. In postmenopausal women, obesity has consistently been shown to increase the rate of breast cancer by 30–50% [22]. Further investigation on breast cancer survival and mortality indicate that adiposity is associated with lower survival rates and higher likelihood of recurrence, irrespective of menopausal status, tumor stage, and treatment [23]. In fact, breast cancer patients with a body mass index (BMI) $\geq$ 40.0 kg/m^2 not only have a mortality rate three times higher compared to those with BMI < 20.5 kg/m^2 but is also associated with poorer differentiation, larger tumor size, and increased lymph node invasion frequency [24].

Recently, the Women's Health Initiative (WHI) clinical trials showed that a higher metabolic syndrome score, which included factors such as high waist circumference and history of T2DM, hypertension, and/or high cholesterol, in postmenopausal breast cancer patients was significantly associated with poorer prognosis and higher mortality rates [20]. In the WHI cohort of 544 women with TNBC, after a median follow-up of 8.3 years, those with three or four risk factors had more than double the risk of breast cancer–specific mortality (HR 2.05; 95% CI 0.94–4.47; trend $p = 0.114$, not statistically significant) and a significantly higher risk of overall mortality (HR 2.13; 95% CI 1.22–3.71; trend $p = 0.006$) [25]. Additionally, a meta-analysis demonstrated that T2DM was associated with an increased risk of both estrogen receptor-positive (ER+) (RR = 1.09, 95% CI: 1.00–1.20) and triple-negative breast cancer (RR = 1.41, 95% CI: 1.01–1.96) [26].

Racial and ethnic disparities further shape the epidemiological landscape of obesity and breast cancer risk, particularly in the United States. Obesity prevalence is disproportionately higher among Black women compared to other racial and ethnic groups, a difference attributed to a complex interplay of genetic, biochemical, behavioral, and socioeconomic factors. Recent analyses demonstrate that these disparities are not only widespread but also persist across age groups and geographic regions. For example, a national cohort study reported a significantly higher prevalence of both general and abdominal obesity among non-Hispanic Black women compared with non-Hispanic White women, findings that have important implications for breast cancer risk and outcomes in this population [27]. Moreover, epidemiologic data indicate that Black women experience higher rates of metabolic syndrome, which may contribute to the observed disparities in breast cancer incidence and prognosis [28]. These findings highlight the need to integrate racial and ethnic considerations into studies examining the metabolism–cancer interface, particularly when evaluating prevention and intervention strategies.

3.2. Obesity-Induced Estrogen Production

One of the principal mechanisms underlying the association between metabolic syndrome and the development of breast cancer is the obesity-driven increase in systemic estrogen levels. In postmenopausal women, adipose tissue becomes the predominant site of estrogen biosynthesis through the activity of the cytochrome P450 enzyme aromatase, which catalyzes the conversion of adrenal androgens, such as androstenedione and testosterone, into estrone and estradiol [29,30]. Elevated estrogen availability enhances estrogen receptor signaling, thereby stimulating breast epithelial proliferation and promoting tumorigenesis [31]. Notably, aromatase is expressed not only in peripheral adipose depots, such as breast, abdominal, thigh, and gluteal fat, but can also be upregulated within the tumor tissue itself, creating a localized estrogen-rich microenvironment that supports malignant cell growth [32].

3.3. Chronic Inflammation in Adipocytes

Beyond estrogen production, obesity induces chronic low-grade inflammation within adipose tissue, which contributes to tumorigenesis through multiple signaling pathways. Obese adipocytes secrete pro-inflammatory cytokines such as tumor necrosis factor-α (TNF-α) and interleukin-6 (IL-6), which activate the "obesity–inflammation–aromatase axis" [33,34]. These cytokines enhance transcription of the CYP19A1 gene, further increasing aromatase expression and thereby amplifying intratumoral estrogen biosynthesis [35]. This inflammatory milieu is accompanied by alterations in adipose-derived endocrine function.

Adipose tissue is now recognized as a metabolically active endocrine organ that secretes a wide range of bioactive molecules, collectively termed adipokines, which exert systemic effects on metabolism, immune function, and tumor biology [36]. Among these adipokines, leptin has been consistently implicated in breast cancer progression. Hyperleptinemia, commonly observed in obesity, promotes cell proliferation, angiogenesis, and resistance to apoptosis, largely through activation of the Janus kinase/signal transducers and activators of transcription (JAK/STAT), phosphatidyl inositol 3 kinase (PI3K)/AKT, and mitogen-activated protein kinase (MAPK) signaling pathways [37,38]. In contrast, the adipokine adiponectin appears to exert protective effects, with anti-proliferative, insulin-sensitizing, and anti-inflammatory properties. Low circulating adiponectin levels have been associated with increased breast cancer risk, more aggressive tumor phenotypes, and poorer clinical outcomes in studies [39,40]. Collectively, these mechanisms highlight the multifaceted contribution of obesity to breast cancer pathogenesis, through both endocrine and paracrine signaling networks that converge on estrogen metabolism, chronic inflammation, and adipokine imbalance.

3.4. *Insulin Resistance and Breast Cancer Risk*

Another hallmark of metabolic syndrome, insulin resistance, has been independently associated with increased breast cancer risk. The mechanistic basis for this relationship is thought to involve bidirectional crosstalk between estrogen signaling and the insulin/insulin-like growth factor-1 (IGF-1) axis [41]. Estrogen receptor α (ERα) activity can be modulated by insulin and IGF-1 through both genomic and non-genomic mechanisms, including phosphorylation of ERα and its coregulators via downstream PI3K/AKT and MAPK signaling cascades [42,43]. Estrogen enhances the expression and activity of key mediators of the insulin/IGF-1 pathway, thereby amplifying proliferative and anti-apoptotic signaling in breast epithelial cells [44]. This synergistic interplay promotes mitogenic signaling, metabolic reprogramming, and resistance to cell death, features that may contribute to the preferential association of T2DM with ER+ tumors. Moreover, hyperinsulinemia and elevated IGF-1 bioavailability in T2DM create a systemic environment that favors tumor initiation and progression via enhanced activation of mitogenic pathways [45]. Altogether, these findings support the hypothesis that metabolic- and hormone-driven signaling converge to increase susceptibility to ER+ breast cancer in women with T2DM.

3.5. *Current Models to Study Metabolic Dysregulation in Breast Cancer*

Complementary in vitro and in vivo systems are often employed to study metabolic dysregulation in breast cancer. Extensively characterized human breast cancer cell lines (i.e., ER+ MCF-7/T-47D; HER2+ SK-BR-3; TNBC MDA-MB-231/MDA-MB-436) enable controlled interrogation of glycolysis, oxidative metabolism, and lipid pathways [46]. To better replicate tumor architecture and nutrient–oxygen gradients, three-dimensional cultures and patient-derived organoid co-culture systems are increasingly employed to model cell–cell and cell–matrix interactions that influence metabolic flux, although variability in derivation, passaging, and microenvironmental fidelity remains a challenge [47]. For translational applications, animal models are commonly utilized. These include syngeneic murine tumors (e.g., 4T1, E0771) in immunocompetent hosts to study immune-metabolic crosstalk, genetically engineered mouse models and patient-derived xenografts to preserve oncogenic drivers and intratumoral heterogeneity, and obesity/insulin-resistance models to examine how adiposity, hyperinsulinemia, and dyslipidemia reprogram tumor and immune compartments [48].

4. IGF-1 Signaling in Breast Cancer

4.1. The IGF-1 Signaling Pathway

Insulin-like growth factor-1 (IGF-1) is a peptide hormone structurally related to insulin that exerts potent anabolic and mitogenic effects [49]. Its production is primarily regulated by growth hormone (GH) and represents a critical mediator of GH's systemic and tissue-specific biological actions [50]. IGF-1 signaling is initiated through binding to the IGF-1 receptor (IGF-1R), a transmembrane tyrosine kinase receptor belonging to the same family as the insulin receptor. Upon ligand binding, IGF-1R undergoes autophosphorylation and recruits adaptor proteins such as insulin receptor substrates and SH2-domain containing proteins, which function as docking molecules to propagate downstream signaling [51]. Activation of IGF-1R triggers multiple intracellular cascades that regulate cell growth, survival, and metabolism.

Two major oncogenic pathways are preferentially activated: the PI3K/AKT/mTOR pathway and the Ras/Raf/MEK/extracellular signal-regulated kinase (ERK) cascade [52]. The PI3K/AKT/mTOR axis promotes protein synthesis, cell proliferation, and metabolic reprogramming, while simultaneously inhibiting apoptosis through phosphorylation of pro-apoptotic proteins and activation of anti-apoptotic members of the Bcl-2 family [49]. In parallel, Ras/Raf/MEK/ERK signaling drives cell cycle progression and enhances mitogenic activity, further contributing to tumor initiation and progression.

4.2. Effect of IGF-1 Signaling on Breast Cancer Biology

IGF-1 has long been implicated in mammary gland biology, where it plays a critical role in ductal morphogenesis, mammary gland development, and maintenance of epithelial cell homeostasis [53]. In breast cancer, aberrant activation of the IGF axis is frequently observed. Both preclinical and clinical studies demonstrate that the IGF-1R is overexpressed and hyperphosphorylated across multiple breast cancer subtypes, thereby providing a mechanistic basis for its oncogenic contribution [54]. Elevated circulating levels of IGF-1 and its primary binding protein, IGFBP-3, have been identified as independent risk factors for both breast cancer development and recurrence in the general population [55]. This association is particularly strong for ER+ tumors, irrespective of menopausal status, suggesting that IGF-1 synergizes with estrogen receptor signaling to drive tumor initiation and progression [55].

In addition, metabolic dysregulation and obesity appear to amplify IGF-1-mediated oncogenic effects. Increased BMI and associated dysmetabolic states have been shown to influence IGF-1 signaling, particularly in human epidermal growth factor receptor 2 (HER2)-positive patients, in which overweight status is correlated with enhanced IGF pathway activity and worse clinical outcomes [56]. IGF-1 may also function as a prognostic biomarker. Several studies have suggested that elevated circulating IGF-1 confers negative prognostic significance, both in unselected breast cancer populations [57] and in women undergoing endocrine therapy, where IGF-1 activation has been linked to therapeutic resistance and disease progression [57,58].

4.3. IGF-1 Dysregulation in Breast Cancer Treatment Resistance

Importantly, dysregulation of the IGF-1/IGF-1R axis has been increasingly implicated in therapy resistance in breast cancer. In ER+ disease, IGF-1R signaling exhibits extensive crosstalk with the ERα pathway, enhancing ligand-independent ER activation and promoting resistance to endocrine therapies such as tamoxifen and aromatase inhibitors [59]. Similarly, in HER2+ breast cancer, IGF-1R cooperates with HER2/ERBB2 signaling to sustain downstream PI3K/AKT and MAPK activation, thereby undermining the efficacy of HER2-targeted agents such as trastuzumab [60]. Beyond receptor crosstalk, IGF-1-mediated activation of survival pathways also contributes to resistance against chemotherapy and radiotherapy by promoting DNA damage repair, inhibiting apoptosis, and enhancing epithelial-to-mesenchymal transition, which fosters metastatic progression [61]. In summary, these findings demonstrate that IGF-1R signaling not only drives tumorigenesis but also serves as a critical mediator of therapeutic resistance across multiple breast cancer subtypes, highlighting the potential of targeting IGF-1R or its downstream pathways as a strategy to overcome resistance and improve patient outcomes.

5. Glucose Metabolism Reprograming in the Tumor Microenvironment

Glucose metabolism reprogramming exerts profound effects not only within breast cancer cells but also across the broader tumor microenvironment, where it influences the function of immune and stromal compartments, including T cells, macrophages, and cancer-associated fibroblasts (CAFs) [62]. Crosstalk among these cell types is largely mediated by key glucose metabolites such as lactate and pyruvate. Elevated intratumoral lactate accumulation and the resulting acidification suppress antitumor immunity, whereas reducing lactate levels has been shown to enhance immunosurveillance and improve responses to immunotherapies [63].

Lactate also functions as a critical immunomodulator, driving the polarization of macrophages from pro-inflammatory M1-like states to immunosuppressive M2-like phenotypes, while exerting context-dependent effects on T cell function [64]. In parallel, CAFs undergo a "reverse Warburg effect," wherein they secrete lactate and pyruvate that are subsequently imported by adjacent cancer cells, fueling tumor proliferation and contributing to multidrug resistance [65,66]. Distinct metabolic programs further delineate T cell subsets: effector T cells rely on glycolysis and glutamine metabolism to support proliferation and differentiation, while regulatory T cells (Tregs) favor oxidative metabolism and display reduced glycolytic activity. This divergence may partly explain why TNBC tumors characterized by heightened glycolytic activity often demonstrate poor responsiveness to immune checkpoint blockade [63]. Together, these findings emphasize the central role of glucose metabolism in shaping tumor–immune–stromal interactions and suggest that targeting metabolic reprogramming may enhance antitumor immunity and improve responsiveness to existing therapies in breast cancer.

6. Targeting Glucose Metabolism in Breast Cancer

6.1. *Metformin*

Metformin, a first-line oral antidiabetic agent belonging to the biguanide class, is widely used for the management of T2DM due to its ability to reduce circulating glucose levels and enhance insulin sensitivity in peripheral tissues [67]. Over the past twenty years, a growing body of research has linked metformin use to lower rates of cancer incidence and mortality, sparking considerable interest in exploring its potential as an anticancer therapy [68,69]. At the systemic level, metformin reduces hepatic gluconeogenesis and consequently lowers circulating insulin concentrations, thereby attenuating activation of insulin- and IGF-1-dependent signaling cascades that are frequently co-opted in tumor growth [70]. At the cellular level, metformin has been shown to activate AMP-activated protein kinase (AMPK), a central metabolic sensor and tumor suppressor [71]. Elevated AMPK activity exerts downstream inhibitory effects on two critical oncogenic signaling axes, the PI3K/AKt/mTOR and Ras/Raf/MEK/ERK pathways, both of which regulate protein synthesis, cell cycle progression, and proliferation [72]. By dampening these anabolic and mitogenic signals, metformin can inhibit tumor cell growth and survival.

6.2. Metformin Preclinical Evidence

In preclinical models of breast cancer, metformin treatment has produced positive effects on dampening tumor progression. In a 2005 study by Anisimov et al., lifelong metformin administration (240 mg/kg body weight) starting at two months of age in transgenic FVB/N female mice with the HER-2/neu mammary cancer gene led to an approximate 8% increase in lifespan [73]. Although metformin did not reduce tumor incidence or the number of tumors (as all mice developed tumors), it significantly delayed tumor onset and reduced tumor size. Similarly, the Schedin laboratory demonstrated that in female Wistar rats (both with intact ovaries and those ovariectomized) fed a high-fat diet and treated with MNU to induce mammary tumors, an 8-week course of metformin (2 mg/mL) significantly lowered tumor volume [74,75]. In the ovariectomized group, metformin also reduced the presence of aromatase-positive macrophages at the tumor margins. Since aromatase converts androgens into estrogens, this finding suggests that metformin may suppress local estrogen production, potentially limiting the growth of ER-positive tumor cells [76]. Notably, metformin's anticancer activity may extend beyond hormone-responsive cancers, as it also inhibited the growth of triple-negative breast tumors derived from MDA-MB-231 cells in BALB/c-nu nude mice [77].

6.3. Metformin Clinical Studies

Several clinical trials have investigated the use of metformin in combination with standard-of-care treatments in breast cancer. The Phase II NCT02028221 trial demonstrated that metformin intake in premenopausal women with a BMI $\geq$ 25 who were at risk for breast cancer resulted in favorable changes in measures of adiposity and borderline breast density reduction compared to placebo [78]. The meaningful reductions in central adiposity, as measured by waist circumference and waist-to-hip ratio, exemplify metformin's potential as a metabolic intervention to modulate obesity-related breast cancer risk. Another study, the Phase II NCT01340300 clinical trial, investigated the effect of exercise $\pm$ metformin in patients with stage I-III breast cancer post adjuvant treatment. The trial reported that both exercise and metformin led to significantly reduced levels of fasting insulin and improvements in BMI compared to the control group (NCT01340300).

MA.32 is a Phase III randomized, double-blind, placebo-controlled international study designed to evaluate whether the addition of metformin to standard adjuvant therapy improved outcomes in patients with high-risk operable breast cancer (NCT01101438). The results demonstrated that addition of metformin to standard adjuvant therapy did not confer a significant improvement in invasive disease–free survival (IDFS) or other key breast cancer outcomes when compared to placebo [79]. Exploratory analyses from the trial raised the possibility of a clinically relevant effect of metformin in specific molecular subgroups. For instance, in HER2+ (ERBB2+) breast cancer, metformin use was associated with longer IDFS and overall survival, particularly among patients harboring at least one C allele of the rs11212617 single nucleotide variant (SNV). This observation is consistent with findings from the METTEN study, which reported a markedly higher pCR rate when metformin was added to neoadjuvant chemotherapy and HER2-targeted therapy in patients carrying the C allele (81.2% vs. 35.3% in non-carriers) [80].

The Phase II METEOR trial evaluated the combination of neoadjuvant metformin with aromatase inhibitor letrozole in postmenopausal women with ER+ breast cancer (NCT01589367) [81]. The trial reported an overall clinical response rate of 61.4% but did not reach statistical significance between the metformin + letrozole vs. placebo + letrozole cohort (66.7% versus 56.4%, $p = 0.193$). Though this difference lacked statistical significance, early reductions in Ki-67 at 4 weeks, particularly in the metformin group, was predictive of clinical response. Recently, the BREAKFAST Phase II trial investigated whether cyclic fasting-mimicking diets (FMD) with or without metformin can enhance the efficiency of neoadjuvant chemotherapy in patients with localized (Stages I–III, cT > 1 cm) TNBC (NCT04248998). Despite early termination due to adoption of chemoimmunotherapy as the new standard-of-care treatment, the trial reported an overall pCR rate of 56.6%, markedly higher than the historical range of 26–39% reported in prior Phase II/III neoadjuvant chemotherapy trials [82]. RNA sequencing of tumor samples uncovered that the patients who achieved pCR exhibited robust early infiltration of activated T cells and natural killer (NK) cells [83]. Concurrently, tumors demonstrated early downregulation of glycolysis and mitochondrial oxidative phosphorylation pathways. These findings suggest that metabolic reprogramming and immune activation underlie the enhanced response observed with FMD ± metformin.

6.4. *SGLT2 Inhibitors*

Sodium–glucose co-transporter 2 inhibitors (SGLT2i), also known as gliflozins, represent a relatively novel class of orally administered glucose-lowering agents that have demonstrated robust efficacy in the management of glycemic control. Mechanistically, SGLT2 blockade reduces renal glucose reuptake, induces glycosuria, and lowers plasma glucose levels through an insulin-independent mechanism, an important advantage in patients with impaired β-cell function or insulin resistance [84].

6.5. *SGLT2i Preclinical Evidence*

In breast cancer, SGLT2 expression has been detected in both primary human breast cancer tissue as well as cell lines via immunohistochemistry and RT-PCR [85,86]. Preclinical studies provide compelling evidence that SGLT2i can exert direct anti-tumor effects. Zhou and colleagues demonstrated that treatment with SGLT2i dapagliflozin and canagliflozin suppressed breast cancer cell proliferation both in vitro and in vivo using nude mouse xenograft models [87]. The inhibitory effects were linked to cell cycle arrest and induction of apoptosis, indicating that SGLT2 blockade disrupts essential tumor cell survival pathways. Furthermore, SGLT2i exposure modulated key signaling cascades, enhancing phosphorylation of AMP-activated protein kinase (AMPK), a central energy-sensing and tumor-suppressive kinase, while simultaneously reducing phosphorylation of 70 kDa ribosomal protein S6 kinase 1 (p70S6K1), a critical downstream effector of the mTOR pathway involved in protein synthesis and cell growth [87].

In another investigation, ipragliflozin was shown to exert a potent inhibitory effect on breast cancer cell proliferation. Specifically, treatment with ipragliflozin at concentrations ranging from 1 to 50 μM significantly and dose-dependently suppressed growth of the human ER+ breast cancer cell line, MCF-7 [88]. Importantly, this anti-proliferative effect was abrogated when SGLT2 expression was silenced via targeted knockdown, providing strong evidence that the observed growth inhibition is mediated through blockade of the SGLT2 transporter. Furthermore, SGLT2i can augment the therapeutic efficiency of cytotoxic chemotherapy. In a study by Akingbesote et al., dapagliflozin reduced tumor glucose uptake and enhanced the cytotoxic effects of paclitaxel by lowering circulating insulin, ultimately prolonging overall survival in the MMTV-PyMT murine model of breast cancer [89]. Interestingly, this benefit was not universally observed across all breast tumor subtypes. The enhanced response was restricted to tumors characterized by mutations occurring upstream of canonical insulin signaling pathways, such as those affecting receptor or proximal signaling components. In contrast, tumors harboring mutations downstream of the insulin signaling cascade, where signaling remains constitutively active irrespective of insulin levels, did not exhibit the same degree of chemosensitization.

6.6. SGLT2i Clinical Studies

Many ongoing clinical trials are investigating the combination of SGLT2i with standard-of-care anticancer therapies, including PI3K inhibitors (i.e., alpelisib), endocrine therapies (i.e., fulvestrant), and conventional chemotherapeutic regimens. The TIFA study is an ongoing multicenter, randomized Phase II clinical trial designed to assess the efficacy of SGLT2i canagliflozin in combination with standard-of-care treatment alpelisib plus fulvestrant in preventing PI3K inhibitor-associated hyperglycemia in patients with metastatic, PIK3CA-mutant, HR+HER2- breast cancer (NCT05090358). The primary endpoint is the proportion of patients who remain free from grade 3–4 hyperglycemia at 12 weeks. Key secondary endpoints extend beyond glycemic safety and include the overall response rate (ORR) and progression-free survival (PFS) at 6 and 12 months, providing insight into whether improved glycemic control translates into enhanced cancer-related efficacy. Furthermore, the study will evaluate adherence to alpelisib, indicating that effective mitigation of hyperglycemia likely leads to sustained drug exposure and optimization of therapeutic benefit.

Another clinical trial, NCT05989347, is a single-arm Phase I study that examines whether dapagliflozin, administered in conjunction with standard neoadjuvant therapies, can favorably alter metabolic markers of insulin resistance in women with early-stage HER2- breast cancer. Specifically, the study will assess changes in fasting plasma glucose and insulin levels as surrogate markers for insulin resistance. Additionally, it explores whether dapagliflozin can be safely integrated into multi-agent chemotherapy regimens without compromising tolerability. SGLT2i are also under investigation for their potential as cardioprotective agents. The multicenter, Phase III PROTECTAA trial aims to evaluate whether dapagliflozin can prevent anthracycline-related cardiotoxicity in non-diabetic patients with stage I–III invasive breast cancer (NCT06304857). The trial not only explores a novel strategy for mitigating a major treatment-limiting toxicity but also broadens the potential application of SGLT2i beyond their established roles in diabetes, heart failure, and chronic kidney disease.

6.7. Glucagon-like Peptide-1 Receptor Agonists

Glucagon-like peptide-1 (GLP-1) is an incretin hormone produced by intestinal L-cells in response to food intake, increasing insulin secretion and inhibiting glucagon release from the pancreas. However, therapeutic use of GLP-1 is limited by its short half-life due to rapid degradation by dipeptidyl peptidase-4 (DPP-4) [90]. GLP-1 receptor agonists (GLP-1RAs) were developed to mimic the action of native GLP-1 and engineered to resist DPP-4 degradation. As a result, they are highly efficacious and widely utilized for the management of T2DM [91]. In addition, GLP-1RAs have pleotropic effects on reducing plasma glucose, inducing weight loss, and immune modulation [92], resulting in their recent FDA approval for the treatment of obesity [93]. Recently, GLP-1RAs have garnered attention for their multifaceted roles beyond glycemic control, including in malignancy through modulating anti-tumor immunity and regulating signaling pathways involved in cancer cell growth and survival. In breast cancer, various GLP-1RAs have been shown to result in a decrease in tumor size, weight, and proliferation index [94].

6.8. GLP-1R Expression in Breast Cancer

Studies have demonstrated that GLP-1R is expressed on human breast cancer cell lines as well as primary breast cancer tissue. Iwaya et al. reported the expression of membrane-localized GLP-1R in human breast cancer tissue via immunohistochemical staining for GLP-1R and three human breast cancer cells lines (MCF-7, MDA-MB-231, and KPL-1) [95]. Among the breast cancer cell lines, MCF-7 cells, which display high estrogen-sensitive proliferation, had the highest *GLP-1R* gene expression as measured by quantitative real-time PCR (RT-PCR). In addition, transfection of MCF-7 cells using a GLP-1R-expressing lentiviral vector containing a Flag epitope tag showed exclusive GLP-1R expression to the breast cancer cell membrane rather than the cytosol or nucleus. Similarly, Tanaka et al. also confirmed that GLP-1R is expressed on MCF-7 cells via immunofluorescence analysis [96]. Another study showed that GLP-1R was widely expressed in many human breast cancer cell lines, including MCF-7, MDA-MB-231, BT483, MDA-MB-468, and ZR751, via RT-PCR and Western blot analyses [97].

6.9. Exendin-4

Exendin-4 (Ex-4) is a GLP-1RA that mimics the incretin hormone GLP-1, enhancing insulin secretion and exhibiting glucose-lowering effects [98]. Beyond its role in diabetes treatment, emerging evidence suggests that Ex-4 may influence tumor cell dynamics, including those of breast cancer. Fidan-Yaylali et al. demonstrated that Ex-4 significantly suppressed the proliferation of MCF-7 breast cancer cells by inducing apoptosis and inhibiting MCF-7 cell migration, invasion, and colony formation [99].

In a similar vein, Iwaya et al. showed that Ex-4 attenuated breast cancer cell proliferation in vitro in MCF-7 cells via inhibition of NF-κB activation and subsequent target gene expression, including VEGF, IL-8, and COX-2 [95]. Treatment with Ex-4 dramatically reduced Ki-67 expression, a marker of cancer cell proliferation and cell cycle progression. Furthermore, Ex-4 treatment inactivated Akt and IκBα phosphorylation in MCF-7 cells. Anti-carcinogenesis activity was also present in vivo, with significantly decreased tumor size observed in Ex-4-treated athymic nude mice transplanted with MCF-7 cells. Nine weeks after subcutaneous transplantation of MCF-7 cells, tumor size was nearly halved in Ex-4-treated mice compared to controls, while body weight and blood glucose levels remained unchanged. This suggests that Ex-4 suppresses tumor growth independently of its systemic metabolic effects.

6.10. Liraglutide

Preclinical studies evaluating the effects of liraglutide on breast cancer cells demonstrate that the second generation GLP-1RA exhibits both anti-proliferative and pro-proliferative properties. In one report, liraglutide demonstrated the ability to suppress proliferation and induce apoptosis in MCF-7 breast cancer cells [100]. Treatment with liraglutide resulted in decreased cell viability and increased apoptotic factors, such as cleaved-caspase-3. These effects were dose- and time-dependent, with significant changes observed at higher concentrations and prolonged exposure times. Mechanistically, liraglutide downregulates microRNA-27a (miR-27a), leading to the upregulation of AMP-activated protein kinase catalytic subunit α2 (AMPKα2), a tumor suppressor gene involved in cell cycle regulation and apoptosis. The modulation of miR-27a and AMPKα2 suggests a pathway through which liraglutide exerts its antiproliferative effects on estrogen receptor-positive breast cancer cells.

Similarly, Alanteet et al. showed that in models simulating obesity-associated breast cancer, liraglutide attenuated tumor cell proliferation by inhibiting the PI3K/Akt/mTOR signaling pathway [101]. Growth of liraglutide-treated MCF-7 cells incubated in conditioned media generated from human adipose-derived stem cells (ADSCs) significantly decreased by approximately 48% when compared to no treatment. When treated with liraglutide, MCF-7 cells from either lean ADSCs or obese ADSCs displayed significantly reduced protein levels of phosphorylated (p)-PI3K, p-Akt, and p-mTOR. Furthermore, the anti-proliferative effects exhibited by liraglutide were similar to those mediated by PI3K and mTOR inhibitors, LY294002 and rapamycin, respectively. Contrasting findings have emerged in studies involving human triple-negative breast cancer (TNBC) cell lines. In one report, a slightly higher concentration of liraglutide accelerated breast cancer progression in vitro and in vivo through the NOX4/ROS/VEGF signaling pathway after activation the GLP-1 receptor [97]. In vitro, liraglutide promoted the proliferation and migration ability of two TNBC cell lines, MDA-MB-231 and MDA-MB-468, by significantly increasing intracellular ATP. This effect was able to be reversed by GLP-1R inhibitor, Exendin.

6.11. Semaglutide

Semaglutide, a second-generation incretin analog, has recently been shown to exhibit anti-tumor effects in a murine model of TNBC. Intraperitoneal administration of semaglutide in BALB/c mice implanted with syngeneic 4T1 TNBC cells resulted in decelerated tumor growth, appearance, and progression [102]. Notably, semaglutide-treated mice had significantly smaller tumor diameters compared to the control group on day 14 after tumor induction ($p = 0.011$). The authors further explored its effects on anti-tumor immunity, demonstrating that semaglutide enhanced the activation and cytotoxicity of CD8+ T cells within the tumor microenvironment, modulated macrophage polarization towards a pro-inflammatory, M1-phenotype, and increased infiltration and activation of B cells, contributing to a more robust immune response. In vitro, semaglutide stimulated the production of pro-inflammatory cytokines TNF-α, IFN-γ, IL-10, and IL-1β after 24 h of incubation compared to untreated cells. This study for the first time demonstrated the effects of semaglutide on antitumor immunity in a murine model of breast carcinoma.

6.12. GLP-1RAs Clinical Studies

While preclinical studies provide evidence for the anticancer effects of GLP-1RAs in breast cancer, clinical data remain limited. The ongoing Phase II clinical trial NCT06518837 aims to evaluate the role of tirzepatide, a dual GLP-1 and Gastric Inhibitory Polypeptide (GIP) receptor agonist, in patients with HR+HER2- breast cancer receiving adjuvant treatment. The primary objective is to determine the proportion of patients achieving a clinically meaningful weight reduction of $\geq$5% from baseline, a benchmark associated with improved metabolic health and potentially enhanced cancer outcomes. Secondary endpoints include 3-year IDFS and distant relapse-free survival, as well as changes in BMI, fat distribution, metabolic biomarkers, and circulating tumor DNA (ctDNA). Safety, tolerability, and feasibility, measured by treatment discontinuation and completion rates, will also be assessed. The incorporation of exploratory biomarkers, such as ctDNA and adiposity measures, allows the trial to provide mechanistic insights into how metabolic interventions may interact with tumor biology.

The TRIM-EBC trial (NCT06517212) is an ongoing study evaluating tirzepatide-induced weight loss as a strategy to improve outcomes in patients with early-stage, high-risk breast cancer. Specifically, the trial is enrolling individuals with HR+HER2- and node-positive (N+) early breast cancer who are overweight or obese and who have evidence of minimal residual disease (MRD), as detected by ctDNA. The central hypothesis is that pharmacologically induced weight loss with tirzepatide will trigger favorable metabolic and hormonal changes that impair the growth and survival of micrometastatic disease, thereby promoting ctDNA clearance and reducing progression to overt metastasis. The primary endpoints are (1) clearance of plasma ctDNA within two years of documented positivity at study entry and (2) the proportion of patients who remain alive and free of distant metastatic disease two years after ctDNA detection. By targeting obesity-related metabolic drivers of breast cancer recurrence, this trial has the potential to establish a new paradigm in which metabolic modulation complements standard oncologic therapies to improve long-term disease control in a high-risk patient population.

A summary of completed and ongoing clinical trials evaluating the effect of anti-diabetic therapies in breast cancer is summarized in Table 1 [78,79,81,82,103–106]. A schematic depicting the intersection of metabolic syndrome and breast cancer development, highlighting key signaling pathways and mechanisms of action of anti-diabetic medications, is shown in Figure 1.

Table 1. List of Clinical Trials Evaluating the Effect of Anti-diabetic Therapies in Breast Cancer.

NCT ID	Reference	Patient Population	Therapy	Phase	Number of Patients Enrolled	Major Clinical Findings	Metabolic Benefit	Status
NCT01589367	Kim et al. (2014) [81]	Postmenopausal ER+ early-stage BC, neoadjuvant setting	Letrozole ± metformin	Phase II	208	Overall clinical response rate of 61.4%, no statistical significance between the metformin + letrozole vs. placebo + letrozole cohort (66.7% versus 56.4%, $p = 0.193$).	N/A	Completed
NCT01980823	—	Newly diagnosed operable breast cancer ("window of opportunity")	Metformin + atorvastatin	Early Phase I	23	Determine if dual combination treatment significantly impacts Ki-67 proliferation index pre-surgery, results pending	N/A	Completed
NCT00930579	Kalinsky et al. (2015) [103]	Newly diagnosed DCIS or stage I-III early invasive BC in non-diabetic women with BMI $\geq$ 25	Metformin	Phase II (pre-surgical)	35	No change in Ki-67 proliferation index	Significant improvement in BMI and reductions in leptin and cholesterol with metformin	Completed
NCT01793948	—	Overweight/ obese women at elevated breast cancer risk	Metformin vs. placebo	Phase 0	24	Biomarker analysis in prevention setting, results pending	N/A	Completed
NCT01340300	Meyerhardt et al. (2019) [104]	Stage I-III breast or colorectal cancer survivors post adjuvant treatment	Exercise ± metformin	Phase II	139	N/A (primary endpoint is metabolic profile)	Both exercise and metformin led to significantly reduced levels of fasting insulin and improvements in BMI compared to the control group	Completed
NCT01101438	Goodwin et al. (2022) [79]	Early-stage BC, adjuvant setting	Metformin vs. placebo	Phase III	3649	No improvement in IDFS; exploratory analyses show that metformin use was associated with longer IDFS and OS in HER2+ BC	N/A	Completed

Table 1. *Cont.*

NCT ID	Reference	Patient Population	Therapy	Phase	Number of Patients Enrolled	Major Clinical Findings	Metabolic Benefit	Status
NCT02028221	Tapia et al. (2021) [78]	Premenopausal women with BMI ≥ 25 at risk for BC	Metformin vs. placebo	Phase II	151	Compared to placebo, metformin did not change % breast density and dense breast volume but led to a numerical but not significant decrease in non-dense breast volume	Favorable changes in measures of adiposity and borderline breast density reduction in metformin group	Completed
NCT01310231	Pimentel et al. (2019) [105]	Metastatic breast cancer in non-diabetic women	Standard chemotherapy + metformin	Phase II	40	No significance found on response rate, PFS, or OS	N/A	Completed
NCT06763328	—	ER- stage I–III invasive BC	Metformin	Phase III	200	Determine if metformin can prevent or reverse insulin resistance in BC patients after chemotherapy, results pending	Results pending	Recruiting
NCT04248998	Ligorio et al. (2025) [82]	Localized TNBC	Cyclic fasting-mimicking diets ± metformin	Phase II	30	Overall pCR rate of 56.6% was markedly higher than the historical range of 26–39% reported in prior Phase II/III neoadjuvant chemotherapy trials	N/A	Active, not recruiting
NCT05023967 (TEAM study)	—	Non-diabetic patients with early-stage breast cancer (prevention/biomarker setting)	Time-restricted eating (nightly fasting) + extended-release metformin vs. control	Phase IIb	120	Determine whether nightly fasting + metformin reduces tumor cell proliferation and related biomarkers (e.g., Ki-67)	Results pending	Recruiting
NCT01042379 (I-SPY Trial)	Yee et al. (2021) [106]	High-risk stage II/III HER2- operable breast cancer (neoadjuvant platform; multiple experimental arms)	One reported arm: Paclitaxel + ganitumab + metformin (PGM) → AC vs. paclitaxel → AC (SOC)	Phase II (adaptive platform)	5000	PGM showed numerically higher pCR in HR-/HER2- patients (32% vs. 21%) but did not meet graduation thresholds; no EFS improvement	N/A	Recruiting; PGM arm completed within platform
NCT05090358	—	HR+HER2-, PIK3CA-mutant metastatic BC on alpelisib + fulvestrant	Canagliflozin vs. diet	Phase II	15	Primary endpoint: proportion of patients who remain free from grade 3–4 hyperglycemia at 12 weeks; secondary endpoints: ORR and PFS at 6 and 12 months. Results pending.	Results pending	Active, not recruiting

Table 1. *Cont.*

NCT ID	Reference	Patient Population	Therapy	Phase	Number of Patients Enrolled	Major Clinical Findings	Metabolic Benefit	Status
NCT05989347	—	Early-stage HER2- BC in hyperinsulinemic women	Dapagliflozin + neoadjuvant therapy	Phase I	20	Assess changes in fasting plasma glucose and insulin levels as surrogate markers for insulin resistance. Results pending.	Results pending	Recruiting
NCT06304857	—	Non-diabetic patients with early-stage invasive BC receiving anthracyclines	Dapagliflozin vs. placebo	Phase III	188	Evaluate whether dapagliflozin can prevent anthracycline-related cardiotoxicity	Results pending	Recruiting
NCT05025735	—	PIK3CA-mutant metastatic HR+HER2- BC	Dapagliflozin + alpelisib + fulvestrant	Phase II	25	Evaluate whether the addition of dapagliflozin to alpelisib and fulvestrant leads to significant reduction in all-grade hyperglycemia	Results pending	Recruiting
NCT06518837	—	HR+HER2- early-stage BC patients receiving adjuvant treatment	Tirzepatide, a dual GLP-1/GIP receptor agonist	Phase II	40	Primary objective is to determine the proportion of patients achieving a clinically meaningful weight reduction of ≥5% from baseline; secondary endpoints include 3-year IDFS and changes in BMI, fat distribution, metabolic biomarkers, and ctDNA	Results pending	Recruiting
NCT06517212	—	Overweight or obese patients with early-stage, HR+HER2-, node-positive breast cancer and detectable minimal residual disease via ctDNA	Tirzepatide	Phase II	48	Aims to assess whether tirzepatide-induced weight loss leads to metabolic, hormonal, and immunologic changes that result in clearance of ctDNA and reduce progression to overt metastasis	Results pending	Recruiting

Abbreviations: ER = estrogen receptor; BC = breast cancer; DCIS: ductal carcinoma in situ; BMI = body mass index; IDFS = invasive disease-free survival; OS = overall survival; PFS = progression-free survival; pCR = pathological complete response; HR = hormone positive; HER2 = human epidermal growth factor receptor 2; PIK3CA = phosphatidylinositol-4,5-bisphosphate 3-kinase catalytic subunit alpha; ctDNA = circulating tumor DNA; AC = Adriamycin/Cytoxan; SOC = standard of care; EFS = event-free survival.

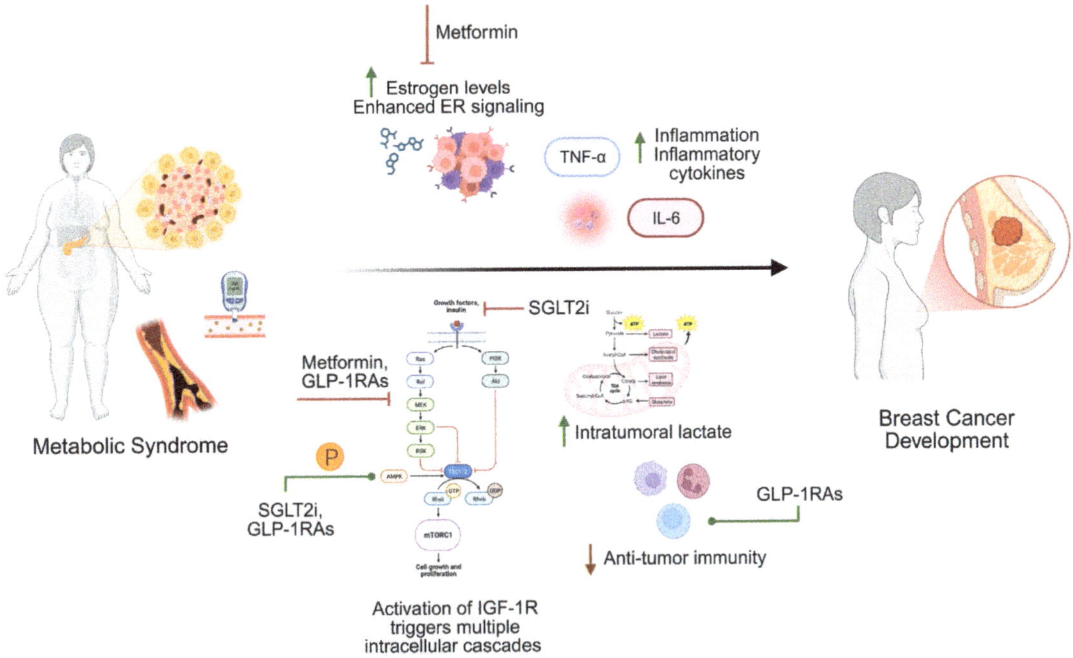

Figure 1. Schematic Illustrating the Intersection of Metabolic Syndrome and Breast Cancer Development.

Key components of metabolic syndrome, including obesity, insulin resistance, dyslipidemia, and chronic inflammation, promote tumor initiation and progression through alterations in insulin/IGF-1 signaling and major oncogenic pathways PI3K/AKT/mTOR and Ras/Raf/MEK/ERK, estrogen production, glucose metabolism reprogramming, and immune modulation. The diagram also highlights proposed mechanisms by which antidiabetic agents, such as GLP-1RAs, SGLT2i, and metformin, may counteract these processes to inhibit tumor growth and improve therapeutic response. Abbreviations: ER = estrogen receptor; SGLT2i = sodium–glucose co-transporter 2 inhibitors; GLP-1RAs = glucagon-like peptide-1 receptor agonists; *p* = phosphorylation; IGF-1R = insulin-like growth factor 1 receptor; TNF-α = tumor necrosis factor-α; IL-6 = interleukin-6. Created in BioRender.com.

7. Limitations of Metabolic Therapy in Breast Cancer

Despite promising preclinical data, the therapeutic application of metabolic therapies such as GLP-1RAs and SGLT2 inhibitors in breast cancer remains preliminary. Few oncology-specific clinical trials have been conducted, and existing safety and efficacy data do not yet justify broad clinical application. One potential limitation to consider is their side effect profile. GLP-1RAs frequently cause gastrointestinal adverse effects and warrant careful monitoring, particularly when combined with chemotherapy, which can further worsen dehydration and impair oral intake [107]. Moreover, GLP-1RAs may have off-target effects, variable bioavailability in tumor tissues, and potential interactions with other therapies that require careful evaluation [108]. SGLT2 inhibitors can increase the risk of genitourinary infections, which complicates peri-treatment management and requires careful patient selection and monitoring [109]. The heterogeneity of patient metabolic phenotypes and breast cancer subtypes also introduces substantial complexity. Future studies are needed to evaluate metabolic therapies in well-designed clinical trials before definitive conclusions can be established.

8. Conclusions

Metabolic dysregulation is increasingly recognized as a driver of breast cancer pathogenesis, influencing not only tumor-intrinsic signaling but also shaping the immune microenvironment through metabolism-immune interactions. Aberrant glucose and lipid metabolism can alter immune cell composition and function, such as T cell exhaustion, macrophage polarization, and fibroblast activation, to support tumor progression and immune evasion. Recognizing this metabolic-immune axis is critical for understanding disease biology and identifying new therapeutic targets.

Although preclinical studies provide compelling evidence that anti-diabetic and anti-obesity agents can modulate these pathways and potentially enhance antitumor immunity and therapeutic response, clinical validation remains limited and mixed. Large, randomized clinical trials of agents such as metformin have yielded inconsistent results, highlighting the challenges in translating metabolic interventions from the bench to the bedside. Moreover, the well-recognized gap between preclinical efficacy and successful oncologic drug development warrants cautious interpretation of these findings. Future research should prioritize identifying patient subgroups most likely to benefit, integrating metabolic strategies with standard and immunotherapeutic approaches and leveraging biomarkers such as circulating metabolites and immune signatures to optimize clinical trial design. Rigorous translational and clinical studies will be essential to determine whether targeting the metabolic-immune axis can meaningfully improve outcomes for patients with breast cancer.

Author Contributions: Conceptualization, Y.Y. and S.L.S.; data curation and writing—original draft preparation, L.C.; writing—review and editing, Y.Y. and S.L.S.; visualization, L.C.; funding acquisition, Y.Y. and S.L.S. All authors have read and agreed to the published version of the manuscript.

Funding: This research was funded by Cedars-Sinai Cancer Developmental Funds.

Institutional Review Board Statement: Not applicable.

Informed Consent Statement: Not applicable.

Data Availability Statement: Not applicable.

Conflicts of Interest: The authors declare no conflicts of interest.

References

1. Sung, H.; Ferlay, J.; Siegel, R.L.; Laversanne, M.; Soerjomataram, I.; Jemal, A.; Bray, F. Global Cancer Statistics 2020: GLOBOCAN Estimates of Incidence and Mortality Worldwide for 36 Cancers in 185 Countries. *CA Cancer J. Clin.* **2021**, *71*, 209–249. [CrossRef]
2. Obeagu, E.I.; Obeagu, G.U. Breast cancer: A review of risk factors and diagnosis. *Medicine* **2024**, *103*, e36905. [CrossRef]
3. Pascual-Morena, C.; Garrido-Miguel, M.; Martinez-Garcia, I.; Luceron-Lucas-Torres, M.; Rodriguez-Gutierrez, E.; Berlanga-Macias, C.; Fernandez-Bravo-Rodrigo, J.; Patino-Cardona, S. Association of Dietary Advanced Glycation End Products with Overall and Site-Specific Cancer Risk and Mortality: A Systematic Review and Meta-Analysis. *Nutrients* **2025**, *17*, 1638. [CrossRef]
4. Glaviano, A.; Foo, A.S.C.; Lam, H.Y.; Yap, K.C.H.; Jacot, W.; Jones, R.H.; Eng, H.; Nair, M.G.; Makvandi, P.; Geoerger, B.; et al. PI3K/AKT/mTOR signaling transduction pathway and targeted therapies in cancer. *Mol. Cancer* **2023**, *22*, 138. [CrossRef]
5. He, Y.; Sun, M.M.; Zhang, G.G.; Yang, J.; Chen, K.S.; Xu, W.W.; Li, B. Targeting PI3K/Akt signal transduction for cancer therapy. *Signal Transduct. Target. Ther.* **2021**, *6*, 425. [CrossRef]
6. Andre, F.; Ciruelos, E.; Rubovszky, G.; Campone, M.; Loibl, S.; Rugo, H.S.; Iwata, H.; Conte, P.; Mayer, I.A.; Kaufman, B.; et al. Alpelisib for PIK3CA-Mutated, Hormone Receptor-Positive Advanced Breast Cancer. *N. Engl. J. Med.* **2019**, *380*, 1929–1940. [CrossRef] [PubMed]

7. Turner, N.C.; Oliveira, M.; Howell, S.J.; Dalenc, F.; Cortes, J.; Gomez Moreno, H.L.; Hu, X.; Jhaveri, K.; Krivorotko, P.; Loibl, S.; et al. Capivasertib in Hormone Receptor-Positive Advanced Breast Cancer. *N. Engl. J. Med.* **2023**, *388*, 2058–2070. [CrossRef] [PubMed]
8. Baselga, J.; Campone, M.; Piccart, M.; Burris, H.A., 3rd; Rugo, H.S.; Sahmoud, T.; Noguchi, S.; Gnant, M.; Pritchard, K.I.; Lebrun, F.; et al. Everolimus in postmenopausal hormone-receptor-positive advanced breast cancer. *N. Engl. J. Med.* **2012**, *366*, 520–529. [CrossRef]
9. Jhaveri, K.L.; Im, S.A.; Saura, C.; Loibl, S.; Kalinsky, K.; Schmid, P.; Loi, S.; Thanopoulou, E.; Shankar, N.; Jin, Y.; et al. Overall Survival with Inavolisib in PIK3CA-Mutated Advanced Breast Cancer. *N. Engl. J. Med.* **2025**, *393*, 151–161. [CrossRef] [PubMed]
10. Liu, D.; Weintraub, M.A.; Garcia, C.; Goncalves, M.D.; Sisk, A.E.; Casas, A.; Harding, J.J.; Harnicar, S.; Drilon, A.; Jhaveri, K.; et al. Characterization, management, and risk factors of hyperglycemia during PI3K or AKT inhibitor treatment. *Cancer Med.* **2022**, *11*, 1796–1804. [CrossRef]
11. Pati, S.; Irfan, W.; Jameel, A.; Ahmed, S.; Shahid, R.K. Obesity and Cancer: A Current Overview of Epidemiology, Pathogenesis, Outcomes, and Management. *Cancers* **2023**, *15*, 485. [CrossRef]
12. Pliszka, M.; Szablewski, L. Associations between Diabetes Mellitus and Selected Cancers. *Int. J. Mol. Sci.* **2024**, *25*, 7476. [CrossRef]
13. Dehesh, T.; Fadaghi, S.; Seyedi, M.; Abolhadi, E.; Ilaghi, M.; Shams, P.; Ajam, F.; Mosleh-Shirazi, M.A.; Dehesh, P. The relation between obesity and breast cancer risk in women by considering menstruation status and geographical variations: A systematic review and meta-analysis. *BMC Womens Health* **2023**, *23*, 392. [CrossRef]
14. Gaudet, M.M.; Press, M.F.; Haile, R.W.; Lynch, C.F.; Glaser, S.L.; Schildkraut, J.; Gammon, M.D.; Douglas Thompson, W.; Bernstein, J.L. Risk factors by molecular subtypes of breast cancer across a population-based study of women 56 years or younger. *Breast Cancer Res. Treat.* **2011**, *130*, 587–597. [CrossRef]
15. Elreda, L.; Kim, A.; Malik, M. Mitigating Breast Cancer Disparities by Addressing the Obesity Epidemic. *Curr. Breast Cancer Rep.* **2022**, *14*, 168–173. [CrossRef]
16. Sankofi, B.M.; Valencia-Rincon, E.; Sekhri, M.; Ponton-Almodovar, A.L.; Bernard, J.J.; Wellberg, E.A. The impact of poor metabolic health on aggressive breast cancer: Adipose tissue and tumor metabolism. *Front. Endocrinol.* **2023**, *14*, 1217875. [CrossRef] [PubMed]
17. Hopkins, B.D.; Goncalves, M.D.; Cantley, L.C. Obesity and Cancer Mechanisms: Cancer Metabolism. *J. Clin. Oncol.* **2016**, *34*, 4277–4283. [CrossRef]
18. Nwabo Kamdje, A.H.; Seke Etet, P.F.; Kipanyula, M.J.; Vecchio, L.; Tagne Simo, R.; Njamnshi, A.K.; Lukong, K.E.; Mimche, P.N. Insulin-like growth factor-1 signaling in the tumor microenvironment: Carcinogenesis, cancer drug resistance, and therapeutic potential. *Front. Endocrinol.* **2022**, *13*, 927390. [CrossRef] [PubMed]
19. Shum, H.C.E.; Wu, K.; Vadgama, J.; Wu, Y. Potential Therapies Targeting the Metabolic Reprogramming of Diabetes-Associated Breast Cancer. *J. Pers. Med.* **2023**, *13*, 157. [CrossRef] [PubMed]
20. Chlebowski, R.T.; Aragaki, A.K.; Pan, K.; Simon, M.S.; Neuhouser, M.L.; Haque, R.; Rohan, T.E.; Wactawski-Wende, J.; Orchard, T.S.; Mortimer, J.E.; et al. Breast cancer incidence and mortality by metabolic syndrome and obesity: The Women's Health Initiative. *Cancer* **2024**, *130*, 3147–3156. [CrossRef]
21. Neeland, I.J.; Lim, S.; Tchernof, A.; Gastaldelli, A.; Rangaswami, J.; Ndumele, C.E.; Powell-Wiley, T.M.; Despres, J.P. Metabolic syndrome. *Nat. Rev. Dis. Primers* **2024**, *10*, 77. [CrossRef]
22. Nahmias-Blank, D.; Maimon, O.; Meirovitz, A.; Sheva, K.; Peretz-Yablonski, T.; Elkin, M. Excess body weight and postmenopausal breast cancer: Emerging molecular mechanisms and perspectives. *Semin. Cancer Biol.* **2023**, *96*, 26–35. [CrossRef]
23. Pang, Y.; Wei, Y.; Kartsonaki, C. Associations of adiposity and weight change with recurrence and survival in breast cancer patients: A systematic review and meta-analysis. *Breast Cancer* **2022**, *29*, 575–588. [CrossRef]
24. Ewertz, M.; Jensen, M.B.; Gunnarsdottir, K.A.; Hojris, I.; Jakobsen, E.H.; Nielsen, D.; Stenbygaard, L.E.; Tange, U.B.; Cold, S. Effect of obesity on prognosis after early-stage breast cancer. *J. Clin. Oncol.* **2011**, *29*, 25–31. [CrossRef]
25. Yuan, Y.; Pan, K.; Mortimer, J.; Chlebowski, R.T.; Luo, J.; Yan, J.E.; Yost, S.E.; Kroenke, C.H.; Adams-Campbell, L.; Nassir, R.; et al. Metabolic syndrome risk components and mortality after triple-negative breast cancer diagnosis in postmenopausal women in the Women's Health Initiative. *Cancer* **2021**, *127*, 1658–1667. [CrossRef]
26. Xiong, F.; Dai, Q.; Zhang, S.; Bent, S.; Tahir, P.; Van Blarigan, E.L.; Kenfield, S.A.; Chan, J.M.; Schmajuk, G.; Graff, R.E. Diabetes and incidence of breast cancer and its molecular subtypes: A systematic review and meta-analysis. *Diabetes Metab. Res. Rev.* **2024**, *40*, e3709. [CrossRef]
27. Sudan, S.K.; Sharma, A.; Vikramdeo, K.S.; Davis, W.; Deshmukh, S.K.; Poosarla, T.; Holliday, N.P.; Prodduturvar, P.; Nelson, C.; Singh, K.P.; et al. Obesity and Early-Onset Breast Cancer and Specific Molecular Subtype Diagnosis in Black and White Women: NIMHD Social Epigenomics Program. *JAMA Netw. Open* **2024**, *7*, e2421846. [CrossRef]
28. Richardson, L.C.; Henley, S.J.; Miller, J.W.; Massetti, G.; Thomas, C.C. Patterns and Trends in Age-Specific Black-White Differences in Breast Cancer Incidence and Mortality—United States, 1999–2014. *MMWR Morb. Mortal. Wkly. Rep.* **2016**, *65*, 1093–1098. [CrossRef] [PubMed]

29. Kurylowicz, A. Estrogens in Adipose Tissue Physiology and Obesity-Related Dysfunction. *Biomedicines* **2023**, *11*, 690. [CrossRef] [PubMed]
30. Steiner, B.M.; Berry, D.C. The Regulation of Adipose Tissue Health by Estrogens. *Front. Endocrinol.* **2022**, *13*, 889923. [CrossRef] [PubMed]
31. Yager, J.D.; Davidson, N.E. Estrogen carcinogenesis in breast cancer. *N. Engl. J. Med.* **2006**, *354*, 270–282. [CrossRef]
32. Bulun, S.E.; Chen, D.; Moy, I.; Brooks, D.C.; Zhao, H. Aromatase, breast cancer and obesity: A complex interaction. *Trends Endocrinol. Metab.* **2012**, *23*, 83–89. [CrossRef]
33. Bowers, L.W.; Brenner, A.J.; Hursting, S.D.; Tekmal, R.R.; deGraffenried, L.A. Obesity-associated systemic interleukin-6 promotes pre-adipocyte aromatase expression via increased breast cancer cell prostaglandin E2 production. *Breast Cancer Res. Treat.* **2015**, *149*, 49–57. [CrossRef]
34. Howe, L.R.; Subbaramaiah, K.; Hudis, C.A.; Dannenberg, A.J. Molecular pathways: Adipose inflammation as a mediator of obesity-associated cancer. *Clin. Cancer Res.* **2013**, *19*, 6074–6083. [CrossRef]
35. Rothenberger, N.J.; Somasundaram, A.; Stabile, L.P. The Role of the Estrogen Pathway in the Tumor Microenvironment. *Int. J. Mol. Sci.* **2018**, *19*, 611. [CrossRef] [PubMed]
36. Duan, F.; Wu, J.; Chang, J.; Peng, H.; Liu, Z.; Liu, P.; Han, X.; Sun, T.; Shang, D.; Yang, Y.; et al. Deciphering endocrine function of adipose tissue and its significant influences in obesity-related diseases caused by its dysfunction. *Differentiation* **2025**, *141*, 100832. [CrossRef] [PubMed]
37. Garofalo, C.; Surmacz, E. Leptin and cancer. *J. Cell. Physiol.* **2006**, *207*, 12–22. [CrossRef] [PubMed]
38. Lin, T.C.; Hsiao, M. Leptin and Cancer: Updated Functional Roles in Carcinogenesis, Therapeutic Niches, and Developments. *Int. J. Mol. Sci.* **2021**, *22*, 2870. [CrossRef] [PubMed]
39. Macis, D.; Gandini, S.; Guerrieri-Gonzaga, A.; Johansson, H.; Magni, P.; Ruscica, M.; Lazzeroni, M.; Serrano, D.; Cazzaniga, M.; Mora, S.; et al. Prognostic effect of circulating adiponectin in a randomized 2 × 2 trial of low-dose tamoxifen and fenretinide in premenopausal women at risk for breast cancer. *J. Clin. Oncol.* **2012**, *30*, 151–157. [CrossRef]
40. Nehme, R.; Diab-Assaf, M.; Decombat, C.; Delort, L.; Caldefie-Chezet, F. Targeting Adiponectin in Breast Cancer. *Biomedicines* **2022**, *10*, 2958. [CrossRef]
41. Bartella, V.; De Marco, P.; Malaguarnera, R.; Belfiore, A.; Maggiolini, M. New advances on the functional cross-talk between insulin-like growth factor-I and estrogen signaling in cancer. *Cell. Signal.* **2012**, *24*, 1515–1521. [CrossRef]
42. Fuentes, N.; Silveyra, P. Estrogen receptor signaling mechanisms. *Adv. Protein Chem. Struct. Biol.* **2019**, *116*, 135–170. [CrossRef]
43. Garcia-Becerra, R.; Santos, N.; Diaz, L.; Camacho, J. Mechanisms of resistance to endocrine therapy in breast cancer: Focus on signaling pathways, miRNAs and genetically based resistance. *Int. J. Mol. Sci.* **2012**, *14*, 108–145. [CrossRef]
44. Kahlert, S.; Nuedling, S.; van Eickels, M.; Vetter, H.; Meyer, R.; Grohe, C. Estrogen receptor alpha rapidly activates the IGF-1 receptor pathway. *J. Biol. Chem.* **2000**, *275*, 18447–18453. [CrossRef] [PubMed]
45. Hua, H.; Kong, Q.; Yin, J.; Zhang, J.; Jiang, Y. Insulin-like growth factor receptor signaling in tumorigenesis and drug resistance: A challenge for cancer therapy. *J. Hematol. Oncol.* **2020**, *13*, 64. [CrossRef] [PubMed]
46. Gallo, M.; Ferrari, E.; Brugnoli, F.; Terrazzan, A.; Ancona, P.; Volinia, S.; Bertagnolo, V.; Bergamini, C.M.; Spisni, A.; Pertinhez, T.A.; et al. Metabolic Profiling of Breast Cancer Cell Lines: Unique and Shared Metabolites. *Int. J. Mol. Sci.* **2025**, *26*, 969. [CrossRef]
47. Gu, Z.; Wu, Q.; Shang, B.; Zhang, K.; Zhang, W. Organoid co-culture models of the tumor microenvironment promote precision medicine. *Cancer Innov.* **2024**, *3*, e101. [CrossRef]
48. Giles, E.D.; Wellberg, E.A. Preclinical Models to Study Obesity and Breast Cancer in Females: Considerations, Caveats, and Tools. *J. Mammary Gland Biol. Neoplasia* **2020**, *25*, 237–253. [CrossRef]
49. Hakuno, F.; Takahashi, S.I. IGF1 receptor signaling pathways. *J. Mol. Endocrinol.* **2018**, *61*, T69–T86. [CrossRef]
50. Khan, M.Z.; Zugaza, J.L.; Torres Aleman, I. The signaling landscape of insulin-like growth factor 1. *J. Biol. Chem.* **2025**, *301*, 108047. [CrossRef] [PubMed]
51. Boucher, J.; Kleinridders, A.; Kahn, C.R. Insulin receptor signaling in normal and insulin-resistant states. *Cold Spring Harb. Perspect. Biol.* **2014**, *6*, a009191. [CrossRef]
52. Li, R.; Pourpak, A.; Morris, S.W. Inhibition of the insulin-like growth factor-1 receptor (IGF1R) tyrosine kinase as a novel cancer therapy approach. *J. Med. Chem.* **2009**, *52*, 4981–5004. [CrossRef]
53. Christopoulos, P.F.; Msaouel, P.; Koutsilieris, M. The role of the insulin-like growth factor-1 system in breast cancer. *Mol. Cancer* **2015**, *14*, 43. [CrossRef]
54. Law, J.H.; Habibi, G.; Hu, K.; Masoudi, H.; Wang, M.Y.; Stratford, A.L.; Park, E.; Gee, J.M.; Finlay, P.; Jones, H.E.; et al. Phosphorylated insulin-like growth factor-i/insulin receptor is present in all breast cancer subtypes and is related to poor survival. *Cancer Res.* **2008**, *68*, 10238–10246. [CrossRef]
55. Endogenous, H.; Breast Cancer Collaborative, G.; Key, T.J.; Appleby, P.N.; Reeves, G.K.; Roddam, A.W. Insulin-like growth factor 1 (IGF1), IGF binding protein 3 (IGFBP3), and breast cancer risk: Pooled individual data analysis of 17 prospective studies. *Lancet Oncol.* **2010**, *11*, 530–542. [CrossRef] [PubMed]

56. Tong, Y.; Wu, J.; Huang, O.; He, J.; Zhu, L.; Chen, W.; Li, Y.; Chen, X.; Shen, K. IGF-1 Interacted With Obesity in Prognosis Prediction in HER2-Positive Breast Cancer Patients. *Front. Oncol.* **2020**, *10*, 550. [CrossRef] [PubMed]
57. Hartog, H.; Boezen, H.M.; de Jong, M.M.; Schaapveld, M.; Wesseling, J.; van der Graaf, W.T. Prognostic value of insulin-like growth factor 1 and insulin-like growth factor binding protein 3 blood levels in breast cancer. *Breast* **2013**, *22*, 1155–1160. [CrossRef] [PubMed]
58. Duggan, C.; Wang, C.Y.; Neuhouser, M.L.; Xiao, L.; Smith, A.W.; Reding, K.W.; Baumgartner, R.N.; Baumgartner, K.B.; Bernstein, L.; Ballard-Barbash, R.; et al. Associations of insulin-like growth factor and insulin-like growth factor binding protein-3 with mortality in women with breast cancer. *Int. J. Cancer* **2013**, *132*, 1191–1200. [CrossRef]
59. Belachew, E.B.; Sewasew, D.T. Molecular Mechanisms of Endocrine Resistance in Estrogen-Positive Breast Cancer. *Front. Endocrinol.* **2021**, *12*, 599586. [CrossRef]
60. Lu, Y.; Zi, X.; Zhao, Y.; Mascarenhas, D.; Pollak, M. Insulin-like growth factor-I receptor signaling and resistance to trastuzumab (Herceptin). *J. Natl. Cancer Inst.* **2001**, *93*, 1852–1857. [CrossRef]
61. Werner, H.; Bruchim, I. IGF-1 and BRCA1 signalling pathways in familial cancer. *Lancet Oncol.* **2012**, *13*, e537–e544. [CrossRef]
62. Liu, S.; Zhang, X.; Wang, W.; Li, X.; Sun, X.; Zhao, Y.; Wang, Q.; Li, Y.; Hu, F.; Ren, H. Metabolic reprogramming and therapeutic resistance in primary and metastatic breast cancer. *Mol. Cancer* **2024**, *23*, 261. [CrossRef] [PubMed]
63. Lei, P.; Wang, W.; Sheldon, M.; Sun, Y.; Yao, F.; Ma, L. Role of Glucose Metabolic Reprogramming in Breast Cancer Progression and Drug Resistance. *Cancers* **2023**, *15*, 3390. [CrossRef] [PubMed]
64. Jin, X.; Zhang, N.; Yan, T.; Wei, J.; Hao, L.; Sun, C.; Zhao, H.; Jiang, S. Lactate-mediated metabolic reprogramming of tumor-associated macrophages: Implications for tumor progression and therapeutic potential. *Front. Immunol.* **2025**, *16*, 1573039. [CrossRef] [PubMed]
65. Becker, L.M.; O'Connell, J.T.; Vo, A.P.; Cain, M.P.; Tampe, D.; Bizarro, L.; Sugimoto, H.; McGow, A.K.; Asara, J.M.; Lovisa, S.; et al. Epigenetic Reprogramming of Cancer-Associated Fibroblasts Deregulates Glucose Metabolism and Facilitates Progression of Breast Cancer. *Cell Rep.* **2020**, *31*, 107701. [CrossRef]
66. Yu, T.; Yang, G.; Hou, Y.; Tang, X.; Wu, C.; Wu, X.A.; Guo, L.; Zhu, Q.; Luo, H.; Du, Y.E.; et al. Cytoplasmic GPER translocation in cancer-associated fibroblasts mediates cAMP/PKA/CREB/glycolytic axis to confer tumor cells with multidrug resistance. *Oncogene* **2017**, *36*, 2131–2145. [CrossRef]
67. Rena, G.; Hardie, D.G.; Pearson, E.R. The mechanisms of action of metformin. *Diabetologia* **2017**, *60*, 1577–1585. [CrossRef]
68. Gong, Z.; Aragaki, A.K.; Chlebowski, R.T.; Manson, J.E.; Rohan, T.E.; Chen, C.; Vitolins, M.Z.; Tinker, L.F.; LeBlanc, E.S.; Kuller, L.H.; et al. Diabetes, metformin and incidence of and death from invasive cancer in postmenopausal women: Results from the women's health initiative. *Int. J. Cancer* **2016**, *138*, 1915–1927. [CrossRef]
69. Soranna, D.; Scotti, L.; Zambon, A.; Bosetti, C.; Grassi, G.; Catapano, A.; La Vecchia, C.; Mancia, G.; Corrao, G. Cancer risk associated with use of metformin and sulfonylurea in type 2 diabetes: A meta-analysis. *Oncologist* **2012**, *17*, 813–822. [CrossRef]
70. Pollak, M. Insulin and insulin-like growth factor signalling in neoplasia. *Nat. Rev. Cancer* **2008**, *8*, 915–928. [CrossRef]
71. Choi, Y.K.; Park, K.G. Metabolic roles of AMPK and metformin in cancer cells. *Mol. Cells* **2013**, *36*, 279–287. [CrossRef]
72. Asati, V.; Mahapatra, D.K.; Bharti, S.K. PI3K/Akt/mTOR and Ras/Raf/MEK/ERK signaling pathways inhibitors as anticancer agents: Structural and pharmacological perspectives. *Eur. J. Med. Chem.* **2016**, *109*, 314–341. [CrossRef]
73. Anisimov, V.N.; Egormin, P.A.; Bershtein, L.M.; Zabezhinskii, M.A.; Piskunova, T.S.; Popovich, I.G.; Semenchenko, A.V. Metformin decelerates aging and development of mammary tumors in HER-2/neu transgenic mice. *Bull. Exp. Biol. Med.* **2005**, *139*, 721–723. [CrossRef]
74. Checkley, L.A.; Rudolph, M.C.; Wellberg, E.A.; Giles, E.D.; Wahdan-Alaswad, R.S.; Houck, J.A.; Edgerton, S.M.; Thor, A.D.; Schedin, P.; Anderson, S.M.; et al. Metformin Accumulation Correlates with Organic Cation Transporter 2 Protein Expression and Predicts Mammary Tumor Regression In Vivo. *Cancer Prev. Res.* **2017**, *10*, 198–207. [CrossRef]
75. Giles, E.D.; Jindal, S.; Wellberg, E.A.; Schedin, T.; Anderson, S.M.; Thor, A.D.; Edwards, D.P.; MacLean, P.S.; Schedin, P. Metformin inhibits stromal aromatase expression and tumor progression in a rodent model of postmenopausal breast cancer. *Breast Cancer Res.* **2018**, *20*, 50. [CrossRef]
76. Caciolla, J.; Bisi, A.; Belluti, F.; Rampa, A.; Gobbi, S. Reconsidering Aromatase for Breast Cancer Treatment: New Roles for an Old Target. *Molecules* **2020**, *25*, 5351. [CrossRef]
77. Song, J.; Du, J.; Han, L.; Lin, X.; Fan, C.; Chen, G. The Effect of Metformin on Triple-Negative Breast Cancer Cells and Nude Mice. *Altern. Ther. Health Med.* **2023**, *29*, 389–395. [PubMed]
78. Tapia, E.; Villa-Guillen, D.E.; Chalasani, P.; Centuori, S.; Roe, D.J.; Guillen-Rodriguez, J.; Huang, C.; Galons, J.P.; Thomson, C.A.; Altbach, M.; et al. A randomized controlled trial of metformin in women with components of metabolic syndrome: Intervention feasibility and effects on adiposity and breast density. *Breast Cancer Res. Treat.* **2021**, *190*, 69–78. [CrossRef] [PubMed]
79. Goodwin, P.J.; Chen, B.E.; Gelmon, K.A.; Whelan, T.J.; Ennis, M.; Lemieux, J.; Ligibel, J.A.; Hershman, D.L.; Mayer, I.A.; Hobday, T.J.; et al. Effect of Metformin vs Placebo on Invasive Disease-Free Survival in Patients With Breast Cancer: The MA.32 Randomized Clinical Trial. *JAMA* **2022**, *327*, 1963–1973. [CrossRef] [PubMed]

80. Martin-Castillo, B.; Pernas, S.; Dorca, J.; Alvarez, I.; Martinez, S.; Perez-Garcia, J.M.; Batista-Lopez, N.; Rodriguez-Sanchez, C.A.; Amillano, K.; Dominguez, S.; et al. A phase 2 trial of neoadjuvant metformin in combination with trastuzumab and chemotherapy in women with early HER2-positive breast cancer: The METTEN study. *Oncotarget* **2018**, *9*, 35687–35704. [CrossRef]
81. Kim, J.; Lim, W.; Kim, E.K.; Kim, M.K.; Paik, N.S.; Jeong, S.S.; Yoon, J.H.; Park, C.H.; Ahn, S.H.; Kim, L.S.; et al. Phase II randomized trial of neoadjuvant metformin plus letrozole versus placebo plus letrozole for estrogen receptor positive postmenopausal breast cancer (METEOR). *BMC Cancer* **2014**, *14*, 170. [CrossRef]
82. Ligorio, F.; Vingiani, A.; Torelli, T.; Sposetti, C.; Drufuca, L.; Iannelli, F.; Zanenga, L.; Depretto, C.; Folli, S.; Scaperrotta, G.; et al. Early downmodulation of tumor glycolysis predicts response to fasting-mimicking diet in triple-negative breast cancer patients. *Cell Metab.* **2025**, *37*, 330–344.e7. [CrossRef] [PubMed]
83. Goncalves, M.D.; Iyengar, N.M. Fueling metabolic disruption via FMD to boost chemotherapy in TNBC. *Cell Metab.* **2025**, *37*, 305–306. [CrossRef]
84. Tentolouris, A.; Vlachakis, P.; Tzeravini, E.; Eleftheriadou, I.; Tentolouris, N. SGLT2 Inhibitors: A Review of Their Antidiabetic and Cardioprotective Effects. *Int. J. Environ. Res. Public Health* **2019**, *16*, 2965. [CrossRef]
85. Rohrborn, D.; Wronkowitz, N.; Eckel, J. DPP4 in Diabetes. *Front. Immunol.* **2015**, *6*, 386. [CrossRef]
86. Teichgraber, V.; Monasterio, C.; Chaitanya, K.; Boger, R.; Gordon, K.; Dieterle, T.; Jager, D.; Bauer, S. Specific inhibition of fibroblast activation protein (FAP)-alpha prevents tumor progression in vitro. *Adv. Med. Sci.* **2015**, *60*, 264–272. [CrossRef] [PubMed]
87. Zhou, J.; Zhu, J.; Yu, S.J.; Ma, H.L.; Chen, J.; Ding, X.F.; Chen, G.; Liang, Y.; Zhang, Q. Sodium-glucose co-transporter-2 (SGLT-2) inhibition reduces glucose uptake to induce breast cancer cell growth arrest through AMPK/mTOR pathway. *Biomed. Pharmacother.* **2020**, *132*, 110821. [CrossRef]
88. Komatsu, S.; Nomiyama, T.; Numata, T.; Kawanami, T.; Hamaguchi, Y.; Iwaya, C.; Horikawa, T.; Fujimura-Tanaka, Y.; Hamanoue, N.; Motonaga, R.; et al. SGLT2 inhibitor ipragliflozin attenuates breast cancer cell proliferation. *Endocr. J.* **2020**, *67*, 99–106. [CrossRef] [PubMed]
89. Akingbesote, N.D.; Norman, A.; Zhu, W.; Halberstam, A.A.; Zhang, X.; Foldi, J.; Lustberg, M.B.; Perry, R.J. A precision medicine approach to metabolic therapy for breast cancer in mice. *Commun. Biol.* **2022**, *5*, 478. [CrossRef]
90. Mariam, Z.; Niazi, S.K. Glucagon-like peptide agonists: A prospective review. *Endocrinol. Diabetes Metab.* **2024**, *7*, e462. [CrossRef]
91. Hinnen, D. Glucagon-Like Peptide 1 Receptor Agonists for Type 2 Diabetes. *Diabetes Spectr.* **2017**, *30*, 202–210. [CrossRef]
92. Bendotti, G.; Montefusco, L.; Lunati, M.E.; Usuelli, V.; Pastore, I.; Lazzaroni, E.; Assi, E.; Seelam, A.J.; El Essawy, B.; Jang, J.; et al. The anti-inflammatory and immunological properties of GLP-1 Receptor Agonists. *Pharmacol. Res.* **2022**, *182*, 106320. [CrossRef]
93. Wang, J.Y.; Wang, Q.W.; Yang, X.Y.; Yang, W.; Li, D.R.; Jin, J.Y.; Zhang, H.C.; Zhang, X.F. GLP-1 receptor agonists for the treatment of obesity: Role as a promising approach. *Front. Endocrinol.* **2023**, *14*, 1085799. [CrossRef] [PubMed]
94. Ibrahim, S.S.; Ibrahim, R.S.; Arabi, B.; Brockmueller, A.; Shakibaei, M.; Busselberg, D. The effect of GLP-1R agonists on the medical triad of obesity, diabetes, and cancer. *Cancer Metastasis Rev.* **2024**, *43*, 1297–1314. [CrossRef] [PubMed]
95. Iwaya, C.; Nomiyama, T.; Komatsu, S.; Kawanami, T.; Tsutsumi, Y.; Hamaguchi, Y.; Horikawa, T.; Yoshinaga, Y.; Yamashita, S.; Tanaka, T.; et al. Exendin-4, a Glucagonlike Peptide-1 Receptor Agonist, Attenuates Breast Cancer Growth by Inhibiting NF-kappaB Activation. *Endocrinology* **2017**, *158*, 4218–4232. [CrossRef] [PubMed]
96. Tanaka, Y.; Iwaya, C.; Kawanami, T.; Hamaguchi, Y.; Horikawa, T.; Shigeoka, T.; Yanase, T.; Kawanami, D.; Nomiyama, T. Combined treatment with glucagon-like peptide-1 receptor agonist exendin-4 and metformin attenuates breast cancer growth. *Diabetol. Int.* **2022**, *13*, 480–492. [CrossRef]
97. Liu, Z.Z.; Duan, X.X.; Yuan, M.C.; Yu, J.; Hu, X.; Han, X.; Lan, L.; Liu, B.W.; Wang, Y.; Qin, J.F. Glucagon-like peptide-1 receptor activation by liraglutide promotes breast cancer through NOX4/ROS/VEGF pathway. *Life Sci.* **2022**, *294*, 120370. [CrossRef]
98. Kolterman, O.G.; Buse, J.B.; Fineman, M.S.; Gaines, E.; Heintz, S.; Bicsak, T.A.; Taylor, K.; Kim, D.; Aisporna, M.; Wang, Y.; et al. Synthetic exendin-4 (exenatide) significantly reduces postprandial and fasting plasma glucose in subjects with type 2 diabetes. *J. Clin. Endocrinol. Metab.* **2003**, *88*, 3082–3089. [CrossRef]
99. Fidan-Yaylali, G.; Dodurga, Y.; Secme, M.; Elmas, L. Antidiabetic exendin-4 activates apoptotic pathway and inhibits growth of breast cancer cells. *Tumour. Biol.* **2016**, *37*, 2647–2653. [CrossRef]
100. Zhao, W.; Zhang, X.; Zhou, Z.; Sun, B.; Gu, W.; Liu, J.; Zhang, H. Liraglutide inhibits the proliferation and promotes the apoptosis of MCF-7 human breast cancer cells through downregulation of microRNA-27a expression. *Mol. Med. Rep.* **2018**, *17*, 5202–5212. [CrossRef]
101. Alanteet, A.; Attia, H.; Alfayez, M.; Mahmood, A.; Alsaleh, K.; Alsanea, S. Liraglutide attenuates obese-associated breast cancer cell proliferation via inhibiting PI3K/Akt/mTOR signaling pathway. *Saudi Pharm. J.* **2024**, *32*, 101923. [CrossRef] [PubMed]
102. Stanisavljevic, I.; Pavlovic, S.; Simovic Markovic, B.; Jurisevic, M.; Krajnovic, T.; Mijatovic, S.; Spasojevic, M.; Mitrovic, S.; Corovic, I.; Jovanovic, I. Semaglutide decelerates the growth and progression of breast cancer by enhancing the acquired antitumor immunity. *Biomed. Pharmacother.* **2024**, *181*, 117668. [CrossRef] [PubMed]

103. Kalinsky, K.; Crew, K.D.; Refice, S.; Xiao, T.; Wang, A.; Feldman, S.M.; Taback, B.; Ahmad, A.; Cremers, S.; Hibshoosh, H.; et al. Presurgical trial of metformin in overweight and obese patients with newly diagnosed breast cancer. *Cancer Investig.* **2014**, *32*, 150–157. [CrossRef] [PubMed]
104. Meyerhardt, J.A.; Irwin, M.L.; Jones, L.W.; Zhang, S.; Campbell, N.; Brown, J.C.; Pollak, M.; Sorrentino, A.; Cartmel, B.; Harrigan, M.; et al. Randomized Phase II Trial of Exercise, Metformin, or Both on Metabolic Biomarkers in Colorectal and Breast Cancer Survivors. *JNCI Cancer Spectr.* **2020**, *4*, pkz096. [CrossRef]
105. Pimentel, I.; Lohmann, A.E.; Ennis, M.; Dowling, R.J.O.; Cescon, D.; Elser, C.; Potvin, K.R.; Haq, R.; Hamm, C.; Chang, M.C.; et al. A phase II randomized clinical trial of the effect of metformin versus placebo on progression-free survival in women with metastatic breast cancer receiving standard chemotherapy. *Breast* **2019**, *48*, 17–23. [CrossRef]
106. Yee, D.; Isaacs, C.; Wolf, D.M.; Yau, C.; Haluska, P.; Giridhar, K.V.; Forero-Torres, A.; Jo Chien, A.; Wallace, A.M.; Pusztai, L.; et al. Ganitumab and metformin plus standard neoadjuvant therapy in stage 2/3 breast cancer. *NPJ Breast Cancer* **2021**, *7*, 131. [CrossRef]
107. Sodhi, M.; Rezaeianzadeh, R.; Kezouh, A.; Etminan, M. Risk of Gastrointestinal Adverse Events Associated with Glucagon-like Peptide-1 Receptor Agonists for Weight Loss. *JAMA* **2023**, *330*, 1795–1797. [CrossRef]
108. Parsons, K.; Montalvo, M.; Fischbach, N.; Taylor, M.; Alfaro, S.; Lustberg, M. The Impact and Safety of GLP-1 Agents and Breast Cancer. *Cancer Med.* **2025**, *14*, e70932. [CrossRef]
109. Confederat, L.G.; Dragostin, O.M.; Condurache, M.I. SGLT2 Inhibitors and the Risk of Urogenital Infections: A Concise Review. *J. Clin. Med.* **2025**, *14*, 1960. [CrossRef]

life

MDPI

Review

Breast Cancer Treatments: Drugs Targeting the PI3K/AKT/mTOR Pathway, TNBC Therapy and Future Directions: A Review

Klaudia Dynarowicz [1], Dorota Bartusik-Aebisher [1], Katarzyna Koszarska [2], Aleksandra Kotlińska [2] and David Aebisher [3,*]

[1] Department of Biochemistry and General Chemistry, Faculty of Medicine, University of Rzeszów, 35-310 Rzeszów, Poland; kdynarowicz@ur.edu.pl (K.D.); dbartusikaebisher@ur.edu.pl (D.B.-A.)
[2] English Division Science Club, Faculty of Medicine, University of Rzeszów, 35-359 Rzeszów, Poland; kk117603@stud.ur.edu.pl (K.K.); ak117604@stud.ur.edu.pl (A.K.)
[3] Department of Photomedicine and Physical Chemistry, Faculty of Medicine, University of Rzeszów, 35-359 Rzeszów, Poland
* Correspondence: daebisher@ur.edu.pl

Abstract

Breast cancer affects women at an increasingly younger age, with genetic predispositions and other factors contributing to its second-highest cancer mortality rate. The diversity of pharmacological treatment stems from its heterogeneity, which favors a more precise approach to each subtype. Despite the extensive advances in medicine in recent decades, the problem of treating cancer patients remains significant. The problem with modern therapeutic methods is low effectiveness, emerging side effects, difficulty in eliminating all cancer cells, and the quite common use of monotherapy and the associated drug resistance, which may lead to disease progression. The aim of this review is to present the latest therapeutic strategies (combination therapies) used in the treatment of breast cancer. PubMed databases and clinical data from ClinicalTrials.gov were used for this purpose. The review included characteristics of the latest clinical trials from the last year (2024–2025), which present currently recruiting studies of breast cancer treatment with immunotherapy. The review also presented characteristics of clinical trials from the last 5 years (2020–2025) using nanoparticles as an adjunct to breast cancer treatment. Articles published between 2016 and August 2025 (excluding articles that describe the first use of a given drug) were included in the review. The review analyzed drugs targeting molecular targets, including intracellular pathways responsible for cell cycle regulation, as well as new directions such as nanotechnology in treatment breast cancer.

Keywords: breast cancer therapy; PI3K/AKT/mTOR pathway; TNBC therapy; nanomedicine; nanoaqualip technology

1. Introduction

Despite extensive technological advances in medical fields such as oncology, radiology, and laboratory diagnostics, cancer is still undoubtedly the leading cause of death worldwide and represents not only a health problem but also a social and demographic one [1–3]. According to the 2024 reports of the National Cancer Institute (NCI) Surveillance, Epidemiology, and End Results (SEER) program and the National Program of Cancer Registries (NPCR), breast cancer currently accounts for approximately 32% of all malignancies in women worldwide [4], ranking first in this group. In terms of mortality, it is the second most common cancer, after lung cancer, and accounts for 15% of all cancer deaths (lung cancer accounts for approximately 21%). The mortality rate among all people diagnosed with breast cancer is approximately 15%, and the incidence rate has been slowly increasing by about 0.6% annually since the mid-2000s [4]. There are various therapeutic approaches to treating breast cancer. These depend on the tumor type, the patient's condition, and the treatment setting [5–7]. The most popular methods include surgery, radiotherapy, chemotherapy, and hormone therapy (Figure 1).

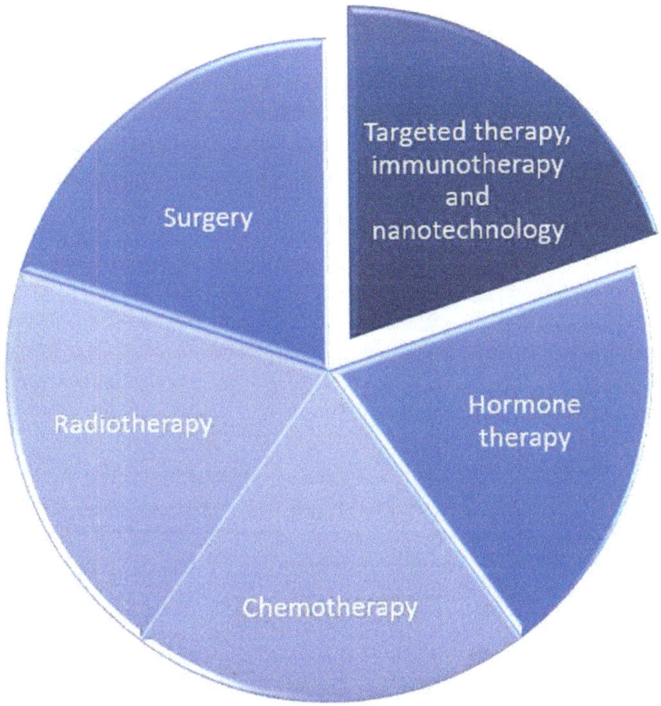

Figure 1. Types of therapeutic methods in the treatment of breast cancer.

Recent advances in understanding the molecular biology of breast cancer have led to the development of targeted therapies and immunotherapies, which aim to exploit the specific genetic and immunological features of different cancer subtypes [8,9]. In recent years, drugs targeting the PI3K/AKT/mTOR pathway have become one of these therapeutic tools [10]. New drugs are being approved by the FDA each year. One of the main sources of knowledge about the current therapeutic status of breast cancer is registered clinical trials. The aim of this review is therefore to characterize the latest drugs targeting the PI3K/AKT/mTOR pathway and to present the characteristics of the latest registered clinical trials using both immunotherapy and nanoparticles. Taking into account the latest treatment trends, progress in breast cancer nanotherapy and Nanoaqualip technology is also presented. The ClinicalTrials.gov database was used to present the characteristics of clinical trials. The following studies using immunotherapy were selected: condition: breast cancer, intervention/treatment: Immune Checkpoint Inhibitors, date range: 1 January 2024–1 January 2025. This review presents those already recruiting patients. The following studies using nanoparticles were selected: condition: breast cancer, intervention/treatment: nanoparticles, study phases: 2 and 3, date range: 1 January 2020–1 August 2025 (a wider range of years was chosen rather than a single year to represent more registered clinical trials). The review presents all eight trials, both those that have not yet started and those currently recruiting patients. This review presents the latest FDA-approved drugs for the treatment of breast cancer in the last two years. Articles cited in the review are from the Pubmed database from the years 2016–2025 (excluding articles that describe the first use of a given drug), selected based on the following keywords: drugs targeting PI3K/AKT/mTOR pathway breast cancer, TNBC therapy, nanoparticles in breast cancer. Full-text articles written in Polish or English, after abstract review, were included in the review.

2. Selected Therapeutic Approaches in Breast Cancer

2.1. Targeting PI3K/AKT/mTOR Pathway

The Ras/MAPK and PI3K/AKT/mTOR signaling pathways play a key role in cell cycle regulation by affecting cyclin D (mainly D1), which forms a complex with CDK4/6 (serine–threonine kinases) phosphorylates the Rb protein [11–14]. Then, in this pathway, the Rb-activated transcription factor E2F activates genes necessary for the cell to enter the S phase of the cell cycle. Studies show that mutations in the PI3K/AKT/mTOR pathway are present in even over 70% of breast cancer cases, which is why drugs targeting its elements are widely used, especially in combined treatment in patients with resistance to hormonal therapy, achieving better effects than drugs from these groups used separately [12].

Drugs targeting cell cycle control pathways have found wide application in oncology, not only in the treatment of breast cancer but also many others, e.g., lung cancer, pancreatic cancer or melanoma [15–17]. New drugs targeting the PI3K pathway in HR+, HER2- breast cancer include Buparlisib, Pictilisib, Alpelisib, Taselisib, and Inavolisib (Figures 2 and 3). Clinical trials are currently evaluating the efficacy of these drugs, with a view to their approval and use on a larger scale. The buparlisib clinical trial was a multicenter, phase III study designed to determine the efficacy and safety of buparlisib plus fulvestrant versus placebo plus fulvestrant [18–21]. The study population consisted of postmenopausal women with breast cancer (HR-positive, HER2-negative) treated with an aromatase inhibitor. The results clearly demonstrated that median progression-free survival was significantly longer in the buparlisib group compared to placebo. Despite satisfactory results, significant or severe adverse events also occurred, primarily elevated aspartate aminotransferase levels, hyperglycemia, hypertension, and fatigue. It is important to emphasize that these effects occurred in both patients receiving buparlisib and those receiving placebo. Several patients also died, but the cause was metastatic breast cancer, and in one case, heart failure. As the authors present, the efficacy of buparlisib supports the use of PI3K inhibitors in combination with endocrine therapy in breast cancer patients. Pictilisib is used in clinical trials to treat breast tumors with various characteristics [22–27]. Registered clinical trials are evaluating the use of pictilisib in combination with fulvestrant in patients with advanced breast cancer whose tumors harbor a PIK3CA mutation. Unfortunately, the results of the clinical trial were unsatisfactory, as the use of pictilisib in combination with fulvestrant did not significantly improve progression-free survival. Furthermore, significant toxicity occurred, significantly reducing the effectiveness of the treatment. Alpelisib is a drug primarily used in the treatment of advanced breast cancer [28–33]. Clinical trials demonstrate the combination of alpelisib with fulvestrant or letrozole in patients with HR-positive, HER2-negative breast cancer with PIK3CA mutation (s) in their tumors. Interpretation of the results indicates that the activity of alpelisib with fulvestrant was relatively good with low toxicity. The results are satisfactory, which may lead to expansion of the ongoing study and further verification of the results. Taselisib is a drug used in clinical trials for the treatment of lymphoma, breast cancer, ovarian cancer, solid tumors, and other conditions. In clinical trials, taselisib is administered in combination with fulvestrant [34,35]. The control group is treated with a placebo. The results were more than satisfactory. Positive antitumor effects were observed in patients receiving taselisib–fulvestrant therapy. As the authors state, these results are significant for better understanding the therapeutic impact of this type of combination therapy. These types of clinical trials highlight the importance of combined, targeted therapies [34,35]. Inavolisib is an innovative drug being studied in the treatment of various types of cancer. It belongs to a class of drugs called PI3K inhibitors, which target a specific pathway in cancer cells that promotes their growth and survival [36,37]. In the clinical trial, this drug was administered to patients with locally advanced or metastatic breast cancer with a PIK3CA mutation. The efficacy of inavolisib in combination with palbociclib and fulvestrant was assessed compared to placebo. Median overall survival was higher in patients receiving inavolisib compared to those receiving placebo treatment. Unfortunately, the occurrence of significant side effects forced the authors to discontinue the study and treatment [36,37]. Adverse events observed included hyperglycemia, stomatitis or mucositis, diarrhea, dry eye syndrome, and blurred vision. The current clinical trial is positive, as inavolisib treatment improved overall survival. However, emerging side effects require further research in this area.

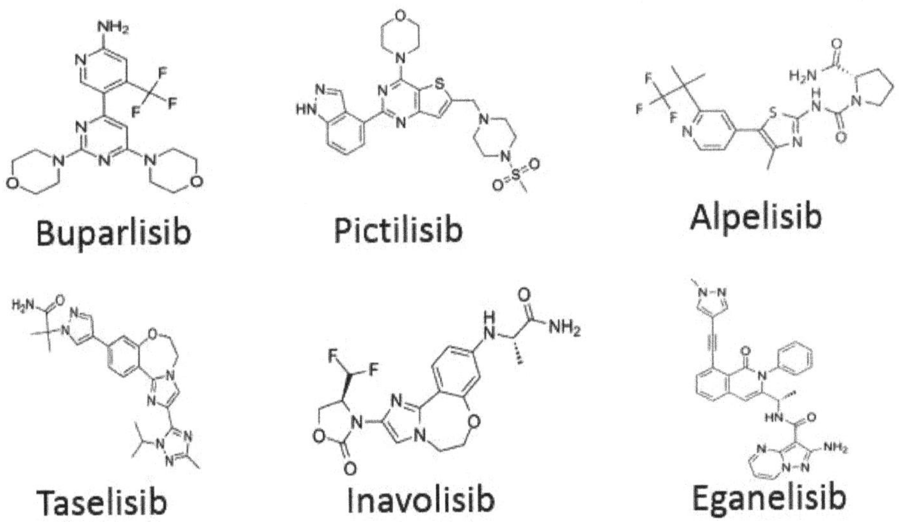

Figure 2. New drugs in clinical trials targeting PI3K pathway.

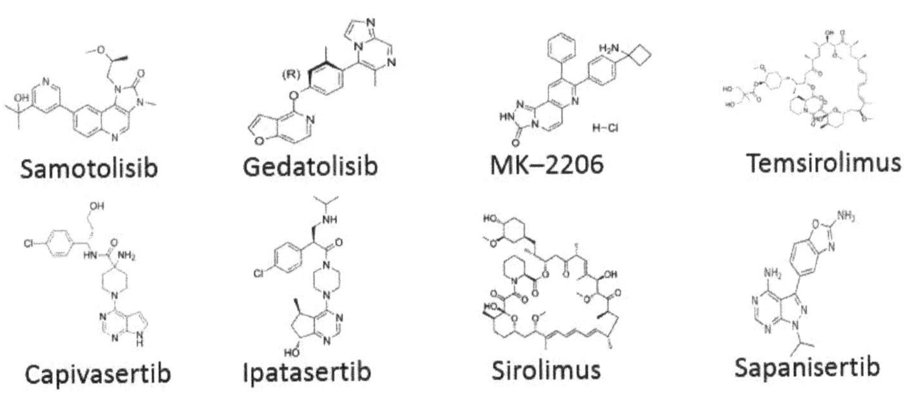

Figure 3. New drugs in clinical trials targeting AKT/mTOR pathway.

For TNBC, a newly analyzed drug is Eganelisib. Eganelisib is a highly selective small-molecule drug with potential immunomodulatory and anticancer effects [38,39]. After administration, eganelisib prevents the activation of signaling pathways, which may lead to reduced cell proliferation in cancer cells. A clinical trial is currently being registered to determine the effectiveness of this drug in patients with TNBC. The results are not yet available, but they may demonstrate promising anticancer activity [38,39].

Newly approved drugs targeting the AKT/mTOR pathway in ER+, HR+, and HER2- breast cancers include Gedatolisib, MK-2206, Capivasertib, Sirolimus, Temsirolimus, Sapanisertib, and Vistusertib. Clinical trials are currently underway to assess the effectiveness of these drugs, with a view to their approval and use on a larger scale.

Gedatolisib is an experimental drug being studied for its potential in treating various types of cancer [40–43]. This drug belongs to a class of drugs called PI3K/mTOR inhibitors, meaning it works by blocking specific proteins involved in the growth and survival of cancer cells. Clinical trials for Gedatolisib are multicenter, open-label studies in patients with metastatic breast cancer [40–43]. The proposed clinical trials aim to determine the maximum tolerated dose of gedatolisib with other combination drugs and to estimate the objective response rate. MK2206, a clinically advanced AKT inhibitor, is used to treat patients with breast cancer [44]. Although MK-2206 is not currently being developed further, this class of drugs continues to generate clinical interest. MK-2206 is estimated to have improved survival after treatment completion. It is important to note that the drug itself is characterized by some toxicity. The most significant grade 3–4 toxicity was rash. Tools such as magnetic resonance imaging and molecular characterization of the results are used to analyze the efficacy of MK-2206 administration. Capivasertib specifically targets and inhibits protein kinase B. By blocking the protein, capivasertib aims to slow or stop the growth of cancer cells. The Phase III clinical trial is evaluating the efficacy of capivasertib in combination with fulvestrant versus placebo in combination with fulvestrant in the treatment of patients with HR+/HER2- metastatic breast cancer after relapse or progression [45,46]. The main finding from the clinical trials was that patients who took capivasertib plus fulvestrant lived longer without relapse compared to those treated with placebo plus fulvestrant. It's worth noting that this result was observed in all participants recruited for the study. This phenomenon was also observed in participants whose tumors had detectable genetic changes in genes called PIK3CA and AKT1. The most common side effects were diarrhea and rash, which the authors say was as expected. Sirolimus, also known by the trade names Rapamune and Rapamycin, is an immunosuppressant drug [47,48]. It was originally approved by the U.S. Food and Drug Administration to prevent kidney transplant rejection. However, researchers are investigating its potential use in treating various other medical conditions. The clinical trials included patients with metastatic breast cancer who were given sirolimus in combination with hormone therapy. Progression-free survival rates in patients treated with sirolimus and everolimus were similar. As in every study, side effects occurred. The most common adverse events were lipid metabolism disorders and stomatitis. However, all recorded side effects were mild. It is worth noting that patients with alterations in the PI3K/AKT/mTOR pathway responded better to sirolimus than those without alterations. Based on the conducted clinical trials, sirolimus is a potentially effective treatment option for patients with hormone receptor-positive advanced breast cancer. Temsirolimus is a derivative and prodrug of sirolimus used in the treatment of renal cell carcinoma. It is an mTOR kinase inhibitor. It is worth noting that temsirolimus is not used in patients with breast cancer. However, clinical trials are being registered that incorporate this drug into the treatment protocol. However, the results are currently unsatisfactory. No overall improvement in the primary endpoint was observed. According to the authors, adding temsirolimus to letrozole did not improve treatment in patients with advanced breast cancer. Further clinical trials may be needed to confirm or refute this finding [49].

Sapanisertib is an experimental mTOR kinase inhibitor used in clinical trials to inhibit the activity of mTORC1 and mTORC2 proteins in the treatment of various types of cancer. In clinical trials, the drug is typically analyzed in combination with fulvestrant [50,51]. Results indicate that median progression-free survival was longer when fulvestrant was administered concurrently with sapanisertib. This data was collected from patients with advanced or metastatic ER+/HER2- breast cancer. However, increased toxicity was observed when these drugs were combined. The most common adverse events were nausea, vomiting, and hyperglycemia, which occurred more frequently with combination therapy. Clinical trials are also underway testing combination therapy combining fulvestrant with vistusertib. However, the results are unsatisfactory, as no significant improvement in outcomes, such as an extension of progression-free survival, has been observed [52,53].

TNBC Drugs

For TNBC, the following drugs are used: Samotolisib and Ipatasertib.

Samotolisib is a potent dual PI3K/mTOR inhibitor. When combined with prexasertib, it inhibits cell proliferation in TNBC cell lines [54]. It's worth noting that no dose-limiting toxicity was observed, but the studies were not free of treatment-related adverse events. Leukopenia, thrombocytopenia, and nausea were common. Despite the challenges, the results are satisfactory. While the combination of the two drugs was associated with toxicity, synergistic antitumor activity was observed. Ipatasertib is a drug used in clinical trials for the treatment of cancer, tumors, solid tumors, breast cancer, and gastric cancer, among others. In TNBC, this drug was used in combination with paclitaxel compared to placebo [55–58]. The results were satisfactory, as the median progression-free survival was 1.3 months longer than placebo in patients. However, the studies were not free of adverse events. Diarrhea and neutropenia were relatively common. Although these side effects were bothersome and common, these are the first results confirming the effectiveness of targeted therapy in TNBC. The use of ipatasertib in patients diagnosed with TNBC requires further research.

Is it possible to clearly and easily compare therapies and drugs targeting the PI3K vs. AKT pathways? Each drug is different. Although the responses are positive, they are almost always associated with toxicity, causing side effects to a greater or lesser extent. When considering treatment and drug selection, it is worth analyzing already registered clinical trials and considering the potential benefits and side effects of the proposed therapy.

Resistance to PI3K/AKT/mTOR pathway inhibitors results from both inherited and acquired mutations that restore the activity of this pathway. Biomarker-based patient stratification utilizes genetic and protein markers, such as PI3K and AKT mutations, elevated phosphorylated protein levels, and cancer stem cell markers, to identify individuals most likely to respond to specific therapies. This approach allows for increased treatment efficacy and supports the selection of therapeutic strategies, including combination therapies. Cancer tumors can activate parallel or bypassing signaling pathways, compensating for inhibition of the PI3K/AKT/mTOR axis. Furthermore, blocking this pathway can suppress protective autophagy, which may contribute to the development of resistance. Identifying patients predisposed to respond to PI3K/AKT/mTOR inhibitors is crucial for optimizing treatment. Biomarker profiling enables precise patient selection for both monotherapy and combination therapies to overcome resistance. Biomarkers can predict which patients are most likely to respond to treatment and identify those at risk of developing resistance, allowing for proactive adaptation of treatment strategies [59–61].

2.2. Anti-HER2 Therapy

In the case of HER2-positive tumors, the first-line treatment is humanized anti-HER2 monoclonal antibodies (mAbs): trastuzumab (Herceptin), introduced in 1998, and pertuzumab (Perjeta), introduced in 2012 as one of the first immunotherapies approved for the treatment of breast cancer [62]. They are designed to bind to the receptor, preventing its activation, while limiting HER2 overexpression, which inhibits the intracellular signaling pathway. Trastuzumab has been shown to inhibit cancer cell proliferation by promoting cell cycle arrest in the G1 phase, but does not affect receptor dimerization, while pertuzumab additionally inhibits heterodimerization with other ERBB family proteins (mainly HER3). In addition to the direct anti-cancer effect, indirect mechanisms are also activated: through the engagement of Fcγ receptors (FCGR), antibody-dependent cellular cytotoxicity (ADCC) or antibody-dependent cellular phagocytosis (ADCP) is stimulated, a strategy that has been recognized in recent years as key to the synergistic mechanism of action (MAO) [63]. Additionally, the indirect effect of mAbs is achieved by complement activation via the classical pathway—complement-dependent cytotoxicity (CDC) and complement-dependent phagocytosis of cancer cells (CDCP) occur [64]. Due to differences in the mechanism of action (taking into account the differences in the HER2 epitopes with which both antibodies bind) and the low efficacy demonstrated in clinical studies in monotherapy in order to maximize the therapeutic effect (enhanced antitumor activity, also against disseminated tumors), these two drugs are usually used simultaneously, especially in the case of an insufficient response in monotherapy with trastuzumab, although the basis for the increased efficacy of this combination has not yet been fully explained [65,66]. In addition, in recent years, many researchers have drawn attention to the use of so-called antibody–drug conjugates (ADCs), which, in addition to mAb, additionally contain cytostatics active against a given type of cancer cells, and thanks to the precise internalization of monoclonal antibodies, they can be delivered directly to the patient's cells, thus minimizing the side effects of systemic toxicity of classic chemotherapy [67]. By using such a combination, the amount of the drug (its concentration) that reaches the tumor directly is increased. Focusing on this drug delivery model, biomolecular ADC complexes combining trastuzumab with cytotoxic drugs targeting microtubule inhibition have been developed: Ado–Trastuzumab emtansine (T–DM1; Kadcyla; registered in 2013) and topoisomerase inhibition: Fam–Trastuzumab deruxtecan (T–DXd; Enhertu registered in 2020). Ghanashyam Biswas et al. have proven that TDM–1 is more effective than trastuzumab in preventing relapses in the long term, prolongs progression-free survival (PFS) and improves survival rates in patients with HER2(+) breast cancer, while preclinical studies are promising, in which T–Dxd shows activity against tumors resistant to T–DM1 [68–71].

Tyrosine kinase inhibitors (TKIs) targeting the intracellular domain of the HER2 receptor have also been used in HER2-positive breast cancer. By competing with ATP, they limit the phosphorylation of tyrosine residues, which inhibits signaling to the cell nucleus [72]. TKIs block enzymatic activity, which prevents excessive activation of the PI3K/Akt and MAPK pathways that determine cancer cell survival and proliferation [73]. In HER2 (+) breast cancer, they have been registered for about a decade and include lapatinib [74](Tykerb; 2007), neratinib [75], pyrotinib [74] , and tucatinib [76] , while afatinib has not yet been registered for this indication. Lapatinib is a dual inhibitor, inhibiting both EGFR and HER2. Its efficacy has been demonstrated mainly in mBC in combination with chemotherapy (it extended overall survival from 20.5 months for paclitaxel in monotherapy to 27.8 months in combination therapy) [74]. In the case of triple-positive BC (HR+, HER2+) it is used in combination with AI (letrozole). Similarly to tucatinib, it is an inhibitor with reversible action, unlike neratinib, pyrotinib and afatinib, which bind to the target receptor in an irreversible manner. TKIs are drugs with proven good synergistic effect, therefore they often appear in combination treatment regimens to increase the efficacy of therapy. Studies on further therapeutic combinations and assessment of their efficacy are also conducted [75,76].

2.3. Immunotherapy

Immunotherapy uses the patient's own immune system and, in addition to the aforementioned monoclonal antibodies in the treatment of HER2-positive cancer, has found its application to an even wider extent in the case of TNBC therapy, because here it has the potential to significantly affect the quality of life of patients by reducing the share of classic chemotherapy [77]. In addition to the aforementioned mAbs targeting the HER2 receptor, checkpoint inhibitors (ICIs) are used here. These are PD-1 and PDL-1 (Programmed cell death—protein and ligand 1) inhibitors, because cancer cells, in order to avoid recognition and destruction by the T lymphocytes of the patient's immune system, have developed an adaptive mechanism in which the ligand expressed on their surface binds to the PD-1 receptor on the cells of the immune system and in this way, mimicking physiological mechanisms, inhibits the proliferation of T lymphocytes and promotes their depletion—it silences the immune response, which allows the tumor to escape from the host's own mechanisms [78]. In recent years, PARP inhibitors have become a breakthrough in the treatment of TNBC, particularly in patients with BRCA1/2 mutations. Their mechanism of action is based on the concept of synthetic lethality, in which cancer cells deficient in homologous recombination repair are selectively attacked, leading to their death. According to Jie et al., clinical trials have demonstrated the effectiveness of PARP inhibitors, such as olaparib and talazoparib, in improving progression-free survival in patients with germline BRCA mutations. Currently, there are projects aimed at using PARP inhibitors in TNBC without BRCA mutations. Another example of future research is the use of inhibitors in combination with other therapies, such as chemotherapy. The inclusion of PARP inhibitors in TNBC therapy requires further research, particularly clinical trials [79].

The main principle of action of PARP inhibitors is to block the repair of single-strand breaks in DNA, leading to the accumulation of double-strand breaks that are lethal to cells with defective HRR mechanisms, such as cells with BRCA mutations [80]. Five PARP inhibitors have been developed to date: olaparib, niraparib, rucaparib, talazoparib, and veliparib. These drugs differ in their ability to bind PARP to DNA and in their ability to inhibit its catalytic activity. They have demonstrated efficacy in the treatment of triple-negative breast cancer (TNBC) in patients with BRCA mutations, exacerbating DNA damage and leading to the death of cancer cells with defective DNA repair mechanisms. TNBC is a particularly aggressive subtype of breast cancer, affecting 10–15% of patients [81]. Compared to other types of cancer, TNBC has a three-fold higher risk of recurrence. It is characterized by the absence of estrogen, progesterone, and HER2 receptors, which for years limited treatment options to chemotherapy alone. However, new therapies are now available, such as immunotherapy (e.g., pembrolizumab), PARP inhibitors, and novel antibody–drug conjugates (e.g., sacituzumab govitecan), which have significantly improved the prognosis for patients with metastatic TNBC [82]. Immunotherapy has proven particularly important because, unlike other subtypes of breast cancer, TNBC lacks "classical" therapeutic targets. This treatment is usually better tolerated than chemotherapy, although it can lead to immune-related side effects. In most cases, they can be controlled with corticosteroids, but more serious toxicity develops in approximately 14% of patients, requiring careful consideration of the benefits and risks, especially when treated early in the disease course [82]. In the last year alone, 10 clinical trials were registered that evaluated the effectiveness of immunotherapy in breast cancer. Table 1 presents 3 clinical trials registered in the last year (2024–2025) and already underway.

Table 1. Registered clinical trials from the last year in which immunotherapy was selected in breast cancer treatment. Based on ClinicalTrials.gov.

Countries	Study Type	Phase	Official Title	Detailed Description	Interventions	Years and Status
Beijing, China	Interventional	Not Applicable	Safety Assessment of Concurrent Radiotherapy and Novel Systemic Therapy for Breast Cancer	The aim of the study is to evaluate the safety of combination therapy in patients with lymphatic drainage within the chest wall/breast requiring capecitabine, a CDK 4/6 inhibitor, HER2-targeted therapy, or immunotherapy.	Radiation HER2 inhibitors, CDK4/6 inhibitors, PARP inhibitors, ICIs	2024–2027
Jacksonville, FL, USA	Observational	Not Applicable	Observational Basket Trial to Collect Tissue to Train and Validate a Live Tumor Diagnostic Platform	This study is being done to collect tissue samples to test how accurately a tumor response platform, Elephas, can predict clinical response across multiple types of immunotherapies, chemoimmunotherapy and tumor types.	Patients will receive standard treatment with checkpoint inhibitors and undergo standard tumor assessment during screening and follow-up.	2024–2027
This study has locations in Greece	Observational	Not applicable	Mechanisms of Response and Resistance to Innovative Treatments in Patients With Locally Advanced or Metastatic Breast Cancer	The study aims to utilize modern methods such as DNA sequencing, analysis of circulating genetic material, digital imaging, and radiological analysis, as well as data from patient medical records. Based on this, a machine learning algorithm will be developed that will predict whether a patient will respond well to treatment or develop resistance, depending on the genetics and molecular profile of the tumor.	Patients receiving treatment with an Antibody–Drug conjugate (e.g., Trastuzumab–Deruxtecan, Sacituzumab–Govitecan) Patients receiving treatment with an immune checkpoint inhibitor (ICI) (e.g., pembrolizumab) Patients receiving treatment with a PARP inhibitor (e.g., Olaparib)	2024–2029

It is worth noting that each combination therapy and radiotherapy has a molecular effect on cells. Current molecular combination therapy can target specific targets based on the patient's tumor characteristics [83]. To optimize therapeutic outcomes and minimize the likelihood of recurrence after surgery, some combination therapies are currently used as adjuvant or first-line treatment for recurrent or metastatic breast cancer in clinical settings [84]. Due to the specificity of new and precise combination therapies, side effects can often be minimized, for example, by altering the therapeutic approach throughout treatment, or by changing the drug regimen, known as combination therapy. With any combination, there are risks to treatment effectiveness due to the use of multiple drugs. This can lead to problems such as inflammation, renal or liver failure, thrombocytopenia, and brain metastases. As with any new treatment, the potential for adverse effects and toxicity must be considered for each combination. A promising potential solution to this problem is the use of molecular docking system to aid in the development of the most effective combination treatment regimens. However, an increase in the number of extended clinical trials will improve the efficacy of therapy for patients with various subtypes of breast cancer [85]. Radiotherapy is increasingly used in combination therapies, supported by clinical trials. Radiotherapy damages cancer cells by destroying their DNA. This can occur directly, through ionizing radiation, or indirectly, when radiation stimulates nearby molecules that produce free radicals. These free radicals create reactive oxygen species (ROS), which stress and damage the cell. Cancer cells sometimes defend themselves by activating survival mechanisms, which can allow tumor recurrence. How well a tumor responds to treatment depends on its radiosensitivity (ease of destruction) or radioresistance (resistance to damage). Radiation alters the way cancer cells grow and divide, engaging them in cell death processes such as apoptosis. Depending on the patient's condition, different treatment regimens are selected to achieve the best possible outcome [86].

2.4. TNBC Cancer Therapy

The exceptional genomic heterogeneity of TNBC, including its molecular subtypes, along with the lack of expression of hormone receptors and HER2, limits the possibilities of targeted therapy, as none of the hormonal therapy options or monoclonal antibodies used in other cancers show efficacy in this case [87]. So far, the basic treatment regimen has been mastectomy or sparing surgery (depending on the stage of the cancer), radiotherapy and neoadjuvant or adjuvant chemotherapy, which has its many limitations. The main aspect is the intrinsic or acquired resistance to chemotherapeutic agents, which translates into a high relapse rate. Focusing on the molecular stratification of TNBC subtypes could bring benefits in the field of targeted treatment and the development of immunotherapy. A conventional chemotherapy regimen includes the following options: Anthracyclines (doxorubicin, epirubicin), Regimens with anthracycline and cyclophosphamide:AC: doxorubicin and cyclophosphamide and ACT: doxorubicin, cyclophosphamide, paclitaxel, Taxane regimens (paclitaxel, docetaxel), TC: docetaxel and cyclophosphamide, TFEC (docetaxel, 5-fluorouracil, epirubicin, cyclophosphamide), Other Schemes: CMF (cyclophosphamide, methotrexate, 5-fluorouracil), Antimetabolites (gemcitabine or capecitabine), microtubule-inhibiting drugs (vinorelbine), and Antitubulin drugs other than taxanes: eribulin, ixabepilone.

The choice of drug is individualized depending on the size of the tumor, the lymph node involvement status, the general performance status or the presence of comorbidities, but anthracyclines in combination with a taxane are most often used, as it was proven in a meta-analysis of 86 randomized studies on 100,000 patients that adding anthracyclines to the taxane therapy regimen reduced the risk of relapse by 14% and the annual mortality by 12% [88]. Adding any other drug is considered a new treatment regimen. Despite early good responses to this type of treatment, there is no positive correlation between early complete remission and overall survival. It has been proven that as many as 90%

of failures in mBC treatment result from the development of multidrug resistance [89]. Cancer cells have developed various resistance mechanisms that allow them to maintain viability after exposure to chemotherapeutic agents. Understanding them can significantly affect the efficacy of therapy and avoid toxic side effects in systemic chemotherapy, as well as the development of new drugs that take into account changes in cancer cells. The most important of them are ABC transporters, overexpression of β-tubulin III, mutations in DNA repair enzymes, changes in genes responsible for apoptosis and signaling pathways ALDH1, GSH/GST (glutathione/glutathione-S-transferase), and NF-κB.

Over the past two years, the U.S. Food and Drug Administration (FDA) has approved several new drugs and therapeutic indications for the treatment of breast cancer (Table 2). The most important include inavolisib in combination with palbociclib and fulvestrant for advanced breast cancer [90]; datopotamab deruxtecan for patients with unresectable or metastatic hormone receptor-positive (HR+) and HER2-negative breast cancer [91]; and capivasertib in combination with fulvestrant for patients with mutations in the PIK3CA, AKT1, or PTEN genes in HR+/HER2- breast cancer [92].

Inavolisib (novel PI3K inhibitor): Approved for the treatment of advanced breast cancer, in combination with palbociclib and fulvestrant. Datopotamab deruxtecan (novel antibody–drug conjugate): Approved for the treatment of unresectable or metastatic HR+, HER2- breast cancer. Capivasertib (novel inhibitor): Approved in combination with fulvestrant for patients with HR-positive, HER2-negative breast cancer with a PIK3CA/AKT1/PTEN mutation [90–92].

The failure rate of currently used immunotherapies in unselected populations of patients with TNBC remains high, and response rates to checkpoint inhibitor monotherapy often do not exceed 10%. However, significant improvements are being observed with combination therapies, such as the combination of immunotherapy with chemotherapy, which is currently undergoing intensive clinical trials. Combining immunotherapy and chemotherapy can lead to synergistic antitumor effects, making this approach promising in the treatment of TNBC. The effectiveness of immunotherapy is strongly dependent on the expression of PD-L1—a biomarker present on both tumor cells and immune cells—which predicts potential treatment response. Using immunotherapy as a first-line therapy, especially in previously untreated patients, may yield better results than its application after multiple prior lines of therapy [93,94].

Table 2. New drugs and their indications approved by the FDA in breast cancer treatment in the last 2 years (as of early 2025).

Drugs	Structure	Characteristics	Indications
Capivasertib	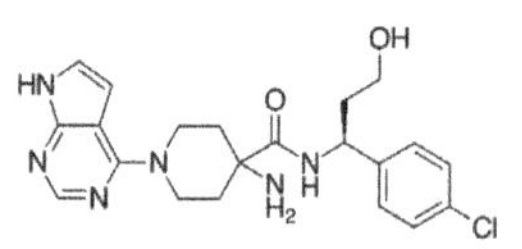	Capivasertib is a kinase inhibitor drug that works by blocking specific proteins that play a key role in the development and survival of cancer cells. This substance selectively inhibits the activity of the AKT protein (also known as protein kinase), a key component of the PI3K/AKT/mTOR signaling pathway. Overactivity of this pathway in cancer promotes uncontrolled cell growth and survival. Inhibition of AKT by capivasertib aims to limit or stop cancer cell proliferation.	Approved for the treatment of hormone receptor-positive, HER2-negative breast cancer with an abnormal PIK3CA gene

Table 2. *Cont.*

Drugs	Structure	Characteristics	Indications
Datopotamab deruxtecan-dlnk	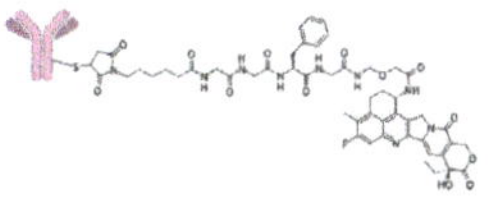	This drug binds to Trop-2 and internalizes the receptor, enabling transport of the active drug into the cell. The release of exatecan inhibits cell replication, leading to apoptosis. The process of exatecan penetration into neighboring cells triggers a series of processes that ultimately result in cell death.	Approved for the treatment of: hormone receptor-positive (HR+) and HER2-negative (HER2-) breast cancer. It is also being studied for the treatment of other types of cancer.
Inavolisib	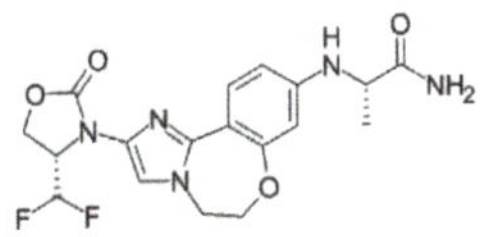	Inavolisib is a potent and selective inhibitor of the PI3K-alpha enzyme (phosphoinositide 3-kinase alpha). It is designed to inhibit the PI3K pathway through HER2-dependent degradation. It aims to inhibit tumor growth in patients whose cancers are caused by PI3K mutations.	Approved for the treatment of hormone receptor-positive, HER2-negative breast cancer with an abnormal PIK3CA gene. It is also being studied for the treatment of other types of cancer
Kisqali Femara Co-Pack (Ribociclib Succinate and Letrozole)	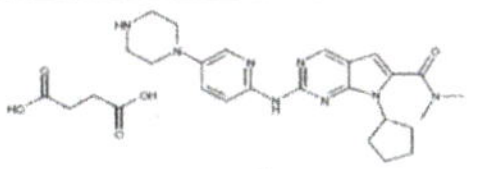 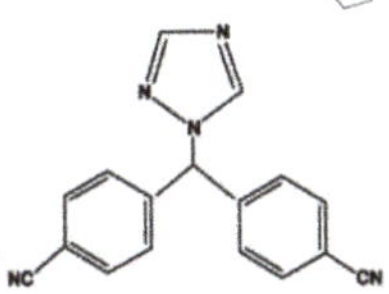	It inhibits the growth of cancer cells, which are ultimately destroyed.	Approved for the treatment of hormone receptor-positive (HR+) and HER2-negative (HER2-) breast cancer. Used for stage II or III breast cancer.

2.5. Limitations and Challenges

Compared to traditional cancer treatments such as surgery, radiotherapy, and chemotherapy, immunotherapy is significantly more effective in prolonging patient survival. This is due to its ability to target multiple stages and targets of tumor development, while selectively attacking tumor tissue and limiting damage to healthy cells. However, cancer cells can develop mechanisms that allow them to escape immune surveillance, promoting metastasis and relapse—two key causes of death in cancer patients. Immunotherapy utilizes a variety of strategies, including stimulating innate and adaptive immunity, overcoming immunosuppressive mechanisms, and strengthening the body's existing defenses. As a result, immunotherapeutics not only enhance the immune response at the primary tumor site during treatment but also induce systemic and long-lasting protective effects, helping to prevent metastasis and tumor recurrence. To date, cancer immunotherapy has been used in the treatment of many advanced malignancies, but its development faces significant barriers. One of the main problems is toxicity—its occurrence and severity can limit further progress and widespread implementation of these methods, as confirmed by numerous clinical trials. Therefore, a key challenge remains developing ways to reduce the risk of adverse events. Another important issue in immuno-oncology is the variable patient response to treatment. While some patients demonstrate significant improvement, others do not respond to immunotherapy at all. Although this phenomenon was not observed in the analyzed studies, this problem is well documented and requires further research to improve the effectiveness of therapy in a broader group of patients [95].

Another significant problem is the lack of selective toxicity, which means that many therapies affect not only cancer cells but also healthy tissues. This imprecision not only leads to reduced therapeutic efficacy but also increases the risk of damage to healthy tissues. Consequently, lower drug doses are often necessary, further limiting treatment effectiveness. An additional challenge is the tendency of conventional therapies to accumulate in healthy tissues at levels significantly higher than within the tumor. This results in suboptimal drug concentrations in cancer cells, meaning they are insufficient to achieve the desired therapeutic effect. As a result, treatment effectiveness is further reduced [96].

3. Future Research Directions in Breast Cancer Treatment—Nanomedicine

Although each year new drugs are approved and used in the treatment of breast cancer, none are 100% effective or guaranteed to cure the cancer. Therefore, developing new methods and treatment plans can increase survival, inhibit tumor growth, and even prevent metastasis [97]. The use of nanoparticle-based systems has recently attracted considerable interest in breast cancer treatment [98]. Nanomedicine offers opportunities that can improve the efficacy of administered drugs, enable better tissue penetration of the therapeutic system, and make the tumor more sensitive to a specific form of treatment [99].

These specially designed molecules have found wide application in various fields of medicine, including oncology, both in terms of diagnostics, as contrast agents (in combination with fluorophores or radioisotopes), and treatment, as carriers for drugs (chemotherapeutics, siRNA, radiosensitizers, immunostimulants) [100]. Thanks to this, nanoparticles can play an important role in oncology by killing cells through various mechanisms: generating reactive oxygen species (ROS), down- and up-regulation in protein production pathways, immunological interactions, and inhibition of transcription [101]. Nanoparticles, due to their size (1–100 nm) and ability to overcome biological barriers, enable precise, intracellular drug delivery and controlled release, which ensures high bioavailability of therapeutic compounds, while limiting the doses of systemically used chemotherapeutics with their simultaneously increased local concentration within the tumor [102]. Their main advantages are the reduction in systemic toxicity of conventional chemotherapy, wide possibilities in creating nanoformulas depending on the needs and precise targeting of cancer cells, and their destruction using electrical, magnetic or optical properties. Additionally, they are bioresponsive structures that can respond to changes inside the body such as pH or temperature changes, which improves the controllability, selectivity and effectiveness of the therapy [103]. Depending on the desired properties, nanoparticles can be made of polymers, lipids, polysaccharides of metal proteins, metal oxides or carbon, which constitute the core of the molecule and functional groups—one or more. They can be polymers, dendrimers, or nanotubes or take liposomal or micellar forms [104]. Two main mechanisms of drug delivery by nanoparticles have been characterized: passive and active. Passive transport, using the specificity of the tumor microenvironment: new, dynamically emerging, defective and leaky blood vessels and lack of or limited lymphatic drainage determine the effect of increased permeability and retention (EPR) [105].

However, this is a heterogeneous phenomenon within different tumors and even within different regions of the same tumor, therefore multifunctional nanoparticles are used, which have the ability to escape from blood vessels independently of EPR, generating local hypertension, loosening intercellular connections of the endothelium or inducing new damage within the tumor blood vessels [106]. In active transport, nanoparticles use surface antigens specific for target cells, different for each type of tissue, which enables selective delivery of the drug to the tumor area [107]. The first doxorubicin-based nanoparticle was registered 30 years ago. To date, liposomes, polymers, micelles, and carbon nanotubes formulated with the doxorubicin nanoparticle have been developed. Combining the nanocomposite with the drug demonstrates superior results compared to standard treatment with the drug alone, for example, in mouse models [108]. The nanodiamond–doxorubicin complex increases apoptosis and inhibits tumor growth and lung metastasis of breast cancer [109]. Nanomedicine enables the broad development of anticancer therapy based not only on chemotherapeutics, but also on therapy directed at siRNA (small interfering RNA), which, taking part in the RNA interference process (RNAi), is able to silence the expression of selected oncogenes by compatible binding to specific mRNA and its degradation [110]. Such a specific treatment deprives the cancer cell of a number of proteins responsible for growth, metastasis, or generating mechanisms of resistance to the applied treatment. This is a promising field of research, but the siRNA delivery system is a challenge, as these are molecules that have their grip directly inside the cell [111].

Although the first siRNA-based drug was registered in 2018 for the treatment of hereditary transthyretin amyloidosis (hATTR), followed by three others for different indications, research on molecules targeting breast cancer is ongoing. Currently, the biggest challenge is to deliver siRNA to the tumor in vivo, because these molecules have a short half-life (about 10 min) and are poorly taken up by cells due to their high molecular weight and the anionic charge of the phosphodiester backbone [112]. They are also rapidly degraded by endonucleases contained in serum and quickly removed by glomerular filtration. Therefore, nanocarriers are used, which protect siRNA from undesirable interactions, metabolism or degradation and have a beneficial effect on the targeted delivery of molecules. Viral and non-viral vectors are used in siRNA delivery, which are becoming increasingly popular due to their potentially lower toxicity and elimination of insertional mutagenesis. Non-viral vectors include positively charged vectors such as cationic cell-penetrating peptides, cationic polymers, dendrimers, conjugation of siRNA with small molecules such as cholesterol or bile acids (bioconjugates) and lipid-based nanocarriers (micelles, liposomes, microemulsions and solid lipid nanoparticles) [113]. The use of nanoparticles enables the delivery of many thousands of siRNA molecules to the cell, while conjugates using antibodies only 1–10 molecules [114]. In addition, their size in the range of 30–200 nm potentially prolongs systemic clearance, because it is not filtered in the renal tubules but is not retained in the reticuloendothelial system of the liver and spleen (additional support in the construction of NC is the use of PEG). NCs are used that combine siRNA targeting a gene that determines the resistance mechanism with a chemotherapeutic agent, in order to be combined with deposition in the cell to obtain a synergistic cytotoxic effect [115]. NC containing siRNA, depending on the targeting of individual genes, may affect specific features of the tumor: metastasis—PTPN22 phosphatase, NF- κB p65 subunit, Twist transcription factor, VEGF, Lcn2 glycoprotein, PLK1 kinase, drug resistance: Pgp glycoprotein, multidrug resistance protein 1 (MRP- 1), autophagy-related protein 7 (ATG7), CXCR4 chemokine receptor, MTDH metadherin, HER2, immune evasion: VEGF, macrophage migration inhibitory factor (MIF), CCR2 receptor, PITPNM3 protein, PD- 1 or PD- L1 and CD73 [116].

Nanoaqualip Technology

Particles developed using "NanoAqualip" technology can increase drug availability by improving permeability and retention in damaged tumor blood vessels. These particles eliminate the need for corticosteroid premedication due to the significantly lower risk of hypersensitivity compared to conventional drug administration (Figure 4). Application of these particles provides improved response rates and better tolerability in locally advanced or metastatic breast cancer compared to conventional drug administration. There is increasing evidence that this technology is effective and safe for local therapy in advanced breast cancer [117,118]. Another example of the use of nanomedicine in breast cancer treatment in recent years has been the so-called Nanoaqualip technology [119,120].

This technology involves the formation of a lipid-based nanomedicine in an aqueous environment. This eliminates the involvement of toxic drug solvents, primarily ethanol, preventing systemic toxicity and many other serious adverse effects. This eliminates the need for premedication with glucocorticosteroids (GCs) before drug administration [119].

Both NPLS and NDLS are currently undergoing clinical trials and demonstrate safety and high efficacy—pCR and CR response rates, as well as overall survival, remain comparable to the previously used formulation, docetaxel [117]. These drugs appear promising, particularly in patients with advanced or metastatic cancer, in patients who would prefer to avoid GCS, and in those who, for some reason, are at increased risk of hypersensitivity reactions.

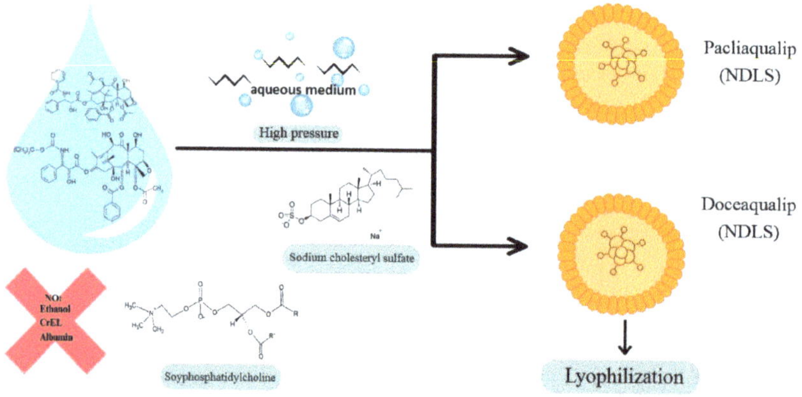

Figure 4. Nanoaqualip technology. The particles are uniform, approximately 100 nm in size, and are produced in an aqueous environment under high pressure using soy phosphatidylcholine and sodium cholesterol sulfate. This suspension is then lyophilized. Dilution is performed in 5% dextrose solution. Nanosomal lipid suspensions of paclitaxel (Pacliaqualip; NPLS) and docetaxel (Doceaqualip; NDLS) have been developed using this technology. The goal was to develop formulations with a superior safety profile to conventional taxanes. CrEL—Cremophor EL—formulation vehicle.

Experiments using nanoparticles as a supportive therapy for breast cancer are already in advanced clinical trials. In the last five years alone, seven clinical trials have been officially registered in Phase II or Phase III on the ClinicalTrials.gov website. Table 3 presents ongoing clinical trials registered over the past five years.

A major challenge facing nanomedicine is the interaction of nanoparticles with biological systems and their long-term cytotoxicity, which is still rarely studied and reported in current reports. The impact of inorganic nanoparticles is of particular concern, as ions released upon their degradation can disrupt the metabolism and proliferation of cancer cells [121].

Furthermore, the Nanoaqualip technique faces significant translational challenges, including difficulties in scaling production due to the complexity of the nanomaterials, high manufacturing costs, unclear and evolving regulatory guidelines for new nanomedicines, and the need for rigorous analytical methods to ensure consistent characterization of nanomaterials for safety and efficacy. Nanomaterial production is inherently more complex than conventional drug formulations, which hinders the development of standardized large-scale processes. The lack of uniform production standards across the industry limits the ability to obtain reproducible, high-quality batches, and the need for costly and extensive safety testing further increases the costs and lengthens the approval process for nanomedicines such as Nanoaqualip [117–120].

Despite promising results in preclinical models, the clinical implementation of nanomedicines is slow and often challenging. General protocols are lacking, characterization of materials and biological mechanisms remains insufficient, and the statistical analyses used do not always ensure complete reliability of results. Furthermore, heterogeneity in research models, limited data sharing, and flaws in study design hinder the progression of nanomedicines to late-stage clinical trials.

Table 3. Registered clinical trials from the last 5 years (2020–2025) in which nanoparticles were included in breast cancer treatment. Based on ClinicalTrials.gov.

Countries	Study Type	Phase	Official Title	Detailed Description	Interventions	Years and Status
Shanghai, China	Interventional	2	A Phase II Study to Explore the Safety, Tolerability, and Preliminary Antitumor Activity of Sitravatinib Plus Tislelizumab or Combination With Nab-paclitaxel in Patients With Locally Recurrent or Metastatic Triple Negative Breast Cancer (TNBC)	The aim of the study is to evaluate the efficacy of sitravatinib in combination with tislelizumab and nab-paclitaxel in patients with previously untreated metastatic breast cancer (TNBC) or with recurrence/metastasis after surgery.	Drug: Sitravatinib Drug: Tislelizumab Drug: Nab-paclitaxel	2021–2024
Gothenburg, Sweden	Interventional	2	Sentinel Lymph Node Localisation With an Ultra-low Dose of Superparamagnetic Iron Oxide Nanoparticles in Patients With Breast Cancer	The main objective is to demonstrate that the use of superparamagnetic iron oxide nanoparticles (SPIO) as a tracer is equally effective in detecting sentinel lymph nodes (SLNs) in breast cancer patients.	Drug: Superparamagnetic Iron Oxide Device: Technetium99	2023–2027
This study has locations in USA, Hong Kong, and Sweden	Interventional	3	Sentinel Lymph Node Biopsy in Ductal Cancer in Situ or Unclear Lesions of the Breast and How to Not do it. An Open-label, Phase 3, Randomised Controlled Trial. (SentiNot 2.0).	The aim of this study is to investigate the use of superparamagnetic iron oxide nanoparticles as a marker for delayed sentinel lymph node dissection in patients for whom initial axillary surgery is considered oncologically unnecessary and should be avoided. This includes patients with a preoperative diagnosis of ductal carcinoma in situ.	Diagnostic Test: Delayed SLND Diagnostic Test: Late SLND	2020–2027
Houston, TX, USA	Interventional	2	A Phase-2, Two-Cohort Trial of Neoadjuvant Nab-Paclitaxel and Alpelisib in Anthracycline Refractory Triple Negative Breast Cancer With PIK3CA or PTEN Alterations	The aim of the study is to evaluate the efficacy of nab-paclitaxel and alpelisib in the treatment of patients with triple-negative breast cancer with mutations in the PIK3CA or PTEN gene that does not respond to anthracycline chemotherapy (anthracycline-resistant).	Drug: Alpelisib Drug: Nab-paclitaxel	2020–2025

Table 3. *Cont.*

Countries	Study Type	Phase	Official Title	Detailed Description	Interventions	Years and Status
Gothenburg, Sweden	Interventional	1, 2	Sentinel Node Localization and Staging with Low Dose Superparamagnetic Iron Oxide-enhanced Magnetic Resonance Imaging and Magnetic Probe in Patients with Breast Cancer	The aim of this study was to determine whether detection of sentinel node status using ultralow dose superparamagnetic iron oxide nanoparticles is feasible.	Drug: Superparamagnetic Iron Oxide	2021–2024
This study has 30 locations in USA	Interventional	2	Randomized Phase 2 Clinical Trial of Nab-Paclitaxel + Durvalumab (MEDI4736) + Tremelimumab + Neoantigen Vaccine Vs. Nab-Paclitaxel + Durvalumab (MEDI4736) + Tremelimumab in Patients With Metastatic Triple Negative Breast Cancer	The Phase II study is examining the efficacy of the drugs nab-paclitaxel, durvalumab, and tremelimumab, in combination with or without a personalized synthetic long-peptide vaccine (neoantigen vaccine), in the treatment of patients with metastatic triple-negative breast cancer.	Procedure: Biopsy Procedure Procedure: Biospecimen Collection Drug: Carboplatin Procedure: Computed Tomography Biological: Durvalumab Drug: Gemcitabine Hydrochloride Procedure: Magnetic Resonance Imaging Drug: Nab-paclitaxel Biological: Personalized Synthetic Long Peptide Vaccine Drug: Poly ICLC Biological: Sacituzumab Govitecan Biological: Tremelimumab	2021–2025
This study has 1008 locations in USA and Puerto Rico	Interventional	2	(CompassHER2-pCR): Preoperative THP and Postoperative HP in Patients Who Achieve a Pathologic Complete Response	The aim of the study is to evaluate the efficacy of paclitaxel, trastuzumab, and pertuzumab in eliminating further chemotherapy after surgery in patients with HER2-positive stage II-IIIa breast cancer who, after preoperative chemotherapy and HER2-targeted therapy, had no cancer detected at the time of surgery (neither in the breast nor in the axillary lymph nodes).	Drug: Docetaxel Procedure: Lumpectomy Procedure: Mastectomy Drug: Nab-paclitaxel Drug: Paclitaxel Biological: Pertuzumab Radiation: Radiation Therapy Biological: Trastuzumab Biological: Trastuzumab Emtansine	2020–2038

Creating interdisciplinary teams that combine materials science, characterization of new technological platforms, and disease models that best reflect clinical conditions, while shaping a regulatory framework consistent with high scientific standards, will generate the data necessary for approval. This will enable the introduction of breakthrough technologies that address unmet needs in healthcare and cancer therapy [122].

Surgery, chemotherapy, and radiotherapy are the most frequently chosen treatment methods for primary tumors. Cancer recurrence, metastasis, and treatment failure necessitate the use of additional methods. Immune-targeted therapies can be a milestone in cancer treatment. Increasingly frequently conducted and registered preclinical studies suggest that the long-term success of cancer therapies depends on immunotherapy. Therefore, cancer immunotherapy is considered an effective treatment method that eliminates both primary tumors and metastases. Nanomedicine is an innovative platform enabling the simultaneous delivery of various immunomodulatory factors directly to the tumor environment or regional lymph nodes. Its primary goal is to precisely reprogram or modulate the immune response through targeted interventions on specific signaling pathways. The development of cancer immunotherapy, primarily using monoclonal antibodies and T-cell-based cell therapies, represents a breakthrough in oncology. However, small-molecule immunotherapeutics are gaining increasing interest. Compared to biologics, they exhibit favorable properties, such as a simpler chemical structure, better tissue penetration, lower production costs, higher stability, and longer shelf life. However, their rapid systemic distribution remains a limitation, potentially leading to off-target side effects. The use of nanotechnology in the formulation of these compounds opens the possibility of increasing their bioavailability and selectivity, facilitating precise targeting of specific immune cell subpopulations while reducing systemic toxicity. Nanoimmunotherapy is challenging, but these two approaches complement each other, improving the overall antitumor response rate [123,124].

It's worth mentioning that, like other methods, this type of therapy has certain limitations. In the case of liposomes, these primarily concern stability and effective drug storage. Furthermore, there are often limitations to large-scale production. Furthermore, the short half-life can be problematic when it comes to delivering the drug over longer distances and its gradual release. Lipid nanoparticles have been reported to have side effects that can cause inflammation. Other limitations may include heterogeneous physicochemical properties, crystallization, low drug loading efficiency, emulsifier toxicity, and the potential cytotoxicity of some polymer components. To enable the further development of nanoparticles as promising tools in cancer therapy, the development of global regulatory standards is essential, requiring close collaboration between international regulatory agencies, academia, and industry. However, scaling up pharmaceutical nanoparticles presents a number of complex challenges. Their complex structure requires a thorough understanding to identify key properties necessary for reproducibility. This step provides the basis for selecting appropriate industrial-scale manufacturing processes and defining critical analytical parameters. Transitioning from basic research to commercial production of therapeutic nanoparticles requires resolving numerous challenges related to chemistry, manufacturing technology, and quality control. Proving the transferability of a given technology to a specialized development center or contract manufacturing organization, where scalable, cost-effective, and well-controlled processes can be implemented, is a key element. The complexity of the therapeutic nanoparticle manufacturing process further complicates the transition from preclinical trials to clinical trials and full commercialization. To meet Good Manufacturing Practice (GMP) requirements, it is recommended to use proven unit operations already used in the pharmaceutical industry, which can accelerate the process of implementing new therapies [125].

4. Conclusions

Breast cancer therapy is undergoing continuous improvement and research into increasingly precise therapeutic solutions. Drugs based on well-known hormonal mechanisms involved in the pathogenesis of this cancer are still being actively developed, but more and more drugs are being implemented that target molecular targets related to carcinogenesis, such as the PI3K/AKT/mTOR pathway. In addition, many current clinical trials compare the use of existing and registered drugs in various combined regimens, sometimes giving better treatment results. The implementation of nanoparticles as a drug carrier system has been groundbreaking, showing great potential to become the basis for diagnostics and treatment thanks to their unique properties ensuring higher efficacy, but also safety for patients. They can find a special place in TNBC therapy, especially in advanced and metastatic forms, where standard treatment regimens do not bring satisfactory results. They enable the transport of not only chemotherapeutics, but also siRNA, which translates into the precision and multidirectionality of the agents used. The number of clinical trials emerging each year indicates that new treatment methods are being sought. Combination therapies and synergistic approaches herald further attempts to increase the effectiveness of breast cancer treatment. This is a decisive and positive impulse, with hope for the future of oncology.

Author Contributions: Conceptualization, K.D., D.B.-A., K.K., A.K. and D.A.; methodology, K.D., D.B.-A., K.K., A.K. and D.A.; formal analysis, K.D., D.B.-A., K.K., A.K. and D.A.; resources, K.D., D.B.-A., K.K., A.K. and D.A.; writing—original draft preparation, K.D., D.B.-A., K.K., A.K. and D.A.; writing—review and editing, K.D., D.B.-A., K.K., A.K. and D.A.; visualization, K.D., D.B.-A., K.K., A.K. and D.A.; supervision, D.A. All authors have read and agreed to the published version of the manuscript.

Funding: This research received no external funding.

Institutional Review Board Statement: Not applicable.

Informed Consent Statement: Not applicable.

Data Availability Statement: Not applicable.

Conflicts of Interest: The authors declare no conflicts of interest.

References

1. Nwosu, N. Cancer: A Disease of Modern Times? *Cureus* **2024**, *16*, e74666. [CrossRef] [PubMed] [PubMed Central]
2. Marino, P.; Mininni, M.; Deiana, G.; Marino, G.; Divella, R.; Bochicchio, I.; Giuliano, A.; Lapadula, S.; Lettini, A.R.; Sanseverino, F. Healthy Lifestyle and Cancer Risk: Modifiable Risk Factors to Prevent Cancer. *Nutrients.* **2024**, *16*, 800. [CrossRef] [PubMed] [PubMed Central]
3. Kotsifaki, A.; Maroulaki, S.; Armakolas, A. Exploring the Immunological Profile in Breast Cancer: Recent Advances in Diagnosis and Prognosis through Circulating Tumor Cells. *Int. J. Mol. Sci.* **2024**, *25*, 4832. [CrossRef]
4. Siegel, R.L.; Giaquinto, A.N.; Jemal, A. Cancer statistics, 2024. *CA Cancer J Clin.* **2024**, *74*, 12–49. [CrossRef] [PubMed]
5. Wang, J.; Wu, S.G. Breast Cancer: An Overview of Current Therapeutic Strategies, Challenge, and Perspectives. *Breast Cancer* **2023**, *15*, 721–730. [CrossRef] [PubMed] [PubMed Central]
6. Xiong, X.; Zheng, L.W.; Ding, Y.; Chen, Y.F.; Cai, Y.W.; Wang, L.P.; Huang, L.; Liu, C.C.; Shao, Z.M.; Yu, K.D. Breast cancer: Pathogenesis and treatments. *Signal Transduct. Target. Ther.* **2025**, *10*, 49. [CrossRef]
7. Rossi, L.; Mazzara, C.; Pagani, O. Diagnosis and Treatment of Breast Cancer in Young Women. *Curr. Treat. Options Oncol.* **2019**, *20*, 86. [CrossRef] [PubMed]
8. Habibi, S.; Bahramian, S.; Zare Jalise, S.; Mehri, S.; Ababzadeh, S.; Kavianpour, M. Novel strategies in breast cancer management: From treatment to long-term remission. *Crit. Rev. Oncol. Hematol.* **2025**, *211*, 104715. [CrossRef] [PubMed]
9. Maqsood, Q.; Khan, M.U.; Fatima, T.; Khalid, S.; Malik, Z.I. Recent Insights Into Breast Cancer: Molecular Pathways, Epigenetic Regulation, and Emerging Targeted Therapies. *Breast Cancer* **2025**, *19*, 1–20. [CrossRef] [PubMed] [PubMed Central]

10. Liu, L.; Graff, S.L.; Wang, Y. New Emerging Therapies Targeting PI3K/AKT/mTOR/PTEN Pathway in Hormonal Receptor-Positive and HER2-Negative Breast Cancer-Current State and Molecular Pathology Perspective. *Cancers* **2024**, *17*, 16. [CrossRef] [PubMed] [PubMed Central]
11. Stanciu, I.M.; Parosanu, A.I.; Orlov-Slavu, C.; Iaciu, I.C.; Popa, A.M.; Olaru, C.M.; Pirlog, C.F.; Vrabie, R.C.; Nitipir, C. Mechanisms of Resistance to CDK4/6 Inhibitors and Predictive Biomarkers of Response in HR+/HER2-Metastatic Breast Cancer-A Review of the Literature. *Diagnostics* **2023**, *13*, 987. [CrossRef] [PubMed] [PubMed Central]
12. Miricescu, D.; Totan, A.; Stanescu-Spinu, I.I.; Badoiu, S.C.; Stefani, C.; Greabu, M. PI3K/AKT/mTOR Signaling Pathway in Breast Cancer: From Molecular Landscape to Clinical Aspects. *Int. J. Mol. Sci.* **2020**, *22*, 173. [CrossRef]
13. Rampioni Vinciguerra, G.L.; Sonego, M.; Segatto, I.; Dall'Acqua, A.; Vecchione, A.; Baldassarre, G.; Belletti, B. CDK4/6 Inhibitors in Combination Therapies: Better in Company Than Alone: A Mini Review. *Front. Oncol.* **2022**, *12*, 891580. [CrossRef] [PubMed] [PubMed Central]
14. Piezzo, M.; Cocco, S.; Caputo, R.; Cianniello, D.; Gioia, G.D.; Lauro, V.D.; Fusco, G.; Martinelli, C.; Nuzzo, F.; Pensabene, M.; et al. Targeting Cell Cycle in Breast Cancer: CDK4/6 Inhibitors. *Int. J. Mol. Sci.* **2020**, *21*, 6479. [CrossRef]
15. Choudhary, N.; Bawari, S.; Burcher, J.T.; Sinha, D.; Tewari, D.; Bishayee, A. Targeting Cell Signaling Pathways in Lung Cancer by Bioactive Phytocompounds. *Cancers* **2023**, *15*, 3980. [CrossRef]
16. Matsuoka, T.; Yashiro, M. Molecular targets for the treatment of pancreatic cancer: Clinical and experimental studies. *World J. Gastroenterol.* **2016**, *22*, 776–789. [CrossRef] [PubMed] [PubMed Central]
17. Guo, W.; Wang, H.; Li, C. Signal pathways of melanoma and targeted therapy. *Signal Transduct. Target. Ther.* **2021**, *6*, 424. [CrossRef] [PubMed] [PubMed Central]
18. ClinicalTrials.gov. Available online: https://clinicaltrials.gov/study/NCT01610284 (accessed on 20 May 2025).
19. Campone, M.; Im, S.A.; Iwata, H.; Clemons, M.; Ito, Y.; Awada, A.; Chia, S.; Jagiełło-Gruszfeld, A.; Pistilli, B.; Tseng, L.M.; et al. Buparlisib plus fulvestrant versus placebo plus fulvestrant for postmenopausal, hormone receptor-positive, human epidermal growth factor receptor 2-negative, advanced breast cancer: Overall survival results from BELLE-2. *Eur. J. Cancer* **2018**, *103*, 147–154. [CrossRef]
20. ClinicalTrials.gov. Available online: https://clinicaltrials.gov/study/NCT01633060 (accessed on 20 May 2025).
21. Di Leo, A.; Johnston, S.; Lee, K.S.; Ciruelos, E.; Lonning, P.E.; Janni, W.; O'Regan, R.; Mouret-Reynier, M.A.; Kalev, D.; Egle, D.; et al. Buparlisib plus fulvestrant in postmenopausal women with hormone-receptor-positive, HER2-negative, advanced breast cancer progressing on or after mTOR inhibition (BELLE-3): A randomised, double-blind, placebo-controlled, phase 3 trial. *Lancet Oncol.* **2018**, *19*, 87–100.
22. Krop, I.E.; Mayer, I.A.; Ganju, V.; Dickler, M.; Johnston, S.; Morales, S.; Yardley, D.A.; Melichar, B.; Forero-Torres, A.; Lee, S.C.; et al. Pictilisib for oestrogen receptor-positive, aromatase inhibitor-resistant, advanced or metastatic breast cancer (FERGI): A randomised, double-blind, placebo-controlled, phase 2 trial. *Lancet Oncol.* **2016**, *17*, 811–821. [CrossRef]
23. ClinicalTrials.gov. Available online: https://clinicaltrials.gov/study/NCT01437566 (accessed on 20 May 2025).
24. Schmid, P.; Pinder, S.E.; Wheatley, D.; Macaskill, J.; Zammit, C.; Hu, J.; Price, R.; Bundred, N.; Hadad, S.; Shia, A.; et al. Phase II Randomized Preoperative Window-of-Opportunity Study of the PI3K Inhibitor Pictilisib Plus Anastrozole Compared with Anastrozole Alone in Patients with Estrogen Receptor-Positive Breast Cancer. *J. Clin. Oncol. Off. J. Am. Soc. Clin. Oncol.* **2016**, *34*, 1987–1994. [CrossRef]
25. Schöffski, P.; Cresta, S.; Mayer, I.A.; Wildiers, H.; Damian, S.; Gendreau, S.; Rooney, I.; Morrissey, K.M.; Spoerke, J.M.; Ng, V.W.; et al. A phase Ib study of pictilisib (GDC-0941) in combination with paclitaxel, with and without bevacizumab or trastuzumab, and with letrozole in advanced breast cancer. *Breast Cancer Res.* **2018**, *20*, 109.
26. ClinicalTrials.gov. Available online: https://clinicaltrials.gov/study/NCT01740336 (accessed on 20 May 2025).
27. Vuylsteke, P.; Huizing, M.; Petrakova, K.; Roylance, R.; Laing, R.; Chan, S.; Abell, F.; Gendreau, S.; Rooney, I.; Apt, D.; et al. Pictilisib PI3Kinase inhibitor (a phosphatidylinositol 3-kinase [PI3K] inhibitor) plus paclitaxel for the treatment of hormone receptor-positive, HER2-negative, locally recurrent, or metastatic breast cancer: Interim analysis of the multicentre, placebo-controlled, phase II randomised PEGGY study. *Ann. Oncol.* **2016**, *27*, 2059–2066. [PubMed]
28. André, F.; Ciruelos, E.; Rubovszky, G.; Campone, M.; Loibl, S.; Rugo, H.S.; Iwata, H.; Conte, P.; Mayer, I.A.; Kaufman, B.; et al. SOLAR-1 Study Group. Alpelisib for PIK3CA-Mutated, Hormone Receptor-Positive Advanced Breast Cancer. *N. Engl. J. Med.* **2019**, *380*, 1929–1940.
29. André, F.; Ciruelos, E.M.; Juric, D.; Loibl, S.; Campone, M.; Mayer, I.A.; Rubovszky, G.; Yamashita, T.; Kaufman, B.; Lu, Y.S.; et al. Alpelisib plus fulvestrant for PIK3CA-mutated, hormone receptor-positive, human epidermal growth factor receptor-2–negative advanced breast cancer: Final overall survival results from SOLAR-1. *Ann. Oncol.* **2021**, *32*, 208–217. [PubMed]
30. Copur, M.S. Alpelisib to treat breast cancer. *Drugs Today (Barc.)* **2020**, *56*, 357–363.
31. ClinicalTrials.gov. Available online: https://clinicaltrials.gov/study/NCT03056755 (accessed on 20 May 2025).
32. ClinicalTrials.gov. Available online: https://clinicaltrials.gov/study/NCT01923168 (accessed on 20 May 2025).

33. Mayer, I.A.; Prat, A.; Egle, D.; Blau, S.; Fidalgo, J.A.P.; Gnant, M.; Fasching, P.A.; Colleoni, M.; Wolff, A.C.; Winer, E.P.; et al. A Phase II Randomized Study of Neoadjuvant Letrozole Plus Alpelisib for Hormone Receptor-Positive, Human Epidermal Growth Factor Receptor 2-Negative Breast Cancer (NEO-ORB). *Clin. Cancer Res. Off. J. Am. Assoc. Cancer Res.* **2019**, *25*, 2975–2987. [CrossRef]
34. Dent, S.; Cortés, J.; Im, Y.H.; Diéras, V.; Harbeck, N.; Krop, I.E.; Wilson, T.R.; Cui, N.; Schimmoller, F.; Hsu, J.Y.; et al. Phase III randomized study of taselisib or placebo with fulvestrant in estrogen receptor-positive, PIK3CA-mutant, HER2-negative, advanced breast cancer: The SANDPIPER trial. *Ann. Oncol.* **2021**, *32*, 197–207. [CrossRef]
35. ClinicalTrials.gov. Available online: https://clinicaltrials.gov/study/NCT02340221 (accessed on 20 May 2025).
36. ClinicalTrials.gov. Available online: https://clinicaltrials.gov/study/NCT04191499 (accessed on 20 May 2025).
37. Arabshomali, A.; Goswami, S.; Masurkar, P.P. Inavolisib for HR-Positive, HER2-Negative Advanced Breast Cancer: Clinical Trials and Patient Access Implication. *Am. J. Clin. Oncol.* **2025**, *48*, 496–500.
38. ClinicalTrials.gov. Available online: https://clinicaltrials.gov/study/NCT03961698 (accessed on 20 May 2025).
39. Hatem, S.; Hargis, J.; Elias, A.; Lee, A.; Swart, R.; Dahkil, S.; Drakaki, A.; Phan, V.; Kass, F.; Cobleigh, M.; et al. Updated efficacy, safety and translational data from MARIO-3, a phase II open-label study evaluating a novel triplet combination of eganelisib (IPI-549), atezolizumab (atezo), and nab-paclitaxel (nab-pac) as first-line (1L) therapy for locally advanced or metastatic triple-negative breast cancer (TNBC). *Cancer Res.* **2022**, *82*, 5-16-02. [CrossRef]
40. ClinicalTrials.gov. Available online: https://clinicaltrials.gov/show/NCT02684032 (accessed on 20 May 2025).
41. Layman, R.M.; Han, H.S.; Rugo, H.S.; Stringer-Reasor, E.M.; Specht, J.M.; Dees, E.C.; Kabos, P.; Suzuki, S.; Mutka, S.C.; Sullivan, B.F.; et al. Gedatolisib in combination with palbociclib and endocrine therapy in women with hormone receptor-positive, HER2-negative advanced breast cancer: Results from the dose expansion groups of an open-label, phase 1b study. *Lancet Oncol.* **2024**, *25*, 474–487.
42. ClinicalTrials.gov. Available online: https://clinicaltrials.gov/study/NCT05501886 (accessed on 20 May 2025).
43. ClinicalTrials.gov. Available online: https://clinicaltrials.gov/study/NCT06757634 (accessed on 20 May 2025).
44. Chien, A.J.; Tripathy, D.; Albain, K.S.; Symmans, W.F.; Rugo, H.S.; Melisko, M.E.; Wallace, A.M.; Schwab, R.; Helsten, T.; Forero-Torres, A.; et al. I-SPY 2 Consortium. *MK*-2206 and Standard Neoadjuvant Chemotherapy Improves Response in Patients with Human Epidermal Growth Factor Receptor 2-Positive and/or Hormone Receptor-Negative Breast Cancers in the I-SPY 2 Trial. *J. Clin. Oncol.* **2020**, *38*, 1059–1069. [CrossRef]
45. ClinicalTrials.gov. Available online: https://www.clinicaltrials.gov/study/NCT04305496 (accessed on 20 May 2025).
46. Turner, N.C.; Oliveira, M.; Howell, S.J.; Dalenc, F.; Cortes, J.; Gomez Moreno, H.L.; Hu, X.; Jhaveri, K.; Krivorotko, P.; Loibl, S.; et al. CAPItello-291 Study Group. Capivasertib in Hormone Receptor-Positive Advanced Breast Cancer. *N. Engl. J. Med.* **2023**, *388*, 2058–2070. [CrossRef]
47. Yi, Z.; Liu, B.; Sun, X.; Rong, G.; Wang, W.; Li, H.; Guan, X.; Li, L.; Zhai, J.; Li, C.; et al. Safety and efficacy of sirolimus combined with endocrine therapy in patients with advanced hormone receptor-positive breast cancer and the exploration of biomarkers. *Breast* **2020**, *52*, 17–22. [CrossRef]
48. ClinicalTrials.gov. Available online: https://clinicaltrials.gov/study/NCT06957379 (accessed on 20 May 2025).
49. Wolff, A.C.; Lazar, A.A.; Bondarenko, I.; Garin, A.M.; Brincat, S.; Chow, L.; Sun, Y.; Neskovic-Konstantinovic, Z.; Guimaraes, R.C.; Fumoleau, P.; et al. Randomized phase III placebo-controlled trial of letrozole plus oral temsirolimus as first-line endocrine therapy in postmenopausal women with locally advanced or metastatic breast cancer. *J. Clin. Oncol.* **2013**, *31*, 195–202. [CrossRef] [PubMed] [PubMed Central]
50. García-Sáenz, J.Á.; Martínez-Jáñez, N.; Cubedo, R.; Jerez, Y.; Lahuerta, A.; González-Santiago, S.; Ferrer, N.; Ramos, M.; Alonso-Romero, J.L.; Antón, A.; et al. Sapanisertib plus Fulvestrant in Postmenopausal Women with Estrogen Receptor-Positive/HER2-Negative Advanced Breast Cancer after Progression on Aromatase Inhibitor. *Clin. Cancer Res.* **2022**, *28*, 1107–1116. [CrossRef]
51. ClinicalTrials.gov. Available online: https://clinicaltrials.gov/study/NCT02756364 (accessed on 20 May 2025).
52. Schmid, P.; Zaiss, M.; Harper-Wynne, C.; Ferreira, M.; Dubey, S.; Chan, S.; Makris, A.; Nemsadze, G.; Brunt, A.M.; Kuemmel, S.; et al. Fulvestrant Plus Vistusertib vs Fulvestrant Plus Everolimus vs Fulvestrant Alone for Women with Hormone Receptor-Positive Metastatic Breast Cancer: The MANTA Phase 2 Randomized Clinical Trial. *JAMA Oncol.* **2019**, *5*, 1556–1564. [CrossRef]
53. ClinicalTrials.gov. Available online: https://clinicaltrials.gov/study/NCT02216786 (accessed on 20 May 2025).
54. Hong, D.S.; Moore, K.N.; Bendell, J.C.; Karp, D.D.; Wang, J.S.; Ulahannan, S.V.; Jones, S.; Wu, W.; Donoho, G.P.; Ding, Y.; et al. Preclinical evaluation and phase ib study of prexasertib, a chk1 inhibitor, and samotolisib (LY3023414), a dual PI3K/mTOR inhibitor. *Clin. Cancer Res.* **2021**, *27*, 1864–1874. [CrossRef]
55. Kim, S.B.; Dent, R.; Im, S.A.; Espié, M.; Blau, S.; Tan, A.R.; Isakoff, S.J.; Oliveira, M.; Saura, C.; Wongchenko, M.J.; et al. Ipatasertib plus paclitaxel versus placebo plus paclitaxel as first-line therapy for metastatic triple-negative breast cancer (LOTUS): A multicentre, randomised, double-blind, placebo-controlled, phase 2 trial. *Lancet Oncol.* **2017**, *18*, 1360–1372.
56. ClinicalTrials.gov. Available online: https://clinicaltrials.gov/study/NCT02162719 (accessed on 20 May 2025).

57. Dent, R.A.; Kim, S.B.; Oliveira, M.; Barrios, C.; O'Shaughnessy, J.; Isakoff, S.J.; Saji, S.; Freitas-Junior, R.; Philco, M.; Bondarenko, I.; et al. Ipatasertib plus Paclitaxel for Patients with PIK3CA/AKT1/PTEN-Altered Locally Advanced Unresectable or Metastatic Triple-Negative Breast Cancer in the IPATunity130 Phase III Trial. *Clin Cancer Res.* **2024**, *30*, 4329–4338. [CrossRef]
58. ClinicalTrials.gov. Available online: https://clinicaltrials.gov/study/NCT03337724 (accessed on 20 May 2025).
59. Glaviano, A.; Foo, A.S.C.; Lam, H.Y.; Yap, K.C.H.; Jacot, W.; Jones, R.H.; Eng, H.; Nair, M.G.; Makvandi, P.; Geoerger, B.; et al. PI3K/AKT/mTOR signaling transduction pathway and targeted therapies in cancer. *Mol. Cancer.* **2023**, *22*, 138. [CrossRef] [PubMed] [PubMed Central]
60. Owonikoko, T.K.; Khuri, F.R. Targeting the PI3K/AKT/mTOR pathway: Biomarkers of success and tribulation. *Am. Soc. Clin. Oncol. Educ. Book* **2013**, *33*, e395–e401. [CrossRef] [PubMed] [PubMed Central]
61. He, Y.; Sun, M.M.; Zhang, G.G.; Yang, J.; Chen, K.S.; Xu, W.W.; Li, B. Targeting PI3K/Akt signal transduction for cancer therapy. *Signal Transduct. Target. Ther.* **2021**, *6*, 425. [CrossRef] [PubMed] [PubMed Central]
62. Keating, G.M. Pertuzumab: In the first-line treatment of HER2-positive metastatic breast cancer. *Drugs* **2012**, *72*, 353–360. [CrossRef] [PubMed]
63. Maadi, H.; Soheilifar, M.H.; Choi, W.S.; Moshtaghian, A.; Wang, Z. Trastuzumab Mechanism of Action; 20 Years of Research to Unravel a Dilemma. *Cancers* **2021**, *13*, 3540. [CrossRef] [PubMed] [PubMed Central]
64. Ovcinnikovs, V.; Dijkman, K.; Zom, G.G.; Beurskens, F.J.; Trouw, L.A. Enhancing complement activation by therapeutic anti-tumor antibodies: Mechanisms, strategies, and engineering approaches. *Semin. Immunol.* **2025**, *77*, 101922. [CrossRef] [PubMed]
65. Tsao, L.C.; Crosby, E.J.; Trotter, T.N.; Wei, J.; Wang, T.; Yang, X.; Summers, A.N.; Lei, G.; Rabiola, C.A.; Chodosh, L.A.; et al. Trastuzumab/pertuzumab combination therapy stimulates antitumor responses through complement-dependent cytotoxicity and phagocytosis. *JCI Insight* **2022**, *7*, e155636. [CrossRef]
66. von Minckwitz, G.; Procter, M.; de Azambuja, E.; Zardavas, D.; Benyunes, M.; Viale, G.; Suter, T.; Arahmani, A.; Rouchet, N.; Clark, E.; et al. APHINITY Steering Committee and Investigators. Adjuvant Pertuzumab and Trastuzumab in Early HER2-Positive Breast Cancer. *N. Engl. J. Med.* **2017**, *377*, 122–131; Erratum in *N. Engl. J. Med.* **2017**, *377*, 702; https://doi.org/10.1056/NEJMx170011; Erratum in *N. Engl. J. Med.* **2018**, *379*, 1585. [CrossRef] [PubMed] [PubMed Central]
67. Sasso, J.M.; Tenchov, R.; Bird, R.; Iyer, K.A.; Ralhan, K.; Rodriguez, Y.; Zhou, Q.A. The Evolving Landscape of Antibody-Drug Conjugates: In Depth Analysis of Recent Research Progress. *Bioconjug Chem.* **2023**, *34*, 1951–2000. [CrossRef] [PubMed] [PubMed Central]
68. Biswas, G.; Subudhi, G.C.; Biswas, S.; Das, B. Long-Term Success with a TDM-1 Biosimilar in Recurrent HER2-Positive Breast Cancer with Multisite Metastases: A Comprehensive Case Study. *Cureus* **2024**, *16*, e67047. [CrossRef]
69. von Arx, C.; De Placido, P.; Caltavituro, A.; Di Rienzo, R.; Buonaiuto, R.; De Laurentiis, M.; Arpino, G.; Puglisi, F.; Giuliano, M.; Del Mastro, L. The evolving therapeutic landscape of trastuzumab-drug conjugates: Future perspectives beyond HER2-positive breast cancer. *Cancer Treat Rev.* **2023**, *113*, 102500.
70. Ogitani, Y.; Aida, T.; Hagihara, K.; Yamaguchi, J.; Ishii, C.; Harada, N.; Soma, M.; Okamoto, H.; Oitate, M.; Arakawa, S.; et al. DS-8201a, A Novel HER2-Targeting ADC with a Novel DNA Topoisomerase I Inhibitor, Demonstrates a Promising Antitumor Efficacy with Differentiation from T-DM1. *Clin. Cancer Res. Off. J. Am. Assoc. Cancer Res.* **2016**, *22*, 5097–5108. [CrossRef]
71. Fabi, A.; De Laurentiis, M.; Caruso, M.; Valle, E.; Moscetti, L.; Santini, D.; Cannita, K.; Carbognin, L.; Ciccarese, M.; Rossello, R.; et al. Efficacy and safety of T-DM1 in the 'common-practice' of HER2+ advanced breast cancer setting: A multicenter study. *Oncotarget* **2017**, *8*, 64481–64489. [CrossRef]
72. Swain, S.M.; Shastry, M.; Hamilton, E. Targeting HER2-positive breast cancer: Advances and future directions. *Nat. Rev. Drug Discov.* **2023**, *22*, 101–126. [CrossRef] [PubMed] [PubMed Central]
73. Pan, L.; Li, J.; Xu, Q.; Gao, Z.; Yang, M.; Wu, X.; Li, X. HER2/PI3K/AKT pathway in HER2-positive breast cancer: A review. *Medicine* **2024**, *103*, e38508. [CrossRef] [PubMed] [PubMed Central]
74. Xuhong, J.C.; Qi, X.W.; Zhang, Y.; Jiang, J. Mechanism, safety and efficacy of three tyrosine kinase inhibitors lapatinib, neratinib and pyrotinib in HER2-positive breast cancer. *Am. J. Cancer Res.* **2019**, *9*, 2103–2119. [PubMed] [PubMed Central]
75. Li, J.; Wang, H.; Li, J.; Bao, J.; Wu, C. Discovery of a Potential HER2 Inhibitor from Natural Products for the Treatment of HER2-Positive Breast Cancer. *Int. J. Mol. Sci.* **2016**, *17*, 1055. [CrossRef]
76. Yang, X.; Wu, D.; Yuan, S. Tyrosine Kinase Inhibitors in the Combination Therapy of HER2 Positive Breast Cancer. *Technol. Cancer Res. Treat.* **2020**, *19*, 1–14. [CrossRef]
77. Yang, T.; Kang, L.; Li, D.; Song, Y. Immunotherapy for HER-2 positive breast cancer. *Front. Oncol.* **2023**, *13*, 1097983. [CrossRef] [PubMed] [PubMed Central]
78. Abaza, A.; Sid Idris, F.; Anis Shaikh, H.; Vahora, I.; Moparthi, K.P.; Al Rushaidi, M.T.; Muddam, M.R.; Obajeun, O.A.; Jaramillo, A.P.; Khan, S. Programmed Cell Death Protein 1 (PD-1) and Programmed Cell Death Ligand 1 (PD-L1) Immunotherapy: A Promising Breakthrough in Cancer Therapeutics. *Cureus* **2023**, *15*, e44582. [CrossRef] [PubMed] [PubMed Central]
79. Jie, H.; Ma, W.; Huang, C. Diagnosis, Prognosis, and Treatment of Triple-Negative Breast Cancer: A Review. *Breast Cancer* **2025**, *17*, 265–274. [CrossRef] [PubMed] [PubMed Central]

80. Jain, A.; Barge, A.; Parris, C.N. Combination strategies with PARP inhibitors in BRCA-mutated triple-negative breast cancer: Overcoming resistance mechanisms. *Oncogene* **2025**, *44*, 193–207; Erratum in *Oncogene* **2025**, *44*, 1063. https://doi.org/10.1038/s41388-025-03364-6. [CrossRef] [PubMed] [PubMed Central]
81. Zhao, J.; Tang, B.; Shen, P.; Zeng, H.; Wei, Q. Empowering PARP inhibition through rational combination: Mechanisms of PARP inhibitors and combinations with a focus on the treatment of metastatic castration-resistant prostate cancer. *Crit. Rev. Oncol. Hematol.* **2025**, *210*, 104698. [CrossRef] [PubMed]
82. Quintana, A.; Saini, K.S.; Vidal, L.; Peg, V.; Slebe, F.; Loibl, S.; Curigliano, G.; Schmid, P.; Cortes, J. Window of opportunity trials with immune checkpoint inhibitors in triple-negative breast cancer. *ESMO Open* **2024**, *9*, 103713. [CrossRef]
83. Anand, U.; Dey, A.; Chandel, A.K.S.; Sanyal, R.; Mishra, A.; Pandey, D.K.; De Falco, V.; Upadhyay, A.; Kandimalla, R.; Chaudhary, A.; et al. Cancer chemotherapy and beyond: Current status, drug candidates, associated risks and progress in targeted therapeutics. *Genes Dis.* **2022**, *10*, 1367–1401. [CrossRef]
84. Bayat Mokhtari, R.; Homayouni, T.S.; Baluch, N.; Morgatskaya, E.; Kumar, S.; Das, B.; Yeger, H. Combination therapy in combating cancer. *Oncotarget* **2017**, *8*, 38022–38043. [CrossRef]
85. Wang, Y.; Minden, A. Current Molecular Combination Therapies Used for the Treatment of Breast Cancer. *Int. J. Mol. Sci.* **2022**, *23*, 11046. [CrossRef]
86. Liu, Y.P.; Zheng, C.C.; Huang, Y.N.; He, M.L.; Xu, W.W.; Li, B. Molecular mechanisms of chemo- and radiotherapy resistance and the potential implications for cancer treatment. *MedComm* **2021**, *2*, 315–340. [CrossRef] [PubMed] [PubMed Central]
87. Asleh, K.; Riaz, N.; Nielsen, T.O. Heterogeneity of triple negative breast cancer: Current advances in subtyping and treatment implications. *J. Exp. Clin. Cancer Res.* **2022**, *41*, 265. [CrossRef] [PubMed] [PubMed Central]
88. Early Breast Cancer Trialists' Collaborative Group (EBCTCG). Anthracycline-containing and taxane-containing chemotherapy for early-stage operable breast cancer: A patient-level meta-analysis of 100,000 women from 86 randomised trials. *Lancet* **2023**, *401*, 1277–1292. [CrossRef]
89. Yu, J.; Mu, Q.; Fung, M.; Xu, X.; Zhu, L.; Ho, R.J.Y. Challenges and opportunities in metastatic breast cancer treatments: Nano-drug combinations delivered preferentially to metastatic cells may enhance therapeutic response. *Pharmacol. Ther.* **2022**, *236*, 108108. [CrossRef]
90. Turner, N.C.; Im, S.A.; Saura, C.; Juric, D.; Loibl, S.; Kalinsky, K.; Schmid, P.; Loi, S.; Sunpaweravong, P.; Musolino, A.; et al. Inavolisib-Based Therapy in *PIK3CA*-Mutated Advanced Breast Cancer. *N. Engl. J. Med.* **2024**, *391*, 1584–1596. [CrossRef] [PubMed]
91. Royce, M.; Shah, M.; Zhang, L.; Cheng, J.; Bonner, M.K.; Pegues, M.; Miller, C.P.; Leu, L.; Price, L.S.L.; Qiu, J.; et al. FDA Approval Summary: Datopotamab deruxtecan-dlnk for treatment of patients with unresectable or metastatic, HR-positive, HER2-negative breast cancer. *Clin Cancer Res.* **2025**, *ahead of print*. [CrossRef] [PubMed] [PubMed Central]
92. Tokunaga, E.; Iwata, H.; Itoh, M.; Taira, T.; Toyama, T.; Mizuno, T.; Osaki, A.; Yanagita, Y.; Nakamura, S.; Nakamura, R.; et al. Capivasertib and fulvestrant for patients with HR-positive/HER2-negative advanced breast cancer: Analysis of the subgroup of patients from Japan in the phase 3 CAPItello-291 trial. *Breast Cancer* **2025**, *32*, 132–143. [CrossRef] [PubMed] [PubMed Central]
93. Li, L.; Zhang, F.; Liu, Z.; Fan, Z. Immunotherapy for Triple-Negative Breast Cancer: Combination Strategies to Improve Outcome. *Cancers* **2023**, *15*, 321. [CrossRef] [PubMed]
94. Qian, K.; Liu, Q. Narrative review on the role of immunotherapy in early triple negative breast cancer: Unveiling opportunities and overcoming challenges. *Transl. Breast Cancer Res.* **2023**, *4*, 16. [CrossRef]
95. Medina, A.; Carballo, J.; González-Marcano, E.; Blanca, I.; Convit, A.F. Breast cancer immunotherapy: Realities and advances. *Cancer Innov.* **2024**, *3*, e140. [CrossRef] [PubMed] [PubMed Central]
96. Oncology. In Clinical Practice. Available online: https://journals.viamedica.pl/oncology_in_clinical_practice/article/view/106843/83567 (accessed on 2 October 2025).
97. Storme, G.A. Breast Cancer: Impact of New Treatments? *Cancers* **2023**, *15*, 2205. [CrossRef] [PubMed] [PubMed Central]
98. Wong, K.E.; Ngai, S.C.; Chan, K.G.; Lee, L.H.; Goh, B.H.; Chuah, L.H. Curcumin Nanoformulations for Colorectal Cancer: A Review. *Front. Pharmacol.* **2019**, *10*, 152. [CrossRef] [PubMed] [PubMed Central]
99. Abdullah, K.M.; Sharma, G.; Singh, A.P.; Siddiqui, J.A. Nanomedicine in Cancer Therapeutics: Current Perspectives from Bench to Bedside. *Mol. Cancer.* **2025**, *24*, 169. [CrossRef] [PubMed] [PubMed Central]
100. Rahimkhoei, V.; Alzaidy, A.H.; Abed, M.J.; Rashki, S.; Salavati-Niasari, M. Advances in inorganic nanoparticles-based drug delivery in targeted breast cancer theranostics. *Adv. Colloid. Interface Sci.* **2024**, *329*, 103204. [CrossRef] [PubMed]
101. Mundekkad, D.; Cho, W.C. Nanoparticles in Clinical Translation for Cancer Therapy. *Int. J. Mol. Sci.* **2022**, *23*, 1685. [CrossRef]
102. Joseph, T.M.; Kar Mahapatra, D.; Esmaeili, A.; Piszczyk, Ł.; Hasanin, M.S.; Kattali, M.; Haponiuk, J.; Thomas, S. Nanoparticles: Taking a Unique Position in Medicine. *Nanomaterials* **2023**, *13*, 574. [CrossRef] [PubMed]
103. Yang, Y.; Zeng, W.; Huang, P.; Zeng, X.; Mei, L. Smart Materials for Drug Delivery and Cancer Therapy. *VIEW* **2021**, *2*, 20200042. [CrossRef]

104. Petrovic, S.; Bita, B.; Barbinta-Patrascu, M.E. Nanoformulations in Pharmaceutical and Biomedical Applications: Green Perspectives. *Int. J. Mol. Sci.* **2024**, *25*, 5842. [CrossRef] [PubMed] [PubMed Central]
105. Subhan, M.A.; Yalamarty, S.S.K.; Filipczak, N.; Parveen, F.; Torchilin, V.P. Recent Advances in Tumor Targeting via EPR Effect for Cancer Treatment. *J. Pers. Med.* **2021**, *11*, 571. [CrossRef]
106. Fan, S.; Wang, W.; Che, W.; Xu, Y.; Jin, C.; Dong, L.; Xia, Q. Nanomedicines Targeting Metabolic Pathways in the Tumor Microenvironment: Future Perspectives and the Role of AI. *Metabolites* **2025**, *15*, 201. [CrossRef] [PubMed] [PubMed Central]
107. van Eerden, R.A.G.; Mathijssen, R.H.J.; Koolen, S.L.W. Recent Clinical Developments of Nanomediated Drug Delivery Systems of Taxanes for the Treatment of Cancer. *Int. J. Nanomed.* **2020**, *15*, 8151–8166. [CrossRef]
108. Gao, G.; Guo, Q.; Zhi, J. Nanodiamond-Based Theranostic Platform for Drug Delivery and Bioimaging. *Small* **2019**, *15*, e1902238. [CrossRef] [PubMed]
109. Mu, Q.; Wang, H.; Zhang, M. Nanoparticles for imaging and treatment of metastatic breast cancer. *Expert Opin. Drug Deliv.* **2017**, *14*, 123–136. [CrossRef] [PubMed] [PubMed Central]
110. Ashrafizadeh, M.; Zarrabi, A.; Hushmandi, K.; Hashemi, F.; Rahmani Moghadam, E.; Raei, M.; Kalantari, M.; Tavakol, S.; Mohammadinejad, R.; Najafi, M.; et al. Progress in Natural Compounds/siRNA Co-delivery Employing Nanovehicles for Cancer Therapy. *ACS Comb. Sci.* **2020**, *22*, 669–700. [CrossRef] [PubMed] [PubMed Central]
111. Sousa, C.; Videira, M. Dual Approaches in Oncology: The Promise of siRNA and Chemotherapy Combinations in Cancer Therapies. *Onco* **2025**, *5*, 2. [CrossRef]
112. Ali Zaidi, S.S.; Fatima, F.; Ali Zaidi, S.A.; Zhou, D.; Deng, W.; Liu, S. Engineering siRNA therapeutics: Challenges and strategies. *J. Nanobiotechnology* **2023**, *21*, 381. [CrossRef] [PubMed] [PubMed Central]
113. Khan, S.; Rehman, U.; Parveen, N.; Kumar, S.; Baboota, S.; Ali, J. siRNA therapeutics: Insights, challenges, remedies and future prospects. *Expert Opin. Drug Deliv.* **2023**, *20*, 1167–1187. [CrossRef] [PubMed]
114. Ngamcherdtrakul, W.; Castro, D.J.; Gu, S.; Morry, J.; Reda, M.; Gray, J.W.; Yantasee, W. Current development of targeted oligonucleotide-based cancer therapies: Perspective on HER2-positive breast cancer treatment. *Cancer Treat. Rev.* **2016**, *45*, 19–29. [CrossRef]
115. Jones, S.K.; Merkel, O.M. Tackling breast cancer chemoresistance with nano-formulated siRNA. *Gene Ther.* **2016**, *23*, 821–828. [CrossRef] [PubMed]
116. Ngamcherdtrakul, W.; Yantasee, W. siRNA therapeutics for breast cancer: Recent efforts in targeting metastasis, drug resistance, and immune evasion. *Transl. Res. J. Lab. Clin. Med.* **2019**, *214*, 105–120. [CrossRef] [PubMed]
117. Badiginchala, R.; Dattatreya, P.S.; Suresh, A.V.S.; Nirni, S.S.; Andra, V.V.; Bunger, D.; Chaturvedi, A. Efficacy and Safety of Nanosomal Docetaxel Lipid Suspension (NDLS) versus Conventional Docetaxel as Neoadjuvant and Adjuvant Therapy for Primary Operable Breast Cancer. *Onco Targets Ther.* **2023**, *16*, 215–225. [CrossRef] [PubMed] [PubMed Central]
118. Hoogevest, P.; Tiemessen, H.; Metselaar, J.M.; Drescher, S.; Fahr, A. The Use of Phospholipids to Make Pharmaceutical form Line Extensions. *Eur. J. Lipid Sci. Technol.* **2021**, *123*, 2000297. [CrossRef]
119. Satheesh, C.T.; Taran, R.; Singh, J.K.; Shrivastav, S.P.; Vithalani, N.K.; Mukherjee, K.K.; Nagarkar, R.V.; Maksud, T.; Mehta, A.O.; Srinivasan, K.; et al. Treatment with nanosomal paclitaxel lipid suspension versus conventional paclitaxel in metastatic breast cancer patients—A multicenter, randomized, comparative, phase II/III clinical study. *Ther. Adv. Med. Oncol.* **2024**, *16*, 17588359241236442. [CrossRef] [PubMed] [PubMed Central]
120. Subramanian, S.; Majumdar, S.K.D.; Biswas, G.; Joshi, N.; Bunger, D.; Khan, M.A.; Ahmad, I. Efficacy and safety of nanosomal docetaxel lipid suspension based chemotherapy in gastric and gastroesophageal junction adenocarcinoma. *Mol. Clin. Oncol.* **2020**, *13*, 14. [CrossRef] [PubMed] [PubMed Central]
121. Wiley. Online Library. Available online: https://onlinelibrary.wiley.com/doi/full/10.1002%2Fmog2.67 (accessed on 2 October 2025).
122. Đorđević, S.; Gonzalez, M.; Conejos-Sánchez, I.; Carreira, B.; Pozzi, S.; Acúrcio, R.C.; Satchi-Fainaro, R.; Florindo, H.F.; Vicent, M.J. Current hurdles to the translation of nanomedicines from bench to the clinic. *Drug Deliv. Transl. Res.* **2022**, *12*, 500–525. [CrossRef] [PubMed] [PubMed Central]
123. Yu, H.J.; De Geest, B.G. Nanomedicine and cancer immunotherapy. *Acta Pharmacol. Sin.* **2020**, *41*, 879–880. [CrossRef] [PubMed] [PubMed Central]
124. Wang, R.; Kumar, P.; Reda, M.; Wallstrum, A.G.; Crumrine, N.A.; Ngamcherdtrakul, W.; Yantasee, W. Nanotechnology Applications in Breast Cancer Immunotherapy. *Small.* **2024**, *20*, e2308639. [CrossRef] [PubMed] [PubMed Central]
125. Fallatah, M.M.; Alradwan, I.; Alfayez, N.; Aodah, A.H.; Alkhrayef, M.; Majrashi, M.; Jamous, Y.F. Nanoparticles for Cancer Immunotherapy: Innovations and Challenges. *Pharmaceuticals* **2025**, *18*, 1086. [CrossRef] [PubMed] [PubMed Central]

life

MDPI

Case Report

Is Sentinel Lymph Node Biopsy Feasible in Multicentric Breast Cancer? A Case Report and Literature Review

Mihaela Camelia Tîrnovanu [1], Elena Cojocaru [2], Vlad Gabriel Tîrnovanu [3], Elena Țarcă [4,*], Loredana Toma [5], Bogdan Florin Toma [2], Sorana Anton [1], Ștefan Dragoș Tîrnovanu [4], Roxana Ana Covali [5], Cipriana Ștefănescu [6] and Irena Cristina Grierosu [6]

1 Department of Mother and Child Medicine-Obstetrics and Gynecology, Faculty of Medicine, "Grigore T. Popa" University of Medicine and Pharmacy, 700115 Iasi, Romania; mihaela.tirnovanu@umfiasi.ro (M.C.T.); sorana.anton@umfiasi.ro (S.A.)

2 Department of Morphofunctional Sciences I, Faculty of Medicine, "Grigore T. Popa" University of Medicine and Pharmacy, 700115 Iasi, Romania; elena2.cojocaru@umfiasi.ro (E.C.); bogdan.f.toma@umfiasi.ro (B.F.T.)

3 St. Josef Hospital, 65189 Wiesbaden, Germany; vlad.tirno@gmail.com

4 Department of Surgery II, Faculty of Medicine, "Grigore T. Popa" University of Medicine and Pharmacy, 700115 Iasi, Romania; stefan-dragos.tirnovanu@d.umfiasi.ro

5 Department of Medical Bioscience, Faculty of Bioengineering, "Grigore T. Popa" University of Medicine and Pharmacy, 700115 Iasi, Romania; loredana-toma@umfiasi.ro (L.T.); ana.covali@umfiasi.ro (R.A.C.)

6 Department of Morphofunctional Sciences II, Faculty of Medicine, "Grigore T. Popa" University of Medicine and Pharmacy, 700115 Iasi, Romania; cipriana.stefanescu@umfiasi.ro (C.Ș.); irena.raileanu@umfiasi.ro (I.C.G.)

* Correspondence: tarca.elena@umfiasi.ro

Abstract

Accurate lymph node staging is crucial for both prognosis (in the event of early-stage disease) and treatment (for local control of disease) in patients with breast cancer. Sentinel lymph node biopsy (SLNB) has been studied in numerous international trials, showing that it allows about 70% of axillary lymph node dissection (ALND) to be avoided and thus significantly reduces the morbidity associated with ALND. SLNB represents a necessary step in the diagnostic algorithm for breast neoplasms because the surgical treatment for breast cancer has become progressively less invasive. We present a case of a 70-year-old woman with multicentric breast cancer (MBC) treated by surgery at "Cuza Vodă" Women's University Hospital, Iassy, Romania. In this case, only the ultrasonography established the diagnosis of left MBC with certainty. Conclusion: The detection of sentinel lymph nodes (SLNs) for MBC must be indicated. In this type of cancer, SLNB is accurate and practical, with sufficient quality control and interdisciplinary collaboration between surgical, nuclear medicine, and pathology units. Lymphoscintigraphy allows the patient to avoid axillary clearance surgery if the sentinel node is negative for metastatic disease. The variability of Ki67, PR, HER2, and ER status supports the idea that all individual foci should be tested in MBC cases to provide the best management and prognosis.

Keywords: multilocular breast cancer; sentinel lymph node biopsy; axillary lymph node dissection; surgical treatment

1. Introduction

Accurate lymph node staging is crucial for both prognosis (in the event of early-stage disease) and treatment (for local control of disease) in patients with breast cancer [1]. Axillary lymph node dissection (ALND) for breast cancer results in a high incidence of postoperative complications (the most important is arm lymphedema) that can reduce quality of life. Furthermore, in early breast cancer, almost 80% of axillary dissections expose no metastasis and, therefore, constitute useless surgery [2].

An important method in surgical oncology is, without a doubt, the sentinel lymph node (SLN) technique, which was first presented by Donald Morton and Alistair Cochran in 1992. Numerous international trials have assessed sentinel lymph node biopsy (SLNB), which has been shown to dramatically lower the morbidity associated with ALND by allowing for the avoidance of roughly 70% of ALND [3–5].

The localized nodes known as the SLN are responsible for the direct drainage of lymph from the main tumor. When the primary breast cancer is in its early stages (I or II), no imaging modality can accurately detect lymph node metastases. However, SLNB is a very dependable technique for screening axillary nodes and detecting metastatic disease, including micro-metastatic illness, in regional lymph nodes. The indications of the method, according to international guidelines, are as follows [6]: (1) breast neoplasm stage for invasive T1/T2 with clinically negative axilla; (2) in situ ductal carcinoma sufficient to require mastectomy or with suspected/proven micro-invasion; (3) patients with clinically negative axillary nodes following neoadjuvant chemotherapy. Contraindications to the method and implicitly to performing the sentinel lymph node biopsy refer to (a) clinically evidenced metastatic lymph node invasion; (b) breast neoplasm stage T3/T4 [7]; (c) inflammatory breast cancer; (d) previous major breast and axillary surgery or mammary and/or axillary radiotherapy because of significant alterations in lymphatic drainage [8]; (e) in situ ductal carcinoma when breast-conserving surgery is planned [9].

Multicentric breast cancer (MBC) has long been considered as a contraindication for SLNB due to concerns with sensitivity and greater false-negative rates [10]. Other reasons for false-negative results for SLN (missing or faint lymph node uptake) include tumor replacement of nodes, low doses of radiopharmaceuticals, poor quality radiopharmaceuticals, insufficient radiocolloid particles, too early or too late imaging times, and advanced patient age. As seen by guidelines from AGO, ASCO, and others, attitudes and recommendations regarding sentinel node biopsy in multifocal/multicentric breast cancer have changed dramatically over the last 14 years, and they currently support the procedure's safety [9].

MBC is defined as two or more separate foci of breast carcinoma that are more than 2 cm apart within the same quadrant; MBC is the presence of separate autonomous foci of carcinoma in different quadrants [11], or a distance bigger than 4 cm. In industrialized countries, the incidence of MBC has been stated to be between 9 and 75% depending on the diagnostic criteria applied, while in Eastern European or less developed countries, the incidence is higher and can affect the average life expectancy [12,13]. The definition, occurrence, biological characteristics, prognostic importance, and subsequent consequences for therapy—particularly surgical therapy—are among the contentious aspects [14,15].

2. Case Report

We present a case of a 70-year-old woman with multilocular breast cancer treated with surgery at "Cuza Vodă" Women's University Hospital Iassy, Romania. Her menarche occurred at the age of 14 years and menopause was surgical at 47 years after total hysterectomy with bilateral salpingo-oophorectomy for uterine fibroma. She had two vaginal deliveries, the first at 20 years old, and she breastfed for only 6 months. We must also mention the 10 abortions in her obstetrical history. The patient had no family history of breast cancer. As comorbidities, the woman had hypertension and grade II obesity with a body mass index of 34.6.

She detected a lump on her left breast 3 months before being admitted to the hospital. She was sent by the general practitioner to mammography. The imaging investigations performed by the patient were mammography, ultrasound exam, and magnetic resonance imaging (MRI). First of all, the digital mammography 2D was carried out in two incidences and performed in the left breast at a depth of 71 mm in the upper internal quadrant with high-intensity opacity, irregular contour, spiculated margins, polylobate, and a size of 16/20 mm with suspicious characteristics of malignancy—BIRADS 5.

The MRI examination with contrast substance was performed in the upper superior quadrant at 10 o'clock for a nodular mass with an irregular contour, with early intense contrast uptake, and wash-out curve kinetics with malignant appearance—BIRADS 6. Also, at the limit of the two upper quadrants, at 12 o'clock an area of architectural distortion of 25/15/8 mm with heterogeneous early contrast uptake with suspicious appearance for malignancy was found. The absence of suspicious adenopathy in the MRI for the left axilla was mentioned.

The core biopsy from the tumor localized in the upper internal superior quadrant specified the diagnosis of invasive breast carcinoma with no special type (NST) grading G3 (Nottingham score 8). The immunohistochemistry (IHC) found estrogen receptors (ERs), which were intensely positive in more than 90% of tumoral cells (Allred score 8); progesterone receptors (PR), which were intensely positive in approximately 30% of tumoral cells (Allred score 6); HER2neu (human epidermal growth factor receptor 2) 2+; and the proliferation marker Ki67, in 40% of the tumor cells. Because HER2 was equivocal (2+) in tumor A, unfortunately, reflex testing (FISH/SISH) was not performed at the external institution, despite our request for this.

Later, after a month, a second ultrasonography established the diagnosis of left multicentric breast cancer. It revealed in the left upper internal quadrant at 10 o'clock, at 5 cm from the nipple, a hypoechoic lump of 19/19/13 mm (anteroposterior/transversal/craniocaudal), with a horizontal axis bigger than the vertical one (Figure 1), with an irregular shape, inhomogeneous, with micro-lobulated margins, horizontal axis, vascular signal present (Figure 2), and posterior acoustic shadowing, and the distance of the tumor from the skin was 6.26 mm (Figure 3). We will name it tumor A. At the limit of the two upper quadrants, at 7 cm from the nipple, there was a second tumor of 16/14 mm (transversal/craniocaudal) with a heterogenous hypoechoic appearance (Figure 4, with a hyperechoic halo, without a vascular signal, and the distance of the tumor from the skin was 7.72 mm (Figure 5). We will name it tumor B. This second tumor corresponds in location to the area of architectural distortion described on the MRI. Therefore, both breast tumors did not comply with the ultrasound rule for malignancy, of "taller (rather) than broader". The distance between breast tumors was 40 mm. The conclusion of the ultrasound exam was multicentric lesions in the upper quadrants of the breast, with characteristics of malignancy—BIRADS 5.

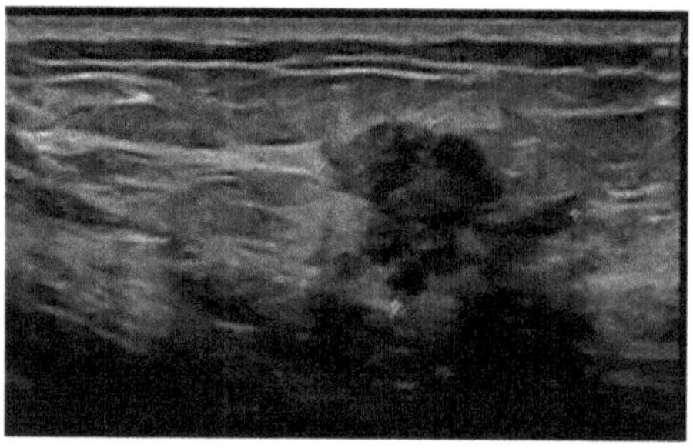

Figure 1. The hypoechoic inhomogeneous lump of 19/19/13 mm from the upper internal with irregular borders—tumor A.

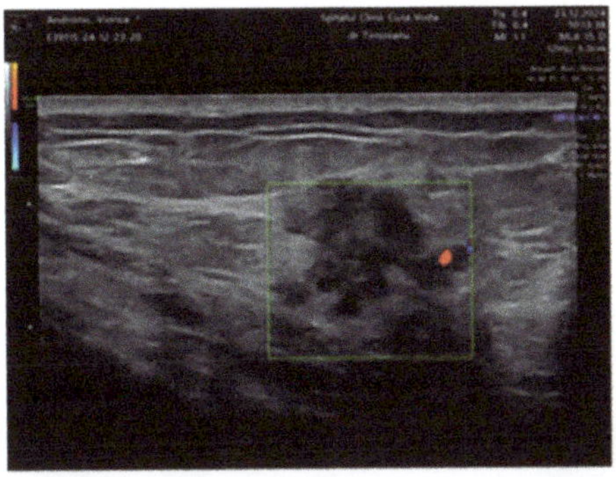

Figure 2. Vascular signal present at the periphery of the tumor A.

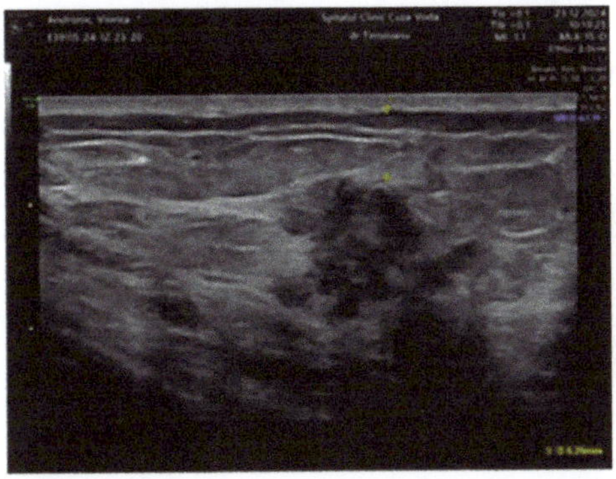

Figure 3. The distance between the bigger tumor and the skin.

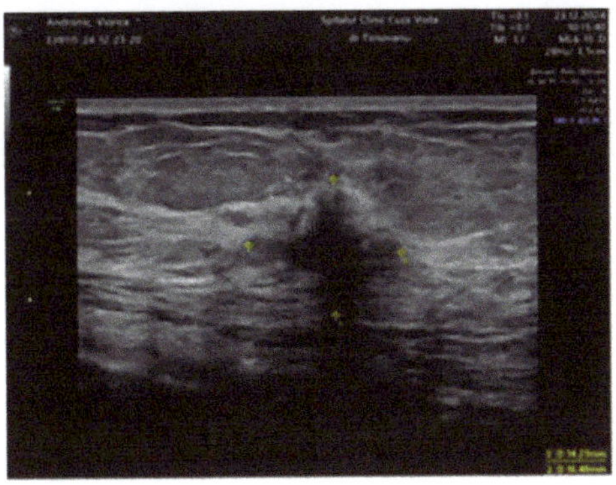

Figure 4. The dimensions of the second lump of 16/14 mm with heterogenous hypoechoic appearance—tumor B.

We reported the presence of lymph node enlargement at the base of the left axilla of 13.6/5.2 mm with a preserved hilum and an increased cortex of 3.86 mm (Figure 6). The preservation of the hilum is a benign sign, but an increased cortex by over 3 mm can be a malignant sign.

The clinical exam upon admission to our hospital found a hard tumor of 3 cm located in the upper internal quadrant, which was imprecisely defined, painless, and mobile on deep and superficial tissues. Another smaller tumor of 1–2 cm was palpated at 12 o'clock on the limit between the upper quadrant, at 4 cm from the areola with the same clinical characteristics. No left axillary adenopathy was palpated. The patient was referred to the Nuclear Medicine Laboratory from the Clinic County Emergency Hospital to identify the SLN, using a General Electric Discovery SPECT-CT (single-photon emission-computed tomography-computed tomography) gamma camera (Figures 7–9).

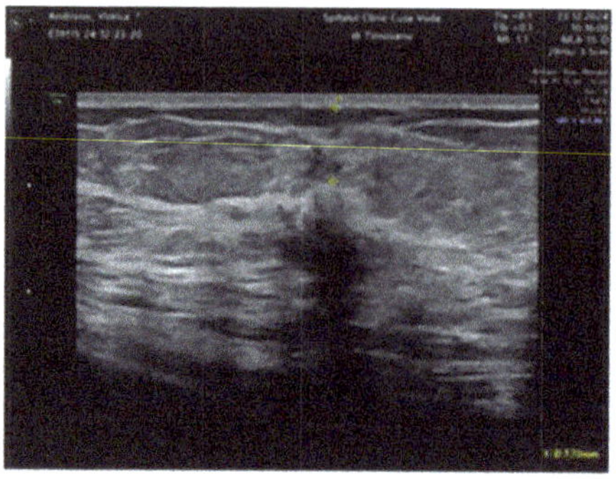

Figure 5. The distance between the second tumor and the skin.

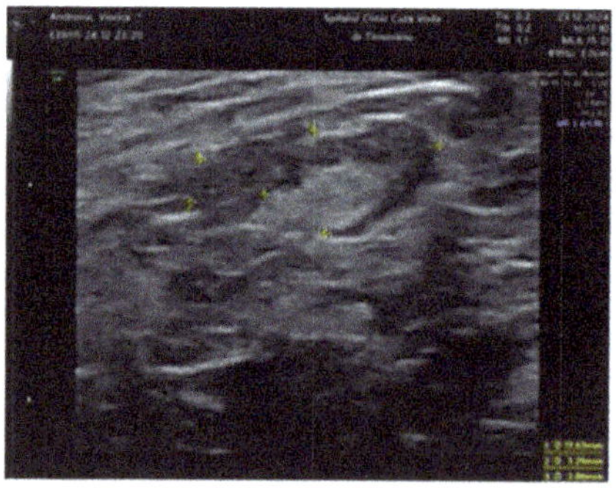

Figure 6. The axillary adenopathy of 13.6/5.2 mm with a preserved hilum and an increased cortex of 3.86 mm.

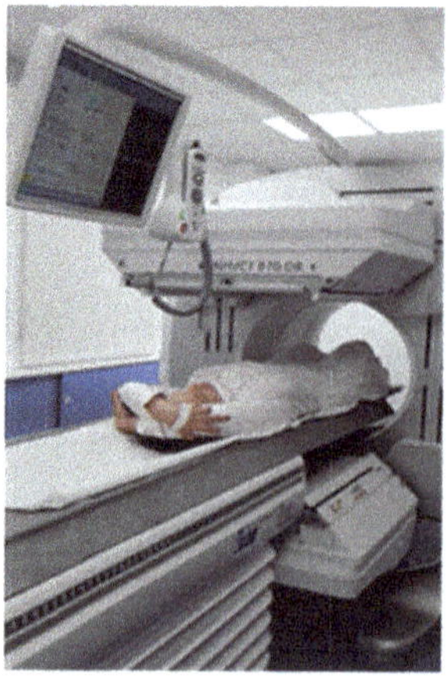

Figure 7. General Electric OPTIMA NM/CT 870DR SPECT-CT gamma camera which precisely locates (due to X-ray low-dose computed tomographic images) the functional abnormalities visualized in scintigraphy. Nuclear Medicine Laboratory, "Sfântul Spiridon" Clinic County Emergency Hospital, Iași.

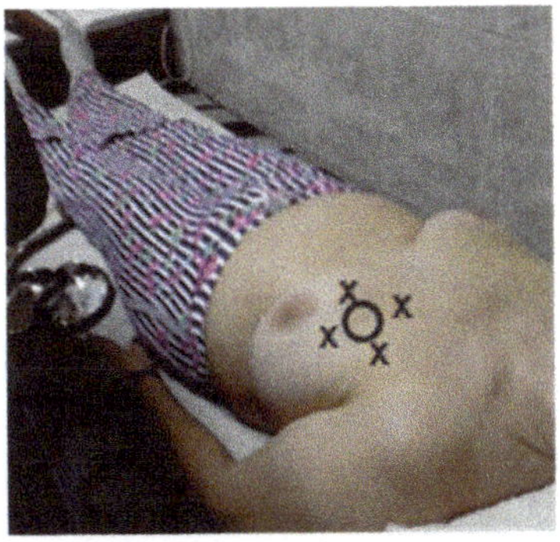

Figure 8. Schematic representation of the SLN technique: breast neoplasm (circle) and the peritumoral injection site of the radiopharmaceutical (marked with x).

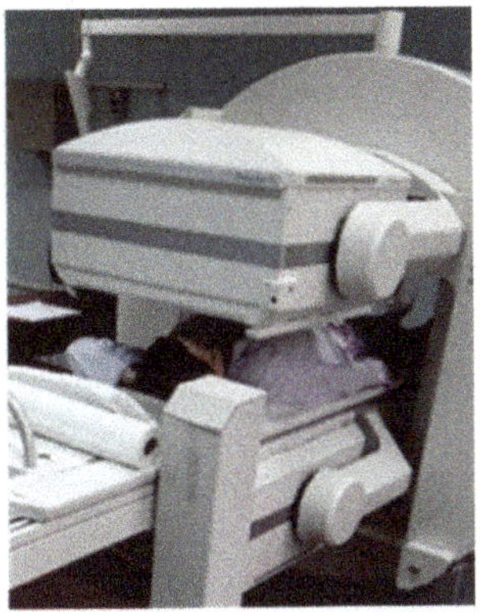

Figure 9. Acquisition of the lymphoscintigraphic image using a classical gamma camera dual detector, Siemens E.CAM from "Sfântul Spiridon" Clinic County Emergency Hospital.

The gamma camera was armed with a low-energy, high-resolution collimator and peaked at 140 KeV, with a 20% window centered over the peak. Shielding of the injection site is typically not needed. ZOOM 1 and 256 × 256 matrix are often used. During the lymphoscintigraphy and SPECT-CT, first of all, two SLNs were identified; one behind the left chondrocostal joint II at the level of the left internal mammary chain (Figures 10–12), and the other at the base of the axillary region (Figures 10 and 13).

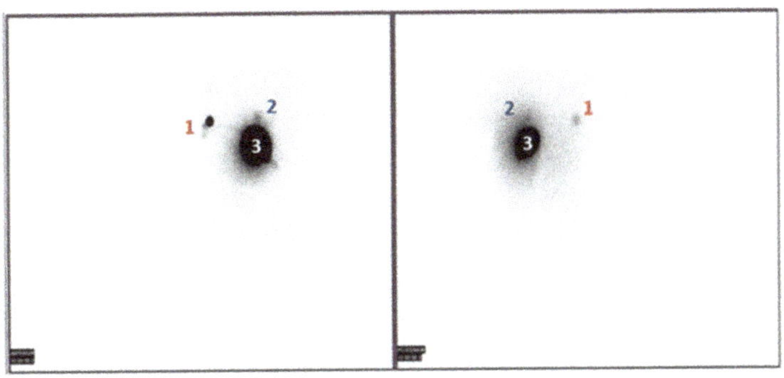

Figure 10. Detection by lymphoscintigraphy of the two SLNs—planar anterior and posterior view (SLN in the left internal mammary chain—1 and left axillary node—2); 3 marks the injected sites.

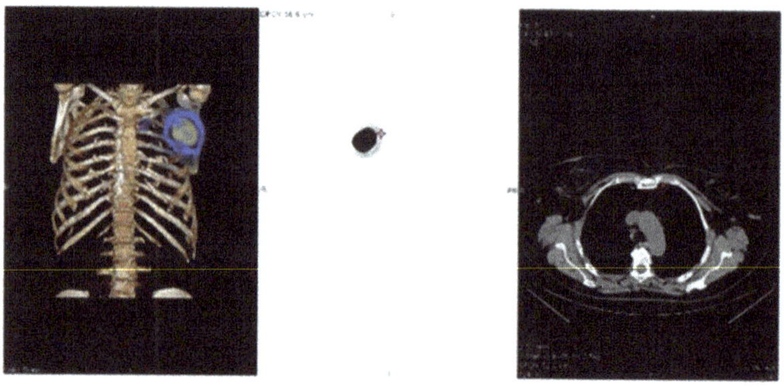

Figure 11. Detection by SPECT-CT of one of the two SLNs located behind the left chondrocostal joint II (bone reconstruction, nuclear medicine, and CT images); the intense area of fixation corresponds to the peritumorally injected radioactivity.

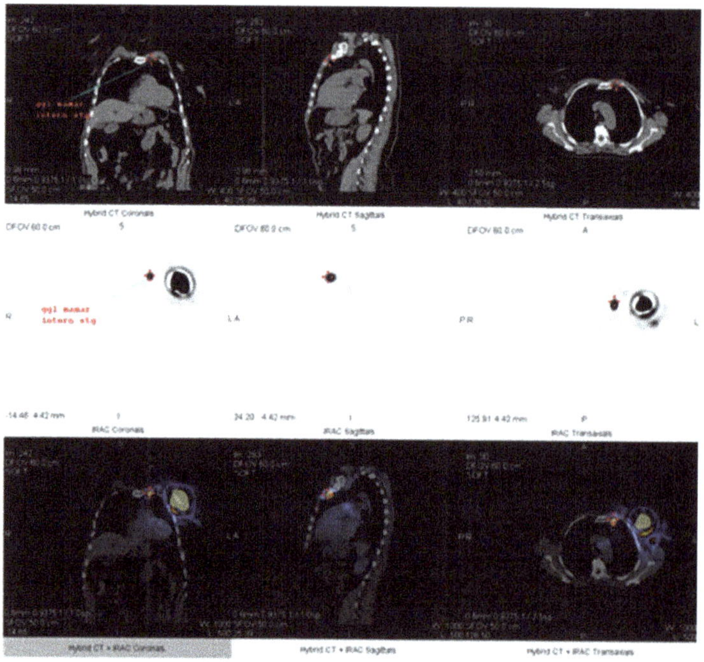

Figure 12. View of SPECT-CT with low dose of the SLN at the level of the left internal mammary chain: coronal, sagittal, and transverse images (CT, nuclear medicine, and fused series of images).

The patient was evaluated by the multidisciplinary team (MDT), which recommended upfront surgery, with no neoadjuvant therapy. The Ki67 of 40% for tumor A was noted, but the overall tumor characteristics and patient age led the team to favor surgical management.

The surgery was performed on the same day with a radiotracer injection after four hours. In the surgery room, we realized a partial mastectomy with the excision of both tumors. The biggest tumor, of 20 mm in concordance with the ultrasound dimension (Figure 14), was confirmed as invasive cancer by core biopsy preoperatively, but the small tumor was sent for a frozen section. The dimension of this second tumor was 17 mm, also in concordance with the ultrasound dimension (Figure 15), and was confirmed as malignant. The distance between the tumors was 45 mm, so the cancer was multicentric.

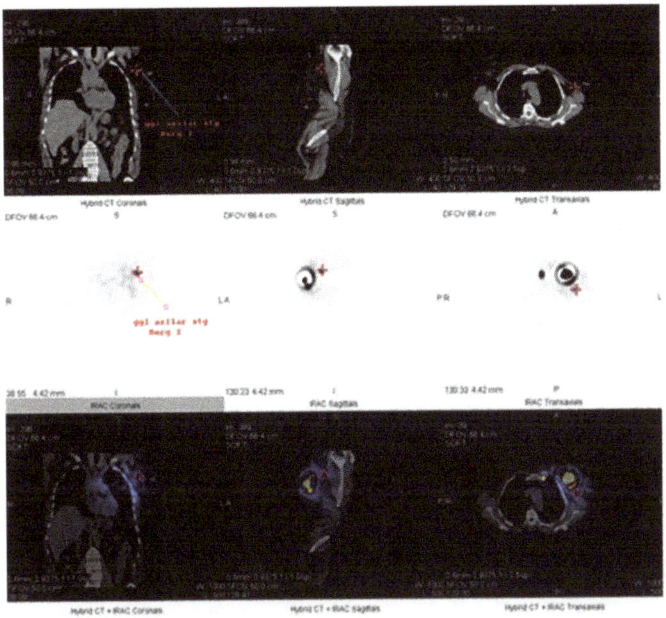

Figure 13. View at SPECT-CT of the SLN in the left axillary region: coronal, sagittal, and transverse images (CT, nuclear medicine, and fused series of images).

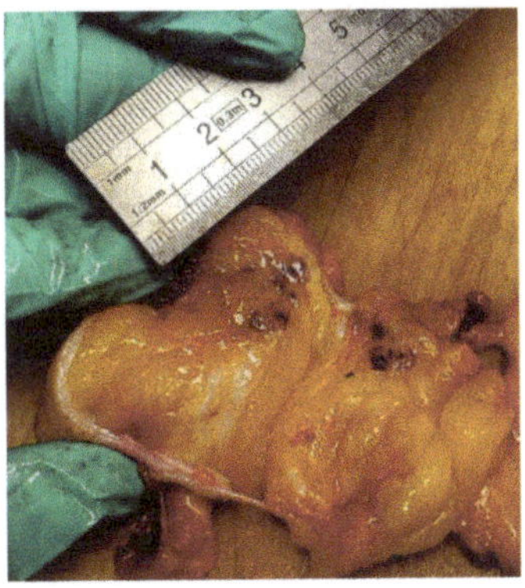

Figure 14. The macroscopical appearance of the first breast tumor (A) of 20 mm on the section.

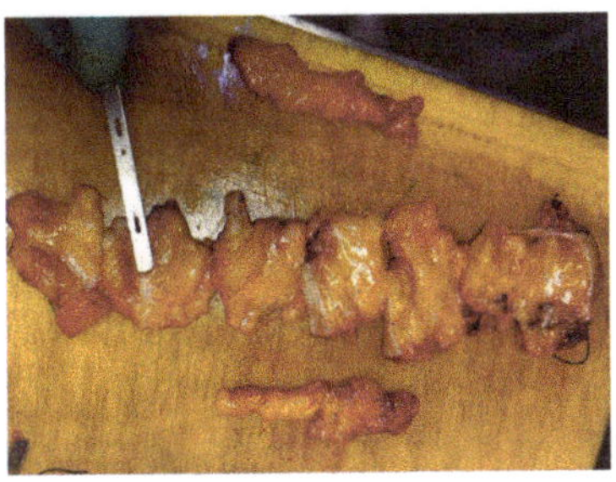

Figure 15. The second tumor (B) of 17 mm was detected after serial sections of the excised breast fragment, and shown by the pathologist.

The second lesion was visualized as an architectural distortion on the MRI and later confirmed by ultrasound and surgery. A core biopsy was not performed initially due to imaging uncertainty, but the lesion was ultimately confirmed intraoperatively.

During surgery, we used a gamma probe with double detection, visual and acoustic, for identification of the SLN (Figure 16). The hottest area was identified by the pitch of the audible signal and the count rate on the digital display. At our patient, we identified the SLN from the left internal mammary chain after partial mastectomy. The intensity of the signal was 358 counters for this SLN. We realized an incision at the base of the left axillary zone of about 4 cm and found one SLN of approximately 2 cm with an intensity of the signal of 1500 counters. The frozen section for the biopsied SLN (Figure 17) specified that it be without metastases.

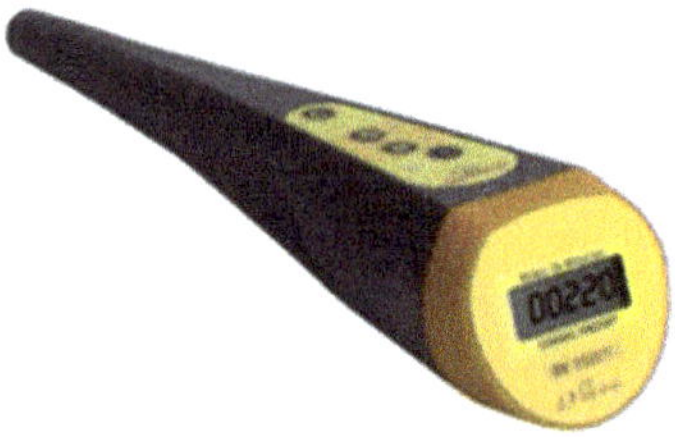

Figure 16. Gamma probe used in our clinic for intraoperative detection of the SLN.

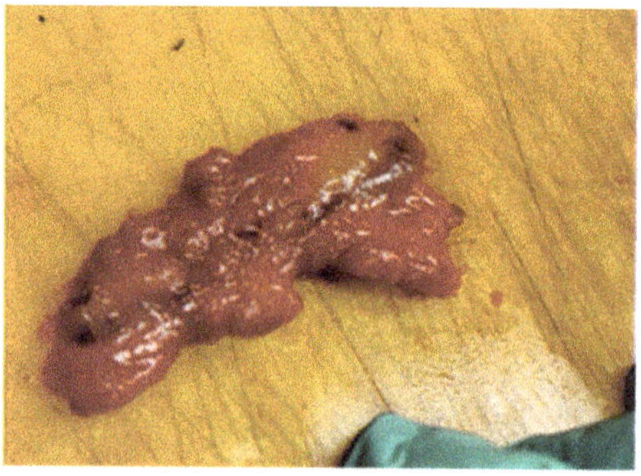

Figure 17. The macroscopical aspect of the axillary SLN in the section before the performance of the frozen section of half of it.

The internal mammary sentinel node was detected preoperatively via lymphoscintigraphy and SPECT-CT, but was not surgically excised due to its location, consistent with institutional limitations and common clinical practice. Only axillary sentinel nodes were excised.

We always check the intensity of the radioactivity of the removed SLN with the Gamma camera, by placing the probe on its surface to confirm the counts. The complete excision of SLNs was confirmed by counting the area after removing the purported sentinel node to see if the counts had fallen to background levels.

The final histological result in paraffin established the malignancy of both tumors—ductal invasive breast cancer NST pT1c(m)N0G3L0V0pN1, but for tumor B, the pathologist requested IHC for e-cadherin to differentiate from a lobular cancer. Microscopically, tumor A at H&E (hematoxylin and eosin) showed tumor cells with solid (Figure 18), trabecular architecture, and focal in nests, as well as cords and cribriform aspect. The tumor cells show marked nuclear pleomorphism (nuclear score 3) (Figure 19) and high mitotic activity (approximately 13/10 hpf/field diameter 0.65). The tumor is associated with foci of tumor necrosis (Figure 20), reduced lymphocytic inflammatory reaction in the periphery of the tumor, and invasion in the adipose tissue (Figure 21), without clear aspects of angioinvasion.

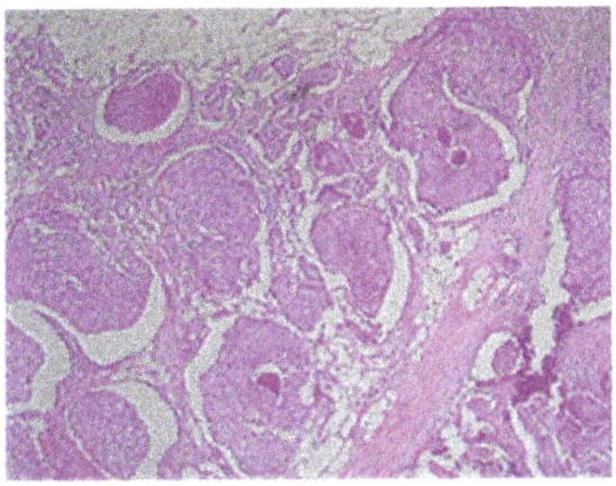

Figure 18. Tumor A—tumoral proliferation with solid architecture and central necrosis HE×4.

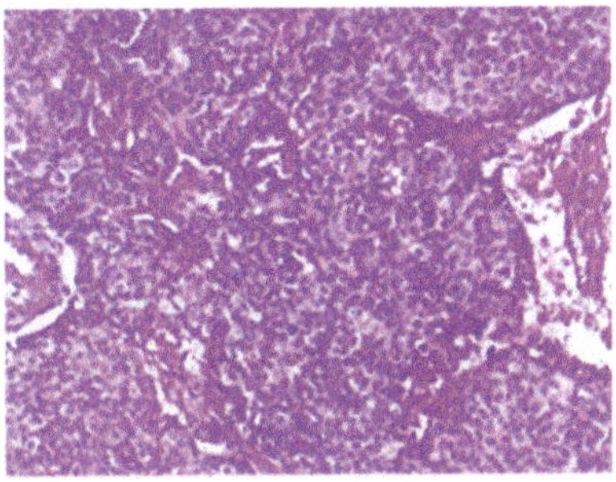

Figure 19. Tumor A with marked nuclear pleomorphism HE×20.

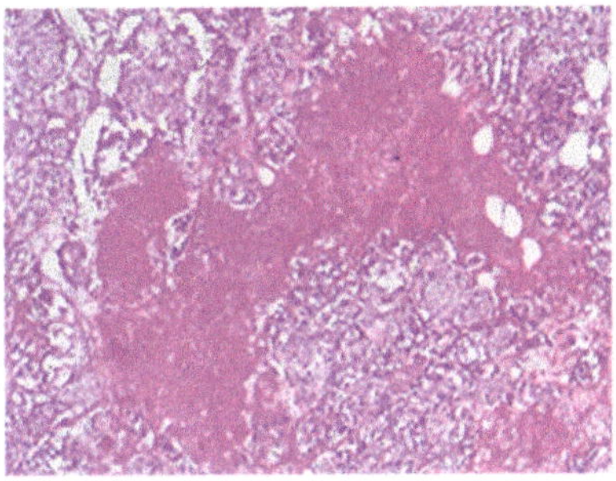

Figure 20. Tumor A—tumoral cells proliferation with central tumoral necrosis HE×10.

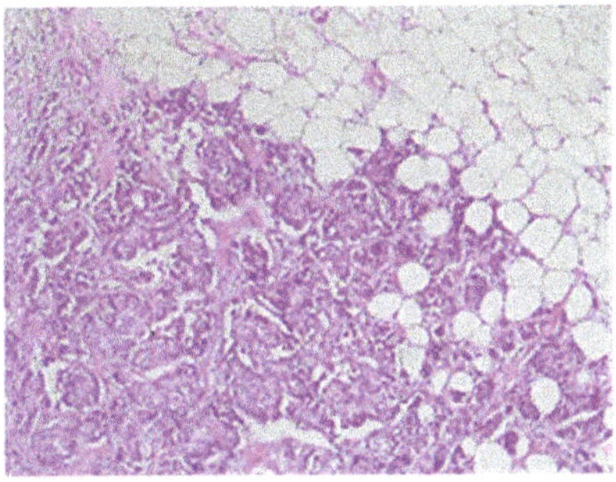

Figure 21. Tumor A tumor infiltration into adipose tissue HE×10.

Tumor B at H&E showed cells with cordonal (Figure 22) and trabecular (Figure 23) proliferation, in clusters or isolated, with moderate-to-marked nuclear pleomorphism (nuclear score 2) and reduced mitotic activity (1–2 mitosis/10 hpf-mitotic score 1). The tumor shows abundant fibrous stroma with reduced lymphocytic inflammatory infiltrate, images of intraductal carcinoma, with adipose and perineural tissue invasion (Figure 24), without clear aspects of angioinvasion. A focus of atypical lobular hyperplasia is visualized at the periphery (Figure 25).

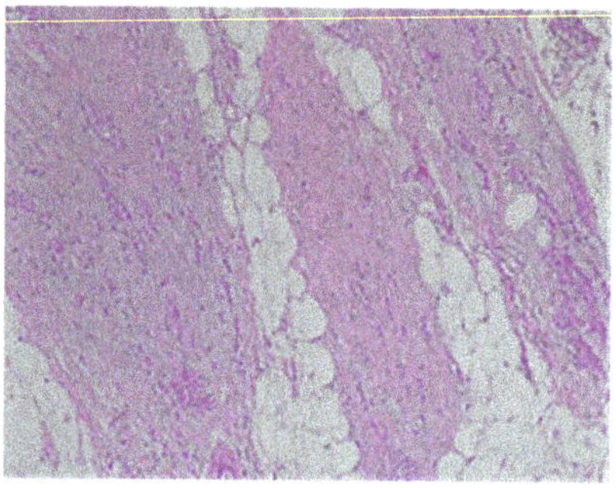

Figure 22. Tumor B cordonal proliferation with sclerohyaline stroma HE × 10.

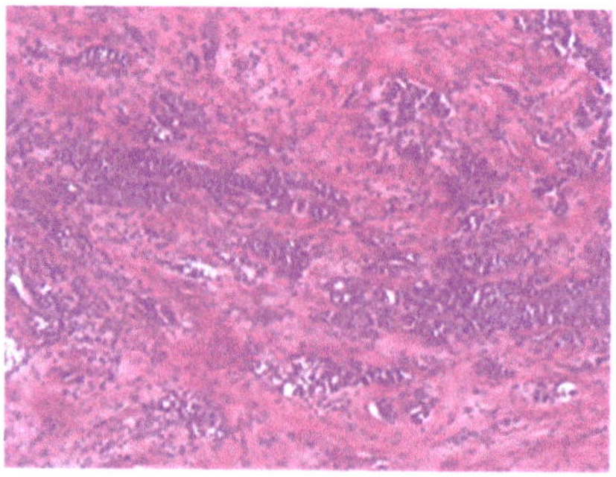

Figure 23. Tumor B trabecular proliferation with fibrous stroma HE × 10.

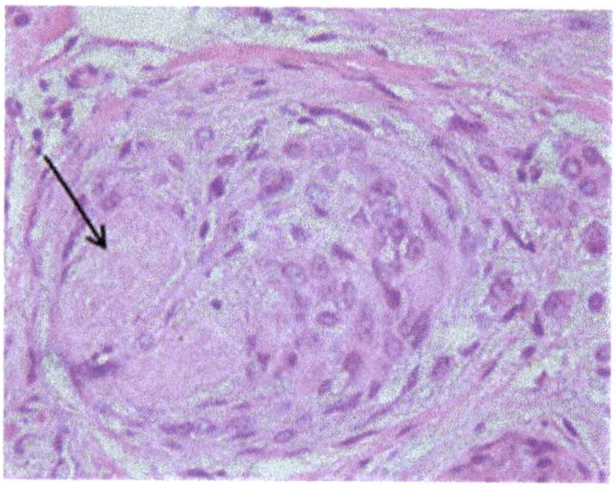

Figure 24. Tumor B—perineural invasion HE × 40.

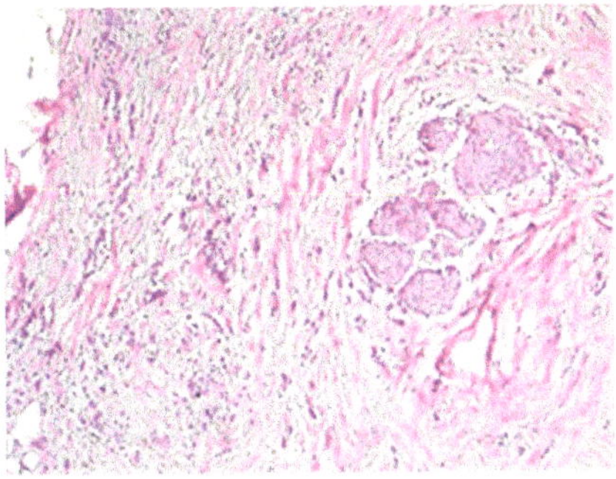

Figure 25. Tumor B—trabecular proliferation on the left, atypical lobular hyperplasia on the right HE×10.

The SLN was without metastases (Figure 26), with reactive changes and lipomatosis. In the neighboring adipose tissue, two more lymph nodes, 1–2 mm without tumor cells, were identified.

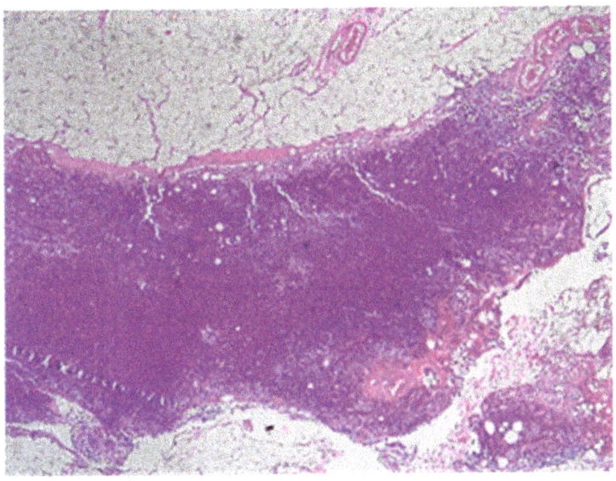

Figure 26. The sentinel lymph node without metastases HE×4.

The IHC for tumor B established that was invasive lobular carcinoma type NOS, with e-cadherin negative in tumoral cells, so we had two different types of cancer in the same breast. Tumor B presents also p120-diffuse cytoplasmic labeling in tumor cells, Ck 5–6-positive, mosaic-type labeling. The carcinoma was intensely positive for ER (90% from tumor cells—Allred score 8), intensely positive for PR (90% from tumor cells—Allred score 8), and HER2-negative. The cell proliferation regulator Ki-67 was positive in approximately 10–15% of tumor cells. Regarding molecular profile, the expression was different in different tumoral foci, because Ki-67 had different values (Table 1). The invasive lobular carcinoma was type Luminal A and the invasive ductal carcinoma was Luminal B.

Table 1. The molecular profile of both tumors.

Markers	Tumor A—Ductal Invasive Carcinoma	Tumor B—Lobular Invasive Carcinoma
ER	90% positive	90% positive
PR	30% positive	90% positive
HER2	2+	Negative
Ki67	40%	10–15%

The patient underwent postoperative bone scintigraphy, which did not detect any bone changes. The osteotropic agent used for skeletal imaging is metastable technetium 99 (99mTc) labeled diphosphonates, and it is dedicated to first-level staging.

The patient received hormone therapy as the only adjuvant treatment, as decided by the oncologist based on her age and comorbidities. We understand that this may seem unexpected in a multicentric case, but the MDT decision was based on comprehensive clinical judgment.

3. Discussion

While some broad aspects of SLN procedures for breast cancer have been agreed upon, not all technical and practical details have been settled on. There are disagreements concerning the radiotracer's particle size, the best injection route, the kind and time of scintigraphy, and intraoperative detection. An ideal tracer administration method for radio-guided SLNB is defined by three key factors: injection site, injected volume, and injected activity. The period between injection and operation is a fourth parameter that should be considered because it has a direct impact on the recommended dosage of radioactivity. Regarding the dimensions of radiocolloids and migration of the radiotracer from the injection site in breast tumors, those of relatively small dimensions, <50 nm, allow for the rapid visualization of lymphatic vessels a few minutes post-injection, but also of a larger number of lymph nodes. In contrast, radiocolloids >300 nm in diameter are preferentially retained in the sentinel node, exhibiting slow migration over time [16]. We exemplify this in Figure 27 [17]. Most physicians argue that the best compromise between rapid lymphatic drainage and ideal retention in SLNs can be found in a radiocolloid with particles that range in size from 100 to 200 nm [9].

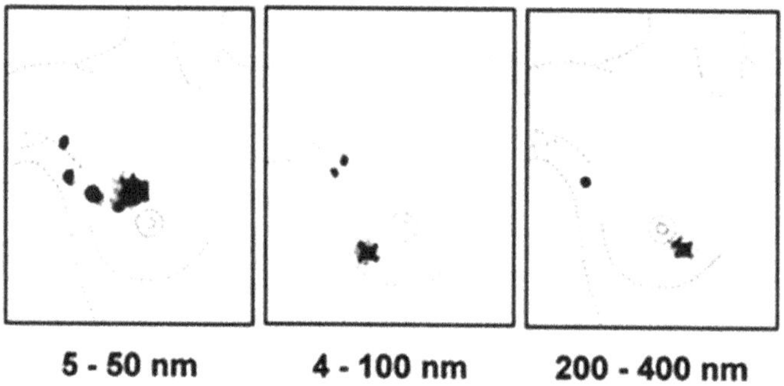

Figure 27. Migration of radiotracer from the injection site in breast tumors to lymph nodes as a function of radiocolloid sizes [17]. https://jnm.snmjournals.org/content/42/8/1198/F5 (accessed on 23 January 2025).

The newest commercially available radiopharmaceutical used for radio-guided SLNB is Tilmanocept (Lymphoseek), composed of a dextran backbone with multiple glucose and mannose residues attached to DTPA for 99mTc labeling. Rapid transit from the primary site to the SLN and selective accumulation in that node, with prolonged retention and limited pass-through to second-echelon nodes, are two potential benefits of 99mTc-Tilmanocept's small molecular size (7.1 nm) and receptor-targeted mannose moieties [18]. These features lead to high SLN visualization and high detection rates in comparison with sulfur colloid tracers [19]. Our colleagues from the Nuclear Medicine Laboratory use 99mTc NanoScan as a radiotracer, which contains human albumin colloidal particles (at least 95% of particles have a diameter $\leq$ 80 nm). This was the radiotracer that was administered to our patient.

Nowadays, depending on the amount of time between scintigraphy and surgery, a total injection of 5 to 30 MBq is usually regarded as enough for surgery scheduled for the same day [6]. According to what is now known, the ipsilateral axilla receives the majority of the multidirectional mammary lymphatic drainage, whereas the internal mammary chain lymph nodes achieve only 3% of the breast's total lymph production [20]. Agreeing with this concept, we detected the location of the largest tumor in the upper internal quadrant in our patient, the SLN with a double location. Some authors, such as in [21], have recommended an intratumoral radiotracer injection for the detection of double lymphatic drainage (Figure 28). In our case, there was a double lymphatic drainage, and an internal mammary chain and ipsilateral axilla, revealed after peritumoral injection in four points.

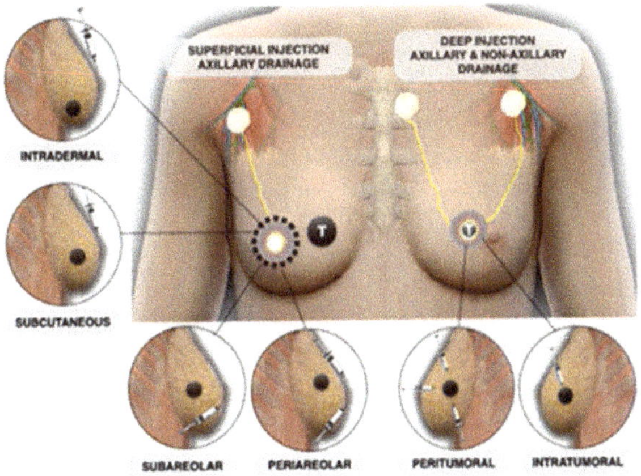

Figure 28. Radiotracer injection sites/modalities for SLN mapping and biopsy. On the left, superficial injections lead to almost exclusively axillary lymphatic drainage. On the right, deep tumor (T)-related injections lead to both axillary and nonaxillary lymphatic drainage [21].

Regarding the imaging procedures for the detection of SLN, preoperative lymphatic mapping has the potential to improve the accuracy of identifying SLN. Additionally, preoperative imaging acts as quality control for the use of the proper tracer, injection failure, radiopharmaceutical failure, and handling of the suitable breast and axilla. It has been demonstrated that SPECT-CT, when used in conjunction with planar imaging, can identify additional SLNs that are not visible on planar images in a significant proportion of patients, altering the drainage territory information in about 17% of cases [22]. The merged SPECT-CT scans give the surgeon valuable pre-operative supplementary information, such as improved localization, a shorter surgical time, and increased method confidence. For our patient, we used both methods of identification of the SLN and we had similar identification sites between planar and SPECT-CT images.

There are other methods for finding the SLN. Advanced tracers such as indocyanine green (ICG), superparamagnetic iron oxide (SPIO), and microbubbles have been discovered [23]. ICG does not involve the nuclear medicine department. Nonetheless, a navigation system for intraoperative fluorescence imaging and the ability to work in low light levels are required. Furthermore, ICG may leak in the surgical field following the resection of the first SLN, which results in the lymphatic vessels section, making it challenging to identify the subsequent SLN. Another noted drawback is the low molecular weight. Because ICG can spread more quickly than blue dye, SLN may needlessly be extensively dissected and removed [24]. The SPIO technique involves injecting a magnetic tracer, which is a contrast agent made of iron oxide crystal nanoparticles coated in carbohydrates. The tracer then migrates into the SLN through the lymph capillaries. The portable magnetometer that performs the detection creates an alternating magnetic field that momentarily magnetizes the SPIO and detects the magnetic response of the particles. After several months, the magnetic tracer is still observable despite gradually fading [25]. Because the magnetic tracer can be given as early as 15 days before surgery or right before the skin incision, this offers a convenient timeframe. The magnetic tracer's brown hue can aid the surgeon in dissecting during surgery. The interpretation of postoperative breast magnetic resonance imaging may be complicated by void artifacts caused by the magnetic tracer's intramammary persistence [26]. Patients who use pacemakers are also ineligible due to the magnetometer's magnetic field, which might cause irregular heartbeats. Similarly, another technological limitation might be caused by metallic surgical equipment that interferes with ferromagnetic signals. Microbubble contrast-enhanced ultrasound imaging is an additional alternate method for SLNB. Following intradermal injection and lymphatic migration, phospholipid-stabilized microbubbles that contain sulfur hexafluoride gas function as sonographic contrast agents [27]. The contrast agent is inexpensive, and SLN can be visualized in real time. This method has a steeper learning curve, is operator-dependent, slower than the others, and necessitates skill in axillary ultrasound examination. It is less sensitive and has a lower detection rate than blue dye. Despite the concern about the risk of false-negative results for multicentric breast cancer, after examination with SLNB, axillary recurrence rates have been reported to be acceptable [10,28,29].

In a prospective multi-institutional trial from the Austrian Sentinel Node Study Group (ASNSG), SLNB evaluated the feasibility and accuracy of the detection of SN in 142 patients with multicentric cancer and compared this with data from 3216 patients with unicentric cancer. They found that SLBN is an indication for MBC without routine axillary lymph node dissection (ALND) [30]. Another study found that the 10-year overall survival for patients treated with sentinel lymph node dissection alone was not worse than the overall survival for patients treated with axillary lymph node dissection among women with T1 or T2 invasive primary breast cancer, no palpable axillary adenopathy, and one or two sentinel lymph nodes with metastases. Based on 10-year results, these results refute the usual use of axillary lymph node dissection in this patient population [31].

A study on 30 patients with MBC identified successfully the SLN in all 30 cases (identification rate 100%). Periareolar injection of radioisotope and blue dye was administered. Axillary metastases were present in 66.7% of cases (20/30). The sensitivity was 100% (20/20) and the false-negative rate was 0% (0/20). The negative predictive value was 100% (10/10), which is encouraging for patients [32]. In some hospital centers, the double labeling technique, a radiotracer, and a blue dye, is used, and the specificity of the technique has increased from 94 to 97% (radioisotopic labeling), to 99% (double labeling) [17]. A study detailed a long experience shows the role of radiocolloids in guiding SLNB as the technique of choice and standard-of-care procedure in patients with breast cancer meeting certain selection criteria [33–37]. In Romania, the use of blue dye is not approved by the national medicine agency, so we can use only the radiotracer technique.

According to a study by Freebody et al., multisite (triple-site) injection is associated with a high axillary sentinel lymph node identification rate. In the study, simultaneous injections of radiotracer at peritumoral, subcutaneous, and subareolar regions had an axillary sentinel lymph node identification rate of 98% in 123 breast lymphoscintigraphy patients [38]. For all cases of breast cancer, we performed lymphoscintigraphy (sentinel lymph node mapping) only by peritumoral injection with an SLN identification rate of 100%.

One study, which performed a meta-analysis of the literature evaluating the feasibility and accuracy of SLNB in multifocal and multicentric breast cancer in a total of 932 patients, established the accuracy rate, sensitivity, specificity, false-negative rate, and negative predictive value for SLN, which were 96.7%, 93.7%, 100%, 6.3%, and 93.5%, respectively. The authors concluded that, in a select subset of patients with multifocal and multicentric breast cancer, SLN biopsy can precisely predict lymph node status with false-negative rates comparable to that of women with unifocal breast cancer [39]. When a multifocal or multicentric breast cancer has an additional relative contraindication to performing SLNB, such as neoadjuvant chemotherapy or T > 5 cm, the false-negative rate increases [40]. Combining the visual evaluation of nodes with intraoperative gamma camera counting and digital palpation through the surgical incision is a good method of lowering false-positive and false-negative outcomes. The current staging system for MBC implies that each tumor that arises independently should be used to estimate the patient's prognosis, based on the size of the largest deposit [41]. The Lang et al. study suggests that all foci of MBC should be considered to estimate the risk of recurrence and metastasis [42].

According to the current guidelines, ER/PR status should be assessed in the sample obtained at the initial core biopsy or the main tumor foci in multifocal breast cancer (more than one distinct tumor foci in a quadrant). We respected this rule for our cases. However, we assessed the receptor status after surgery of both individual foci in MBC and confirmed the variability.

E-cadherin is a crucial molecule for cell adhesion; numerous studies have shown that its down-regulation is linked to the advancement of breast cancer, worse prognoses, and treatment resistance [43–46]. Weissenbacher et al. stated that E-cadherin expression was significantly lower in MBC compared to the unifocal group and suggested the down-regulation of E-cadherin in MBC was causally connected with a worse prognosis [46–48]. In our case, E-cadherin was only analyzed for the lobular carcinoma foci and was negative.

The luminal B subtype is understood as the most aggressive form of ER+ breast cancer cases and often does not show benefits for hormone therapy [49].

Breast cancer is mainly diagnosed in early stages (90–95%); however, 20–30% of these patients become metastatic. Bone is the single most frequent site for bone metastasis in breast cancer patients [50]. Bone metastasis occurred within 3 years in 60.74% of cases after surgery, and after 3 years in 39.26% of cases [51]. On the other hand, 36% of the incidence of bone metastasis is from breast cancer [52], with a tendency of incidence in luminal subtypes (ER+/HER2− 58.52% and in ER+/HER2+ subtype 47.28% of incidence). With an incidence of bone metastases of 34.49% and 36.39%, respectively, the triple-negative (TN) subtype is more impacted by lung involvement (32.09%), whereas the ER−/HER2+ subtype has a larger percentage of liver metastases (31.72%) [53]. Molnar IA et al. show that the luminal A subtype presented a tendency of isolated bone metastases in 59% of cases [54]. Some studies suggest a higher likelihood of lobular breast cancer metastasizing to the bone [55,56].

Breast cancer women older than 65 years developed bone-only metastases 1.5 times more often than younger breast cancer patients did [56]. Marwa and Mubarak show that sixty percent of patients diagnosed with bone metastases were 45 years or younger [57]. Nevertheless, much is still unknown regarding bone metastasis formation. Based on the disruption of multicellular units (osteoblasts, osteoclasts, bone lining cells, and osteocytes) and the release of growth factors (TGF-B, FGF, PDGF, and IGF), the etiopathology of bone metastasis encourages the growth of tumor cells and compromises secondary bone architecture [52]. TGF fuels the further growth of breast cancer cells. Even if breast cancer cells preferentially activate osteoclasts, resulting in osteolytic metastasis, osteoblastic areas are, in general, also present [58]. Breast cancer cells also produce endothelin-1 (ET-1) that activates osteoblasts, resulting in the accumulation of immature mineralized bone (sometimes termed osteoid), and suppresses osteoclast activity [59]. Breast cancer can spread to any bone in the body, but the spine, hips, shoulders, skull, and upper parts of the leg bones near the hips are where it is most often found [60,61].

Reported sensitivity and specificity for bone scintigraphy are 78% and 48%, respectively [62]. Since bone turnover is typically detected by bone scintigraphy, metastases exhibiting a high prevalence of lytic activity may be regarded as a false-negative. An accumulation of agents on bone scans may result from a change in 5–10% of bone, which is not always connected to cancer. However, this can also be a confounding factor, with benign pathologies in such degenerative diseases. Because of this, borderline cases may necessitate a second-level exam. The lack of volumetric evaluation and low spatial resolution (less than 1 cm) are two other drawbacks of bone scans. Breast cancer patients should be monitored with bone scintigraphy, especially in the first 3 years after diagnosis, when they are at an increased risk of bone metastases; even more so if the molecular type of breast cancer is luminal subtype A or B.

4. Conclusions

In the presented case, only the ultrasonography established the diagnosis of left multicentric breast cancer with certainty. The detection of SLN for MBC must be indicated. Given suitable quality control and interdisciplinary cooperation of surgical, nuclear medicine, and pathology units, SLNB is feasible and accurate in this disease entity. Lymphoscintigraphy allows the patient to avoid axillary clearance surgery if the sentinel node is negative for metastatic disease. The conventional radioactivity-dependent technique of SLNB is enshrined in evidence-based algorithms for breast cancer management, a time-proven technique that has been refined and continues to improve by using SPECT-CT detection. SPECT/CT adds value, particularly in cases with negative, inconclusive, or equivocal results, and provides better and more accurate localization of sentinel nodes in patients presenting drainage outside of the ipsilateral axilla, like in our case. However, planar lymphoscintigraphy remains a satisfactory imaging option for imaging departments not equipped with SPECT/CT due to its good patient-based detection rate, a fact confirmed in our patient. The two tumors were histologically different; tumor A was invasive ductal carcinoma type NST, and tumor B was invasive lobular carcinoma type NOS. The variability of Ki67, PR, HER2, and ER status supports the idea that all individual foci should be tested in MBC cases to provide the best management and prognosis.

Author Contributions: Conceptualization, M.C.T., E.C. and I.C.G.; methodology, V.G.T. and E.Ț.; software, L.T.; validation, E.C., E.Ț. and B.F.T.; formal analysis, S.A.; investigation, Ș.D.T.; resources, M.C.T.; data curation, R.A.C.; writing—original draft preparation, M.C.T., Ș.D.T. and C.Ș.; writing—review and editing, E.Ț. and I.C.G.; visualization, L.T.; supervision, V.G.T. and C.Ș.; project administration, S.A.; funding acquisition, R.A.C. All authors have read and agreed to the published version of the manuscript.

Funding: This research received no external funding.

Institutional Review Board Statement: The study was conducted in accordance with the Declaration of Helsinki, and approved by the Ethics Committee of "Cuza Vodă" Women's University Hospital Iassy, Romania (protocol code 378/12.05.2024).

Informed Consent Statement: Written informed consent has been obtained from the patient to publish this paper.

Data Availability Statement: The data presented in this study are available on request from the corresponding author.

Conflicts of Interest: The authors declare no conflicts of interest.

References

1. Zirruda, S.; Veronesi, U. Milestones in Breast Cancer Treatment. *Breast J.* **2015**, *21*, 3–12. [CrossRef] [PubMed]
2. Beek, M.A.; Verheuvel, N.C.; Luiten, E.J.T.; Klompenhouwer, E.G.; Rutten, H.J.T.; Roumen, R.M.H.; Voogd, A.C. Two decades of axillary management in breast cancer. *Br. J. Surg.* **2015**, *102*, 1658–1664. [CrossRef]
3. Krag, D.N.; Anderson, S.J.; Julian, T.B.; Brown, A.M.; Harlow, S.P.; Ashikaga, T.; Weaver, D.L.; Miller, B.J.; Jalovec, L.M.; Frazier, T.G.; et al. Technical outcomes of sentinel-lymph-node resection and conventional axillary-lymph-node dissection in patients with clinically node-negative breast cancer: Results from the NSABP B-32 randomised phase III trial. *Lancet Oncol.* **2007**, *8*, 881–888. [CrossRef]
4. Antes, G.; von Elm, E. The PRISMA Statement—What should be reported about systematic reviews? *Dtsch. Med. Wochenschr.* **2009**, *134*, 1619. [CrossRef]
5. Straver, M.E.; Meijnen, P.; van Tienhoven, G.; van de Velde, C.J.H.; Mansel, R.E.; Bogaerts, J.; Duez, N.; Cataliotti, L.; Klinkenbijl, J.H.G.; Westenberg, H.A.; et al. Sentinel node identification rate and nodal involvement in the EORTC 10981-22023 AMAROS trial. *Ann. Surg. Oncol.* **2010**, *17*, 1854–1861. [CrossRef]
6. Giammarile, F.; Alazraki, N.; Aarsvold, J.N.; Audisio, R.A.; Glass, E.; Grant, S.F.; Kunikowska, J.; Leidenius, M.; Moncayo, V.M.; Uren, R.F.; et al. The EANM and SNMMI practice guideline for lymphoscintigraphy and sentinel node localization in breast cancer. *Eur. J. Nucl. Med. Mol. Imaging* **2013**, *40*, 1932–1947. [CrossRef]
7. Chatterjee, A.; Serniak, N.; Czerniecki, B.J. Sentinel lymph node biopsy in breast cancer: A work in progress. *Cancer J.* **2015**, *21*, 7–10. [CrossRef] [PubMed] [PubMed Central]
8. Mathelin, C.; Lodi, M. Narrative review of sentinel lymph node biopsy in breast cancer: A technique in constant evolution with still numerous unresolved questions. *Chin. Clin. Oncol.* **2021**, *10*, 20. [CrossRef]
9. Lyman, G.H.; Somerfield, M.R.; Bosserman, L.D.; Perkins, C.L.; Weaver, D.L.; Giuliano, A.E. Sentinel Lymph Node Biopsy for Patients With Early-Stage Breast Cancer: American Society of Clinical Oncology Clinical Practice Guideline Update. *J. Clin. Oncol.* **2017**, *35*, 561–564. [CrossRef]
10. Spillane, A.J.; Brennan, M.E. Accuracy of sentinel lymph node biopsy in large and multifocal/multicentric breast carcinoma—A systematic review. *Eur. J. Surg. Oncol.* **2011**, *37*, 371–385. [CrossRef]
11. Cserni, G.; Bori, R.; Sejben, I.; Voros, A.; Kaiser, L.; Hamar, S.; Csorgo, E.; Kulka, J. Unifocal, multifocal and diffuse carcinomas: A reproducibility study of breast cancer distribution. *Breast* **2013**, *22*, 34–38. [CrossRef] [PubMed]
12. Jain, S.; Rezo, A.; Shadbolt, B.; Dahlstrom, J.E. Synchronous multiple ipsilateral breast cancers: Implications for patient management. *Pathology* **2009**, *41*, 57–67. [CrossRef] [PubMed]
13. Țarcă, V.; Țarcă, E.; Moscalu, M. Social and Economic Determinants of Life Expectancy at Birth in Eastern Europe. *Healthcare* **2024**, *12*, 1148. [CrossRef] [PubMed] [PubMed Central]
14. Coombs, N.J.; Boyages, J. Multifocal and multicentric breast cancer: Does each focus matter? *J. Clin. Oncol.* **2005**, *23*, 7497–7502. [CrossRef]
15. Fisher, B. Role of science in the treatment of breast cancer when tumor multicentricity is present. *J. Natl. Cancer Inst.* **2011**, *103*, 1292–1298. [CrossRef]
16. Leidenius, M.H.; Leppanen, E.A.; Krogerus, L.A.; Smitten, K.A. The impact of radiopharmaceutical particle size on the visualization and identification of sentinel nodes in breast cancer. *Nucl. Med. Commun.* **2004**, *25*, 233–238. [CrossRef]
17. Mariani, G.; Moresco, L.; Viale, G.; Villa, G.; Bagnasco, M.; Canavese, G.; Buscombe, J.; Strauss, H.W.; Paganelli, G. Radioguided sentinel lymph node biopsy in breast cancer surgery. *J. Nucl. Med.* **2001**, *42*, 1198–1215.
18. Vera, D.R.; Wallace, A.M.; Hoh, C.K.; Mattrey, R.F. A synthetic macromolecule for sentinel node detection: (99m)Tc-DTPA-mannosyl-dextran. *J. Nucl. Med.* **2001**, *42*, 951–959.
19. Vidal-Sicart, S.; Rioja, M.E.; Prieto, A.; Goñi, E.; Gómez, I.; Albala, M.D.; Lumbreras, L.; León, L.F.; Gómez, J.R.; Campos, F. Sentinel Lymph Node Biopsy in Breast Cancer with (99m)Tc-Tilmanocept: A Multicenter Study on Real-Life Use of a Novel Tracer. *J. Nucl. Med.* **2021**, *62*, 620–627. [CrossRef]
20. Turner-Warwick, R.T. The lymphatics of the breast. *Br. J. Surg.* **1959**, *46*, 574–582. [CrossRef]
21. Giammarile, F.; Vidal-Sicart, S.; Paez, D.; Pellet, O.; Enrique, E.-L.; Mikhail-Lette, M.; Morozova, O.; Camila, N.M.M.; Ivonne, R.S.D.; Bolton, R.C.D.; et al. Sentinel Lymph Node Methods in Breast Cancer. *Semin. Nucl. Med.* **2022**, *52*, 551–560. [CrossRef] [PubMed]
22. Jimenez-Heffernan, A.; Ellmann, A.; Sado, H.; Huić, D.; Bal, C.; Parameswaran, R.; Giammarile, F.; Pruzzo, R.; Kostadinova, I.; Vorster, M.; et al. Results of a Prospective Multicenter International Atomic Energy Agency Sentinel Node Trial on the Value of SPECT/CT Over Planar Imaging in Various Malignancies. *J. Nucl. Med.* **2015**, *56*, 1338–1344. [CrossRef] [PubMed]
23. Goonawardena, J.; Yong, C.; Law, M. Use of indocyanine green fluorescence compared to radioisotope for sentinel lymph node biopsy in early-stage breast cancer: Systematic review and meta-analysis. *Am. J. Surg.* **2020**, *220*, 665–676. [CrossRef]

24. Agrawal, S.K.; Hashlamoun, I.; Karki, B.; Sharma, A.; Arun, I.; Ahmed, R. Diagnostic Performance of Indocyanine Green Plus Methylene Blue Versus Radioisotope Plus Methylene Blue Dye Method for Sentinel Lymph Node Biopsy in Node-Negative Early Breast Cancer. *JCO Glob. Oncol.* **2020**, *6*, 1225–1231. [CrossRef]
25. Karakatsanis, A.; Christiansen, P.M.; Fischer, L.; Hedin, C.; Pistioli, L.; Sund, M.; Rasmussen, N.R.; Jørnsgård, H.; Tegnelius, D.; Eriksson, S.; et al. The Nordic SentiMag trial: A comparison of super paramagnetic iron oxide (SPIO) nanoparticles versus Tc(99) and patent blue in the detection of sentinel node (SN) in patients with breast cancer and a meta-analysis of earlier studies. *Breast Cancer Res. Treat.* **2016**, *157*, 281–294. [CrossRef]
26. Zada, A.; Peek, M.C.; Ahmed, M.; Anninga, B.; Baker, R.; Kusakabe, M.; Sekino, M.; Klaase, J.M.; Haken, B.T.; Douek, M. Meta-analysis of sentinel lymph node biopsy in breast cancer using the magnetic technique. *Br. J. Surg.* **2016**, *103*, 1409–1419. [CrossRef]
27. Cox, K.; Sever, A.; Jones, S.; Weeks, J.; Mills, P.; Devalia, H.; Fish, D.; Jones, P. Validation of a technique using microbubbles and contrast enhanced ultrasound (CEUS) to biopsy sentinel lymph nodes (SLN) in pre operative breast cancer patients with a normal grey-scale axillary ultrasound. *Eur. J. Surg. Oncol.* **2013**, *39*, 760–765. [CrossRef]
28. Huang, X.; Liu, J.Q.; Zhou, Y.D.; Xu, Y.; Chen, C.; Wang, X.; Cao, X.; Yao, R.; Sun, Q. Sentinel lymph node biopsy should be considered for clinically node-negative breast cancer regardless of BRCA1/2 mutation status. *Ann. Transl. Med.* **2020**, *8*, 1183. [CrossRef] [PubMed] [PubMed Central]
29. Gentilini, O.; Veronesi, P.; Botteri, E.; Soggiu, F.; Trifirò, G.; Lissidini, G.; Luini, A. Sentinel lymph node biopsy in multicentric breast cancer: Five-year results in a large series from a single institution. *Ann. Surg. Oncol.* **2011**, *18*, 2879–2884. [CrossRef]
30. Knauer, M.; Konstantiniuk, P.; Haid, A.; Wenzl, E.; Riegler-Keil, M.; Pöstlberger, S.; Reitsamer, R.; Schrenk, P. Multicentric breast cancer: A new indication for sentinel node biopsy—A multi-institutional validation study. *J. Clin. Oncol.* **2006**, *24*, 3374–3380. [CrossRef]
31. Giuliano, A.E.; Ballman, K.V.; McCall, L.; Beitsch, P.D.; Brennan, M.B.; Kelemen, P.R.; Ollila, D.W.; Hansen, N.M.; Whitworth, P.W.; Blumencranz, P.W.; et al. Effect of Axillary Dissection vs No Axillary Dissection on 10-Year Overall Survival Among Women with Invasive Breast Cancer and Sentinel Node Metastasis: The ACOSOG Z0011 (Alliance) Randomized Clinical Trial. *JAMA* **2017**, *318*, 918–926. [CrossRef] [PubMed] [PubMed Central]
32. van la Parra, R.F.D.; de Roos, W.K.; Contant, C.M.E.; Bavelaar-Croon, C.D.L.; Barneveld, P.C.; Bosscha, K. A prospective validation study of sentinel lymph node biopsy in multicentric breast cancer: SMMaC trial. *Eur. J. Surg. Oncol.* **2014**, *40*, 1250–1255. [CrossRef] [PubMed]
33. Park, K.U.; Somerfield, M.R.; Anne, N.; Brackstone, M.; Conlin, A.K.; Couto, H.L.; Dengel, L.T.; Eisen, A.; Harvey, B.E.; Hawley, J.; et al. Sentinel Lymph Node Biopsy in Early-Stage Breast Cancer: ASCO Guideline Update. *J. Clin. Oncol.* **2025**, *43*, 1720–1741. [CrossRef] [PubMed]
34. Charalampoudis, P.; Markopoulos, C.; Kovacs, T. Controversies and recommendations regarding sentinel lymph node biopsy in primary breast cancer: A comprehensive review of current data. *Eur. J. Surg. Oncol.* **2018**, *44*, 5–14. [CrossRef]
35. Ho, C.; Argáez, C. *Sentinel Lymph Node Biopsy for the Management of Breast Cancer: A Review of Guidelines CADTH Rapid Response Reports*; Canadian Agency for Drugs and Technologies in Health: Ottawa, ON, Canada, 2019.
36. Gradishar, W.J.; Anderson, B.O.; Abraham, J.; Aft, R.; Agnese, D.; Allison, K.H.; Blair, S.L.; Burstein, H.J.; Dang, C.; Elias, A.D.; et al. Breast Cancer, Version 3.2020, NCCN Clinical Practice Guidelines in Oncology. *J. Natl. Compr. Cancer Netw.* **2020**, *18*, 452–478. [CrossRef]
37. Fisher, C.S.; Margenthaler, J.A.; Hunt, K.K.; Schwartz, T. The landmark series: Axillary management in breast cancer. *Ann. Surg. Oncol.* **2020**, *27*, 724–729. [CrossRef]
38. Freebody, J.; Fernando, S.; Rossleigh, M.A. Triple-site radiotracer application in breast lymphoscintigraphy and sentinel node discordance. *World J. Nucl. Med.* **2019**, *18*, 127–131. [CrossRef]
39. Moody, L.C.; Chao, C. Indications for Sentinel Lymph Node Biopsy in Multifocal and Multicentric Breast Cancer. *J. Surg. Res.* **2012**, *172*, 227–228. [CrossRef]
40. Han, C.; Yang, L.; Zuo, W. A mini-review on factors and countermeasures associated with false-negative sentinel lymph node biopsies in breast cancer. *Chin. J. Cancer Res.* **2016**, *28*, 370–376. [CrossRef] [PubMed] [PubMed Central]
41. Andea, A.A.; Wallis, T.; Newman, L.A.; Bouwman, D.; Dey, J.; Visscher, D.W. Pathologic analysis of tumor size and lymph node status in multifocal/multicentric breast carcinoma. *Cancer* **2002**, *94*, 1383–1390. [CrossRef]
42. Lang, Z.; Wu, Y.; Li, C.; Li, X.; Wang, X.; Qu, G. Multifocal and Multicentric Breast Carcinoma: A Significantly More Aggressive Tumor than Unifocal Breast Cancer. *Anticancer Res.* **2017**, *37*, 4593–4598. [CrossRef] [PubMed]
43. Baranwal, S.; Alahari, S.K. Molecular mechanisms controlling E-cadherin expression in breast cancer. *Biochem. Biophys. Res. Commun.* **2009**, *384*, 6–11. [CrossRef] [PubMed]
44. Klopp, A.H.; Lacerda, L.; Gupta, A.; Debeb, B.G.; Solley, T.; Li, L.; Spaeth, E.; Xu, W.; Zhang, X.; Lewis, M.T.; et al. Mesenchymal stem cells promote mammosphere formation and decrease E-cadherin in normal and malignant breast cells. *PLoS ONE* **2010**, *5*, e12180. [CrossRef]

45. Stătescu, L.; Trandafir, L.M.; Țarcă, E.; Moscalu, M.; Leon Constantin, M.M.; Butnariu, L.I.; Trandafirescu, M.F.; Tîrnovanu, M.C.; Heredea, R.; Pătrașcu, A.V.; et al. Advancing Cancer Research: Current Knowledge on Cutaneous Neoplasia. *Int. J. Mol. Sci.* **2023**, *24*, 11176. [CrossRef] [PubMed] [PubMed Central]
46. Weissenbacher, T.; Hirte, E.; Kuhn, C.; Janni, W.; Mayr, D.; Karsten, U.; Rack, B.; Friese, K.; Jeschke, U.; Heublein, S.; et al. Multicentric and multifocal versus unifocal breast cancer: Differences in the expression of E-cadherin suggest differences in tumor biology. *BMC Cancer* **2013**, *13*, 361. [CrossRef]
47. Hashmi, A.A.; Aijaz, S.; Khan, S.M.; Mahboob, R.; Irfan, M.; Zafar, N.I.; Nisar, M.; Siddiqui, M.; Edhi, M.M.; Faridi, N.; et al. Prognostic parameters of luminal A and luminal B intrinsic breast cancer subtypes of Pakistani patients. *World J. Surg. Oncol.* **2018**, *16*, 1. [CrossRef]
48. Tsoutsou, P.G.; Vozenin, M.C.; Durham, A.D.; Bourhis, J. How could breast cancer molecular features contribute to locoregional treatment decision making? *Crit. Rev. Oncol. Hematol.* **2017**, *110*, 43–48. [CrossRef]
49. Fragomeni, S.M.; Sciallis, A.; Jeruss, J.S. Molecular Subtypes and Local-Regional Control of Breast Cancer. *Surg. Oncol. Clin. N. Am.* **2018**, *27*, 95–120. [CrossRef]
50. Pulido, C.; Vendrell, I.; Ferreira, A.R.; Casimiro, S.; Mansinho, A.; Alho, I.; Costa, L. Bone metastasis risk factors in breast cancer. *Ecancermedicalscience* **2017**, *11*, 715. [CrossRef]
51. Pan, Y.; Lin, Y.; Mi, C. Clinicopathological characteristics and prognostic risk factors of breast cancer patients with bone metastasis. *Ann. Transl. Med.* **2021**, *9*, 1340. Available online: https://atm.amegroups.org/article/view/77448 (accessed on 27 January 2025). [CrossRef]
52. Hernandez, R.K.; Wade, S.W.; Reich, A.; Pirolli, M.; Liede, A.; Lyman, G.H. Incidence of bone metastases in patients with solid tumors: Analysis of oncology electronic medical records in the United States. *BMC Cancer* **2018**, *18*, 44. [CrossRef] [PubMed]
53. Wu, Q.; Li, J.; Zhu, S.; Wu, J.; Chen, C.; Liu, Q.; Wei, W.; Zhang, Y.; Sun, S. Breast cancer subtypes predict the preferential site of distant metastases: A SEER based study. *Oncotarget* **2017**, *8*, 27990. [CrossRef] [PubMed]
54. Molnár, I.A.; Molnár, B.Á.; Vízkeleti, L.; Fekete, K.; Tamás, J.; Deák, P.; Szundi, C.; Székely, B.; Moldvay, J.; Vári-Kakas, S. Breast carcinoma subtypes show different patterns of metastatic behavior. *Virchows Arch.* **2017**, *470*, 275–283. [CrossRef] [PubMed]
55. Diessner, J.; Wischnewsky, M.; Stüber, T.; Stein, R.; Krockenberger, M.; Häusler, S.; Janni, W.; Kreienberg, R.; Blettner, M.; Schwentner, L.; et al. Evaluation of clinical parameters influencing the development of bone metastasis in breast cancer. *BMC Cancer* **2016**, *16*, 307. [CrossRef]
56. Purushotham, A.; Shamil, E.; Cariati, M.; Agbaje, O.; Muhidin, A.; Gillett, C.; Mera, A.; Sivanadiyan, K.; Harries, M.; Sullivan, R.; et al. Age at diagnosis and distant metastasis in breast cancer—A surprising inverse relationship. *Eur. J. Cancer* **2014**, *50*, 1697–1705. [CrossRef]
57. Ali, M.K.; Mubarak, A. Identification of Risk Factors of Bone Metastasis Among Iraqi Breast Cancer Women. *Int. J. Health Med.* **2024**, *1*, 165–178. [CrossRef]
58. Du, Y.; Cullum, I.; Illidge, T.M.; Ell, P.J. Fusion of metabolic function and morphology: Sequential [18F] fluorodeoxyglucose positron-emission tomography/computed tomography studies yield new insights into the natural history of bone metastases in breast cancer. *J. Clin. Oncol.* **2007**, *25*, 3440–3447. [CrossRef]
59. Yin, J.J.; Mohammad, K.S.; Käkönen, S.M.; Harris, S.; Wu-Wong, J.R.; Wessale, J.L.; Padley, R.J.; Garrett, I.R.; Chirgwin, J.M.; Guise, T.A. A causal role for endothelin-1 in the pathogenesis of osteoblastic bone metastases. *Proc. Natl. Acad. Sci. USA* **2003**, *100*, 10954–10959. [CrossRef]
60. D'Oronzo, S.; Coleman, R.; Brown, J.; Silvestris, F. Metastatic bone disease: Pathogenesis and therapeutic options. *J. Bone Oncol.* **2019**, *15*, 100205. [CrossRef]
61. Vakaet, L.A.L.; Boterberg, T. Pain control by ionizing radiation of bone metastasis. *Int. J. Dev. Biol.* **2004**, *48*, 599–606. [CrossRef]
62. O'Sullivan, G.J.; Carty, F.L.; Cronin, C.G. Imaging of bone metastasis: An update. *World J. Radiol.* **2015**, *7*, 202. [CrossRef]

life

MDPI

Article

Advancing Nipple Reconstruction in Skin-Sparing Mastectomy: The Efficacy of the Long V-Y Flap Technique for Enhanced Size Retention and Symmetry

Woo Seob Kim †, Byung Woo Yoo, Kap Sung Oh, Hyun Woo Shin, Kyu Nam Kim and Junekyu Kim *,†

Department of Plastic and Reconstructive Surgery, Kangbuk Samsung Hospital, School of Medicine, Sungkyunkwan University, Seoul 03181, Republic of Korea

* Correspondence: j7181.kim@samsung.com

† These authors contributed equally to this work.

Abstract

Reconstruction of the nipple is the final step of breast reconstruction following skin-sparing mastectomy (SSM) and plays a critical role in restoring breast symmetry and aesthetic completeness. Various nipple reconstruction techniques have been described, including split transplantation of normal nipples and flap-based procedures; however, long-term maintenance of nipple projection and volume remains challenging due to postoperative tissue loss. This study evaluates the clinical outcomes of a previously described long V-Y flap technique, focusing on its ability to mitigate postoperative nipple projection and volume loss. A retrospective analysis was performed on 27 patients who underwent nipple reconstruction using the long V-Y flap following SSM with tissue expander–implant reconstruction. Nipple projection and volume were measured immediately after surgery and at follow-up beyond six months, and volume loss rates were calculated. Outcomes were compared descriptively with projection loss rates reported for other commonly used flap techniques in the literature. The mean nipple volume loss was 34.23%, and the mean projection loss was 32.79%, while nipple width remained largely unchanged over time. These values were numerically lower than those reported for skate, star, bell, and arrow flaps in prior studies. The long V-Y flap appears to be a practical and reliable option for nipple reconstruction after SSM, particularly for larger nipples, with favorable short- to mid-term maintenance of projection and volume.

Keywords: nipple reconstruction; skin-sparing mastectomy; long V-Y flap technique; enhanced size retention; symmetry; split transplantation

1. Introduction

Skin-sparing mastectomy (SSM) is commonly performed for patients with nipple metastasis from breast cancer. Due to limited skin availability, most patients undergo tissue expansion and implant replacement. Nipple reconstruction is essential for improving breast completeness and achieving symmetry, representing the final step of breast reconstruction. Techniques for nipple reconstruction have been comprehensively categorized, and evidence-based decision-making algorithms have been proposed to guide technique selection based on patient factors, surgical timing, and breast reconstruction characteristics. Nevertheless, no single approach has consistently achieved durable long-term projection. Nipple reconstruction methods are broadly divided into split transplantation of the normal nipple and flap-based techniques [1–6]. Despite the use of various flap designs, long-term outcomes often remain suboptimal due to progressive projection and volume loss. Reported six-month projection loss rates across commonly used nipple reconstruction techniques range from approximately 30% for skate and modified star flaps to over 60% for bell flaps, underscoring the challenge of long-term projection maintenance [7–10].

This study aimed to calculate the long-term volume loss rate of reconstructed nipples, excluding cases of incomplete reconstruction due to flap necrosis or related complications. We analyzed contributing factors to nipple volume loss, explored mitigation strategies, and introduced the long V-Y flap as a streamlined and effective technique for nipple reconstruction [1]. This study offers a systematic evaluation of long-term outcomes of nipple reconstruction using this method, with comparison to other techniques reported in the literature. Despite previous publications describing the long V-Y technique, its long-term volumetric outcomes have not been consistently quantified using standardized measurements, nor sufficiently compared with other established nipple reconstruction methods.

2. Materials and Methods

2.1. Subjects

This study evaluated patients who underwent nipple reconstruction using the long V-Y flap technique after breast reconstruction with tissue expanders and prostheses following SSM between January 2016 and December 2022. Demographics and patient characteristics are listed in Table 1. Written informed consent was obtained from all 27 patients who participated in the study. The eligibility criterion for participation was a successful reconstruction without complications that could lead to incomplete nipple reconstruction. Patients with comorbidities known to impair wound healing, such as diabetes mellitus, vasculopathies or other vascular conditions were excluded to ensure an accurate assessment of the effectiveness of the long V-Y flap technique and to minimize confounding factors in nipple reconstruction outcomes. The study was conducted in accordance with the guidelines of the Declaration of Helsinki and was approved by the Institutional Review Board of Kangbuk Samsung Hospital (IRB no. KBSMC 2023-12-029; date of approval: 15 December 2023).

Table 1. Patient demographics and characteristics.

	Value
No. of patients	27
Method of reconstruction	
DTI	0
TEI	27
Mean age ± SD, year	48.5 ± 7.9
Mean BMI ± SD, kg/m^2	22.6 ± 3.0
Mean follow-up ± SD, month	6.4 ± 5.7
Radiotherapy history	1
Chemotherapy history	1
Mean immediate reconstruction nipple height ± SD, mm	8.19 ± 2.35 (95% CI, 7.25–9.12)
Mean immediate reconstruction nipple width ± SD, mm	10.85 ± 1.90 (95% CI, 10.10–11.60)

2.2. Operation

A centerline was drawn on the patient's torso in a standing position, and the position of the nipple to be reconstructed was determined and marked on the opposite side of the normal nipple. A V-shaped flap was prepared using this point as the base of the long V-Y flap. In case of an existing scar, a flap was created to include the scar on one side; if

https://doi.org/10.3390/life16010088

the flap was distant from the existing scar, it was designed parallel to the scar. The flap width was 130% of the diameter of the normal nipple or approximately 3–5 mm wider. The flap was designed to be longer than wider with a width-to-length ratio of 1:2.5. Instead of a complete triangle, the shape of the flap was more effectively tapered, maintaining an appropriate width over the middle, shaped like half of a fusiform or wine glass. While separating the flap from the pectoralis major muscle or acellular dermal matrix (ADM), the skin and subcutaneous fat layers were raised together. By advancing the flap toward the base, it was folded to achieve the required height, and the triangular tip was positioned between the original base and the tip, where it was sutured in a Y-shape. If the apex of the nipple, where the flap folds sharply, is excised into a rounded shape, it could constrict the central portion of the flap, potentially compromising blood flow. Therefore, upper nipple resection is not recommended. To ensure unimpeded blood circulation to the distal portion of the flap, the apex was folded inwards and sutured to create a round contour (Figure 1). To prevent compression due to external forces, including the underwear, a hole was drilled in the center of the foam dressing to maintain and protect the height, and the sutures were removed approximately two weeks later.

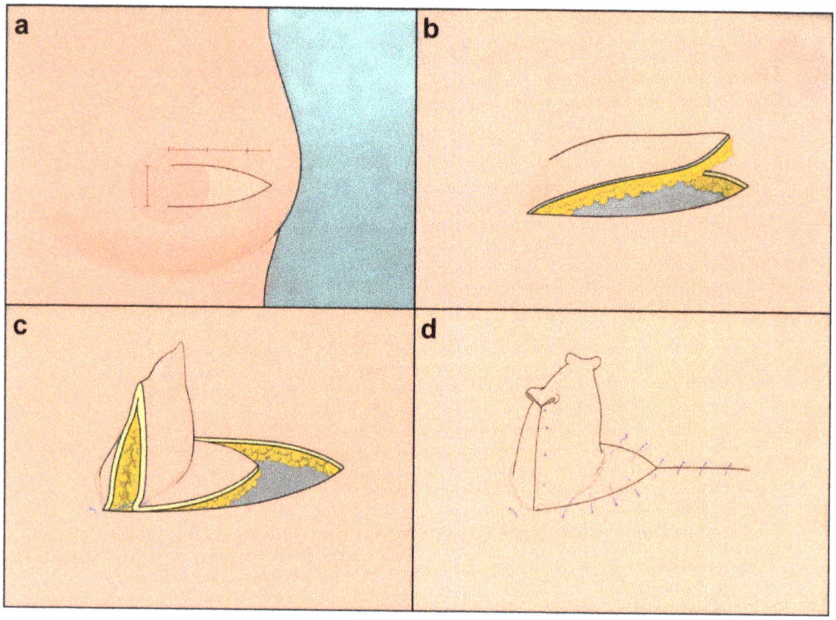

Figure 1. Diagrammatic representation of the long V-Y flap design for nipple reconstruction. (**a**) The flap base is marked at the planned nipple site, designed with a 1:2.5 width-to-length ratio and a half-fusiform shape, tapering in the middle. (**b**) The flap is meticulously separated with concurrent elevation of the skin and subcutaneous fat layers. (**c**) The flap is advanced toward the base and folded to achieve the desired height. (**d**) Depiction of the flap-folding, highlighting the inward manipulation of sharp rectangular tips at the top using a three-point suture, avoiding cuts that could restrict blood flow.

2.3. *Measurement of Nipple Projection and Volume Loss*

In this study, volume retention (also referred to as persistence) is defined as the proportion of the initial nipple volume maintained at follow-up, whereas volume loss represents the complementary reduction from the initial postoperative volume. The height (projection) and width of the reconstructed nipples were measured using a ruler and photographed before, immediately after operation, and during each outpatient visit (2 weeks, 1 month,

https://doi.org/10.3390/life16010088

3 months, 6 months Post-Op) (Figures 2–4). All measurements were performed by the same physician to ensure consistency. The rate of reduction in the projection and volume retention rate (or persistence rate) were calculated as follows:

$$\text{Volume retention rate (Persistence rate) (\%)} = 100 \times [\pi \times \text{height} \times (1/2\ \text{diameter})^2 \text{ at the last measurement}/\pi \times \text{height} \times (1/2\ \text{diameter})^2 \text{ immediately after surgery}]$$

$$\text{Volume loss rate (\%)} = 100 - \text{Volume retention rate (\%)}$$

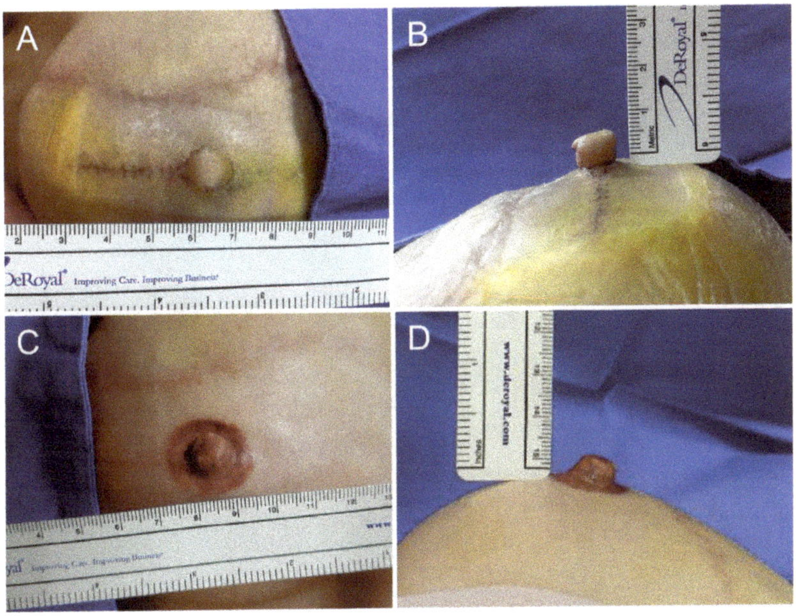

Figure 2. Clinical photographs of a 36-year-old female patient who underwent nipple reconstruction using the Long V-Y Flap Technique. (**A**,**B**) Immediate postoperative appearance showing a nipple width of 11 mm and height of 8 mm. (**C**,**D**) At 3 months postoperatively, the nipple maintained its width at 11 mm with a slight reduction in height to 7 mm.

The losses in projection calculated using this method were statistically compared using STATA (StataCorp. 2021. Stata Statistical Software: Release 17. College Station, TX, USA: StataCorpLLC.), The normality of data distribution was assessed using the D'Agostino-Pearson normality test. The long-term projection losses in nipples reconstructed with the long V-Y flap were compared with those reconstructed using other flaps [7,11–13] using an independent two-sample t-test. A two-sided value of $p < 0.05$ was considered statistically significant in all analyses.

In their thesis on nipple reconstruction using angel flaps, based on a literature review, Wong et al. compiled and organized the results of nipple reconstruction using 15 different flaps, and we selected the data with relatively clear results [11]. Data reported by Shestak et al. (skate flap, modified star flap, and bell flap), Kroll and Hamilton (double opposing tab flap), and Rubino et al. (modified star flap, arrow flap) were used as controls [7,12,13]. Since Kroll and Hamilton's data did not include the standard deviation (SD) for projection loss; for our analysis, we assumed that the SD was equivalent to that observed in our study. This assumption was necessary to calculate the p-value and compare the projection loss between their study and ours [12].

https://doi.org/10.3390/life16010088

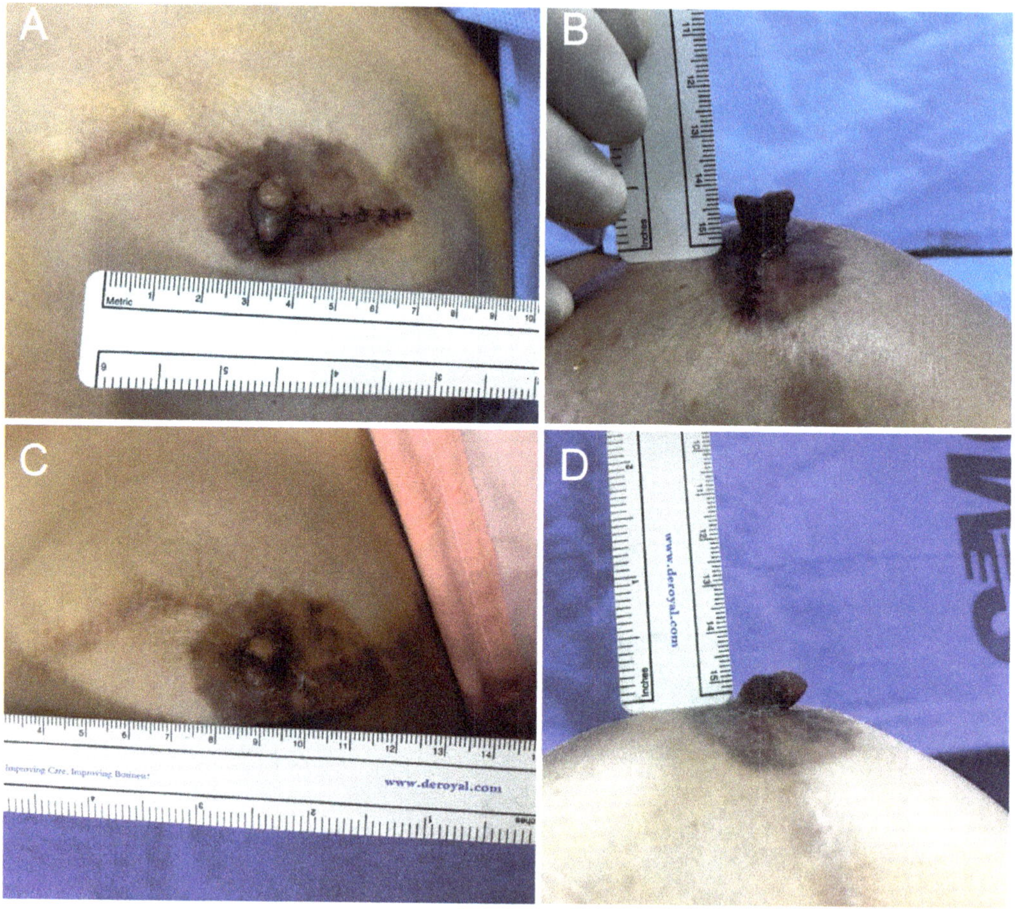

Figure 3. Clinical photographs of a 60-year-old female patient who underwent nipple reconstruction using the Long V-Y Flap Technique. (**A**,**B**) Immediate postoperative views showing a nipple width of 10 mm and height of 11 mm. (**C**,**D**) At 3 months postoperatively, the nipple width was slightly increased to 11 mm while the height decreased to 7 mm.

https://doi.org/10.3390/life16010088

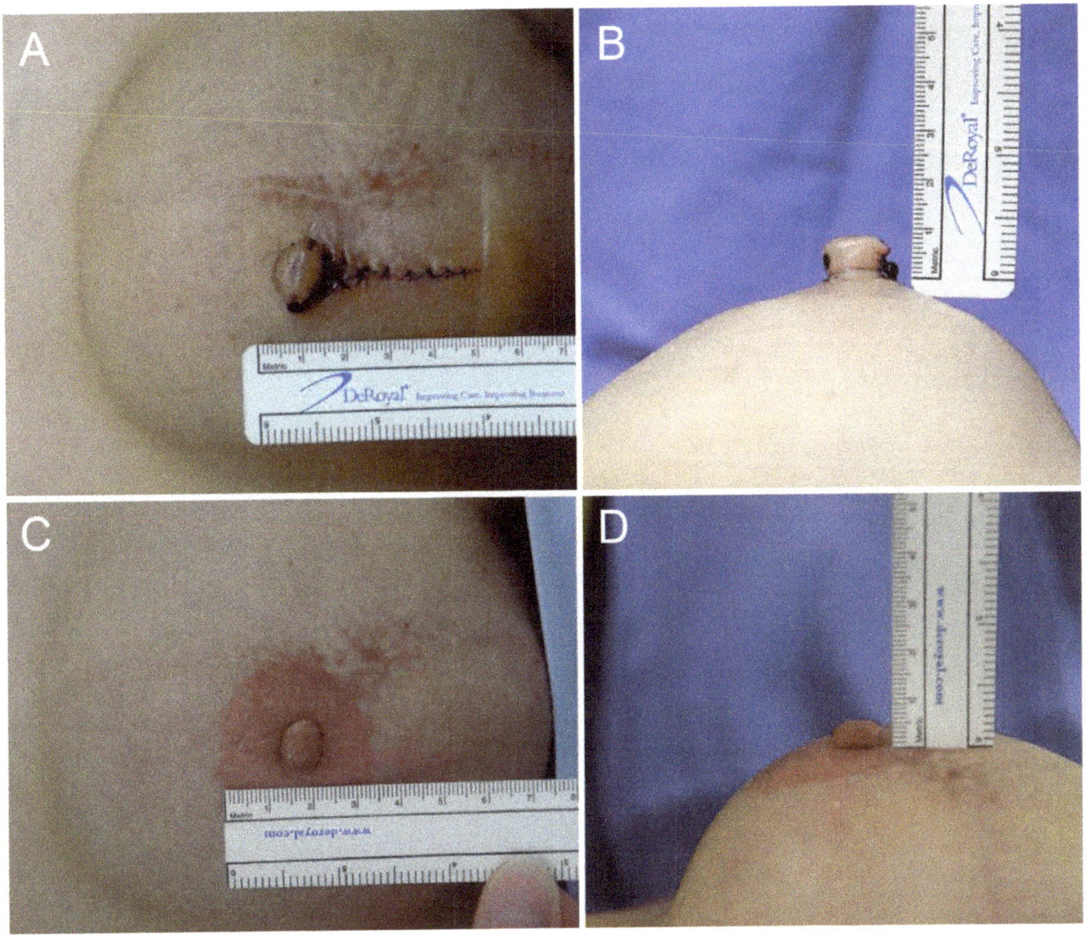

Figure 4. Clinical photographs of a 46-year-old female patient who underwent nipple reconstruction using the Long V-Y Flap Technique. (**A**,**B**) Immediate postoperative views demonstrating a nipple width and height of 8 mm. (**C**,**D**) At 6 months postoperatively, the nipple width had increased slightly to 9 mm while the height decreased to 6 mm.

3. Results

A total of 27 patients were compared and analyzed. They were followed up for an average of 6.4 months. The average height and width of the nipples reconstructed with long V-Y flaps were 8.19 mm and 10.85 mm immediately after the operation and 5.33 mm and 10.67 mm six months later, respectively (Figures 2–4). The average reduction in projection was 32.79% (95% CI, 27.36–38.22), and the average volume loss was 34.23% (95% CI, 28.13–40.33). Hence, the persistence rate of volume was 65.77%.

A literature review revealed that most studies had data only for projection loss and no data for volume loss [7,11–13]; hence, we could only compare the projection loss. Accordingly, comparisons with previously published techniques were performed for exploratory purposes only and should be interpreted with caution. Table 2 presents a comparative analysis of nipple projection loss observed with various flap techniques, illustrating the differential outcomes associated with each method.

https://doi.org/10.3390/life16010088

Table 2. Comparison of nipple projection loss rates between different flaps at the six-month follow-up.

	Long V-Y (N = 27)	Skate (N = 23)	Modified Star (N = 28)	Bell (N = 17)
Nipple projection loss (%) (standard deviation)	32.79 (13.73)	32.96 (12.70)	29.92 (12.68)	62.64 (28.79)

Confidence intervals were not available for historical controls. No correction for multiple comparisons was performed; therefore, reported *p*-values should be interpreted with caution and considered exploratory rather than confirmatory.

For these exploratory comparisons, historical mean values were compared with those of our cohort under the assumption of comparable measurement methods, and standard deviations from our dataset were applied when not reported in the original studies. The projection loss for nipples reconstructed using the long V-Y flap appeared numerically lower (32.79%) than for those that used the skate flap (40.11%, $p = 0.473$) and the modified star flap (40.28%, $p = 0.415$), although these differences were not statistically significant [7].

In contrast, greater projection loss has been reported for the bell flap (69.95%, $p < 0.001$) [7], double opposing tab flap at 10 months (66.00%, $p < 0.001$), modified star flap (69.90%, $p < 0.001$), and arrow flap (50.90%, $p = 0.020$) [12–18]. While exploratory statistical comparisons yielded *p*-values < 0.05 for these techniques ($p < 0.001$, $p < 0.001$, $p < 0.001$, and $p = 0.020$, respectively), these findings should be regarded as descriptive trends rather than definitive inferential evidence due to methodological limitations inherent in historical comparisons.

4. Discussion

In patients with breast cancer undergoing mastectomy for nipple metastasis, nipple reconstruction is often pursued after satisfactory reconstruction of the breast mound. This step is crucial for enhancing the aesthetic completeness of the reconstructed breast and achieving symmetry with the contralateral unaffected breast. Composite grafting involves the transfer of tissue from the healthy nipple and typically exhibits a lower volume loss rate compared to flap-based nipple reconstruction. However, this approach necessitates the sacrifice of healthy tissue and carries a risk of graft failure. Consequently, flap surgery has become the more prevalent choice for nipple reconstruction.

In patients undergoing SSM, the limitation of skin availability poses a significant challenge to achieving successful outcomes from flap surgery, as reusing affected skin often leads to suboptimal results. This difficulty has been corroborated in numerous studies. Various flap surgery techniques have been proposed for nipple reconstruction, including the skate, star, C-V, bell, arrow, and double opposing flaps [1–6,9,14]. Despite these multiple techniques, achieving optimal results remains challenging. Studies by Banducci et al., Bramhall et al., and others have consistently shown a 25–50% reduction in flap size within 2–3 months post-operation and a 40–70% reduction at extended follow-up [8,15]. Given these limitations, achieving a flap volume retention of 60% is often considered acceptable in clinical practice. Notably, the long V-Y flap technique used in this study demonstrated a volume persistence rate of 65.77%, suggesting potential effectiveness in mitigating postoperative nipple volume loss and supporting its role as a promising option for short- to mid-term outcomes.

Compensating for this volume loss by initially creating a larger nipple leads to disproportionate sizes often [16]. For example, to achieve a final size of 6 × 6 mm after expected volume loss, a nipple of about 18 × 18 mm must be constructed, i.e., three times the desired size. With techniques like the star or skate flap, experienced surgeons are acutely aware of how such large flaps can distort the overall shape of the breast mound. Consequently, after six months, patients are often left with a mere 6 × 6 mm-sized flap, falling short of the

https://doi.org/10.3390/life16010088

desired esthetic outcomes. This is because many reconstruction techniques fail to maintain the original size, leading to volume loss, as illustrated in Figure 5 [1,7,17,18].

Figure 5. Nipple reconstruction using the CV flap method. Shown is the nipple reconstruction using the CV flap method illustrating various mechanisms that contribute to the reduction in nipple size post-reconstruction. The red arrows indicate the skin retraction forces triggered by the closure of the donor defect, leading to retraction of the adjacent skin. The blue arrow represents two factors: (a) the number of lobules, where multiple lobules increase scar formation and disrupt the blood supply, potentially resulting in necrosis, fat atrophy, and (b) external pressure forces exerted by clothing or posture.

Contributing factors are discussed below:

(1) Retraction force, i.e., tension required to return to the original skin position

In patients undergoing SSM, after tissue expansion processes, there is considerable tension in the skin over the implant. In this condition, when nipples are created through flap surgery, the surrounding skin must be pulled to close the donor defect at the base of the flap, creating retraction force. The larger the flap, the greater the tension.

(2) Number of lobules

Most flaps are designed to have multiple lobules centered on one pedicle to form the cylindrical shape of the nipple. However, with a higher number of lobules in the flap, two problems affected the volume loss rate of the reconstructed nipple. First, based on the number and complexity of the lobules, the number of scars where lobules touch each other increases during the three-dimensional assembly. Second, as many incisions are made, the possibility of disturbing blood flow to the flap increases, leading to fat necrosis and subcutaneous fat atrophy. These factors can act as vectors that reduce volume and pull the nipple downward.

(3) External pressure

External pressure caused by clothing or posture can cause deformation that crushes the reconstructed nipples. Dressings or external devices can adequately protect this but cannot be ruled out permanently. In addition, the absence of a substructure that reinforces the lower part of the reconstructed nipple has been suggested as a cause. Various methods have been proposed using materials such as ear cartilage, ADM, and subcutaneous inlay grafts to reinforce the inner and lower nipples [12,15,19]. However, these methods have limitations owing to additional surgery and donor site morbidity, foreign body reaction, increased operation time, and high cost.

Although it is impossible to eliminate all these factors by changing the design of the flap, the factors mentioned above can be significantly reduced. We used a 'long V-Y flap'

https://doi.org/10.3390/life16010088

design to reconstruct the nipple [20]. The V-Y advancement flap is a traditional and effective tissue transfer method used to treat soft tissue defects in the face and other areas. There have been few reports of nipple reconstruction using this method, including a study by Riccio et al., who used a very small V-Y flap for autologous tissue breast reconstruction [21]. They designed the tip of the flap to be located at the base of the nipple [21]. In contrast, the long V-Y flap has an implant located just below it; therefore, it relies on the dermal plexus from the skin pedicle without blood flow from deep tissues.

The long V-Y flap is longer than the conventional random flap, with a width-to-length ratio of 1:2 to 1:2.5. Despite this geometry, clinically significant distal ischemia was not observed, possibly for reasons related to the tissue expansion process commonly performed in SSM, which may induce adaptive microvascular changes resembling delayed phenomena.

When the flap is folded, two sharp rectangular tips formed on top of the nipple. Cutting these rounds is not recommended because the middle part of the flap becomes narrow, constricting the blood flow. Instead, it was manipulated such that the pointed part was bent inside using a 3-point suture (Figure 1).

The 'long V-Y flap' has the following advantages. (1) The direct effect of skin tension on the reconstructed nipples is reduced. After the flap donor site is closed, tension is generated at right angles to the scar; however, a flap that moves within the completed nipple is not subjected to this tension. This was evidenced by a lower decrease in the width of the reconstructed papillae, as the tip of the V-Y flap is located in the middle of the donor site when closed, a gradual decrease in tension occurs, reducing the deformation in the breast mound shape. (2) The incision is small, leading to a small scar after assembly. As a result, the change in nipple size from changes in scars, such as contractures on the surface, is reduced. (3) There is only one lobule in the 'long V-Y flap.' Hence, theoretically, there would be no major blood flow disturbances compared to flaps with multiple lobules, reducing the possibility of fat necrosis or atrophy.

Despite these advantages, the long V-Y flap is not immune to deformations arising from external forces, which work in the same manner as in a conventional flap.

To determine whether this theoretical hypothesis translates into a long-term clinical result, we compared nipple size reduction from the long V-Y vs. other flap procedures. Patients were followed up during outpatient visits for an average of six months or longer. The reconstructed nipples were measured and photographed, and the reduction in the projection and volume were calculated. Since most literature focuses on projection loss, this study emphasizes projection comparisons using the long V-Y vs. other flap procedures, and the volume change was used as an additional reference. In line with our hypothesis, the width of the nipple remained stable while height decreased (Figures 2–4).

Published literature reports projection reductions ranging from approximately 40% to over 70% [7–9,11–13], consistent with our prior experience using modified star and C-V flaps. In the present cohort, the long V-Y flap demonstrated a projection reduction rate of 32.79%, suggesting improved resistance to size reduction compared with several conventional techniques.

This study has several limitations that should be acknowledged. First, the study population was highly selective and relatively small, comprising 27 patients from a single institution. Cases with flap necrosis, postoperative complications, and patients with comorbidities such as diabetes mellitus or vascular disease were excluded, despite these conditions being common in the breast reconstruction population. This selection bias may have resulted in more favorable outcomes and limits the generalizability of the findings to routine clinical practice.

Second, the retrospective single-cohort design is inherently subjected to selection bias and unmeasured confounding factors, and the absence of a direct comparator group

https://doi.org/10.3390/life16010088

precludes definitive conclusions regarding the relative performance of the long V-Y flap compared with other nipple reconstruction techniques. Comparisons with previously published techniques were therefore based on heterogeneous historical controls with differing patient populations, follow-up durations, and measurement methods, and should be interpreted as descriptive and exploratory rather than inferential.

Third, patient-reported outcomes, including perceived symmetry, aesthetic satisfaction, and quality of life, were not assessed, despite their increasing importance in the contemporary nipple–areola complex reconstruction literature. Fourth, outcome assessment was limited by methodological constraints, as nipple projection and volume were measured using clinical photographs and ruler-based assessments only, without three-dimensional imaging or objective volumetric analysis. In addition, follow-up duration was heterogeneous across patients (mean 6.4 $\pm$ 5.7 months), raising uncertainty as to whether the reported "six-month" projection measurements reflect a uniform postoperative time point. Given the relatively short follow-up period, the findings should be interpreted as reflecting short- to mid-term outcomes rather than definitive long-term results.

Finally, although the sample size was considered adequate for exploratory analysis, a formal power analysis with clearly defined effect size assumptions and a prespecified primary endpoint was not performed. Accordingly, the statistical findings should be interpreted with caution. Future prospective studies incorporating standardized follow-up intervals, direct comparator groups, objective volumetric assessment, and patient-reported outcome measures are needed to further validate these findings.

Anatomical differences in nipple morphology across populations further affect generalizability. Western women tend to have smaller, wider nipples, with an average reconstructed projection of approximately 3.8 mm and a width of 13 mm [9]. In contrast, we achieved an average reconstructed nipple size of 10 mm in width and 8 mm in height, approximately 20% greater in volume than a nipple measuring 4 mm in height and 13 mm in width. Creating such a large nipple with conventional flaps remains challenging; however, the long V-Y flap demonstrated consistent results. Notably, the average width of the nipples reconstructed with the long V-Y flap was 10.85 mm, and the reduction in width over time was minimal (1.66%), resulting in a high width retention rate of 98.34%.

Kroll and Hamilton emphasized that flap width is a critical factor for preserving long-term projection because wider flaps enhance blood supply and reduce complications such as fat necrosis [12]. Based on these findings, the long V-Y flap technique appears to be a suitable and effective option for reconstructing large nipples while maintaining long-term projection and volume.

The 'long V-Y flap' technique has a shorter operation time due to the simple procedure. Additionally, the base and direction of the flap can be set arbitrarily, allowing the incorporation of the existing scar, leaving only one Y-shaped scar. In addition, the larger the reconstructed nipple, the greater the effect of pulling the anterior skin of the breast, resulting in a flattening deformation. Long V-Y flap surgery may help reduce this deformation, which is often overlooked because they are considered unimportant or unavoidable during nipple reconstruction.

5. Conclusions

In patients undergoing skin-sparing mastectomy, nipple reconstruction using flap techniques tends to result in progressive volume loss, making maintenance of initial projection and volume challenging. In this context, the long V-Y flap technique appears to offer a practical and efficient approach that may help mitigate volume loss, particularly in the reconstruction of relatively large nipples.

https://doi.org/10.3390/life16010088

Further prospective studies with standardized and longer follow-up periods are needed to more clearly define the durability, safety profile, and potential complications associated with the long V-Y flap technique. Comparative studies with larger sample sizes, direct comparator groups, and diverse patient populations could provide more comprehensive insights into the versatility and adaptability of this technique. Additionally, future research exploring the integration of new materials or innovative surgical approaches in conjunction with a long V-Y flap could open new avenues for enhancing aesthetic and functional outcomes. Finally, systematic evaluation of patient-reported outcomes, including aesthetic satisfaction, symmetry, and quality of life, will be essential to fully assess the clinical value of nipple reconstruction techniques and to guide continued improvement of postmastectomy breast reconstruction.

Author Contributions: Conceptualization, J.K.; methodology, J.K.; software, B.W.Y.; validation, J.K. and H.W.S.; formal analysis, W.S.K. and B.W.Y.; investigation, W.S.K., B.W.Y. and H.W.S.; data curation, W.S.K. and B.W.Y.; writing—original draft preparation, W.S.K., B.W.Y. and K.N.K.; writing—review and editing, H.W.S., K.N.K. and J.K.; visualization, B.W.Y.; supervision, K.S.O. and J.K.;W.S.K., B.W.Y., K.S.O., H.W.S. and K.N.K. contributed equally to this work. All authors have read and agreed to the published version of the manuscript.

Funding: This research received no external funding.

Institutional Review Board Statement: The study was conducted in accordance with the guidelines of the Declaration of Helsinki and was approved by the Institutional Review Board of Kangbuk Samsung Hospital (IRB no. KBSMC 2023-12-029; date of approval: 15 December 2023).

Informed Consent Statement: Written informed consent was obtained from all 27 patients who participated in the study.

Data Availability Statement: The data presented in this study are available upon request from the corresponding author. The data are not publicly available due to privacy restrictions.

Conflicts of Interest: The authors declare that they have no conflicts of interest to disclose.

References

1. Gougoutas, A.J.; Said, H.K.; Um, G.; Chapin, A.; Mathes, D.W. Nipple-areola complex reconstruction. *Plast. Reconstr. Surg.* **2018**, *141*, 404e–416e. [CrossRef] [PubMed]
2. Sisti, A.; Grimaldi, L.; Tassinari, J.; Cuomo, R.; Fortezza, L.; Bocchiotti, M.A.; Roviello, F.; D'Aniello, C.; Nisi, G. Nipple-areola complex reconstruction techniques: A literature review. *Eur. J. Surg. Oncol.* **2016**, *42*, 441–465. [CrossRef]
3. Hartrampf, C.R., Jr.; Culbertson, J.H. A dermal-fat flap for nipple reconstruction. *Plast. Reconstr. Surg.* **1984**, *73*, 982–986. [CrossRef] [PubMed]
4. Little, J.W., III. Nipple-areola reconstruction. *Clin. Plast. Surg.* **1984**, *11*, 351–364. [CrossRef]
5. Farhadi, J.; Maksvytyte, G.K.; Schaefer, D.J.; Pierer, G.; Scheufler, O. Reconstruction of the nipple-areola complex: An update. *J. Plast. Reconstr. Aesthet. Surg.* **2006**, *59*, 40–53. [CrossRef]
6. Cronin, E.D.; Humphreys, D.H.; Ruiz-Razura, A. Nipple reconstruction: The S flap. *Plast. Reconstr. Surg.* **1988**, *81*, 783–787. [CrossRef] [PubMed]
7. Shestak, K.C.; Gabriel, A.; Landecker, A.; Peters, S.; Shestak, A.; Kim, J. Assessment of long-term nipple projection: A comparison of three technical techniques. *Plast. Reconstr. Surg.* **2002**, *110*, 780–786. [CrossRef]
8. Banducci, D.R.; Le, T.K.; Hughes, K.C. Long-term follow-up of a modified Anton-Hartrampf nipple reconstruction. *Ann. Plast. Surg.* **1999**, *43*, 467–469. [CrossRef]
9. Losken, A.; Mackay, G.J.; Bostwick, J., 3rd. Nipple reconstruction using the C-V flap technique: A long-term evaluation. *Plast. Reconstr. Surg.* **2001**, *108*, 361–369. [CrossRef]
10. Few, J.W.; Marcus, J.R.; Casas, L.A.; Aitken, M.E.; Redding, J. Long-term predictable nipple projection following reconstruction. *Plast. Reconstr. Surg.* **1999**, *104*, 1321–1324. [CrossRef]
11. Wong, W.W.; Hiersche, M.A.; Martin, M.C. The angel flap for nipple reconstruction. *Can. J. Plast. Surg.* **2013**, *21*, e1–e4. [CrossRef]
12. Kroll, S.S.; Hamilton, S. Nipple reconstruction with the double-opposing-tab flap. *Plast. Reconstr. Surg.* **1989**, *84*, 520–525. [CrossRef] [PubMed]

https://doi.org/10.3390/life16010088

13. Rubino, C.L.D.; Dessy, L.A.; Posadinu, A. A modified technique for nipple reconstruction: The "arrow flap". *Br. J. Plast. Surg.* **2003**, *56*, 247–251. [CrossRef]
14. Eng, J.S. Bell flap nipple reconstruction–a new wrinkle. *Ann. Plast. Surg.* **1996**, *36*, 485–488. [CrossRef]
15. Bramhall, R.J.; Thiruchelvam, P.T.R.; Concepcion, M.; Gui, G.P. Use of acellular dermal matrix (ADM) in nipple reconstruction: The 'central-pillar technique'. *Gland Surg.* **2017**, *6*, 394–398. [CrossRef]
16. Feng, S.; Zhang, B.; Kappos, E.A.; Tremp, M.; Yang, C. Modified S-flap for nipple reconstruction. *Aesthet. Plast. Surg.* **2017**, *41*, 312–317. [CrossRef] [PubMed]
17. Bosch, G.; Ramirez, M. Reconstruction of the nipple: A new technique. *Plast. Reconstr. Surg.* **1984**, *73*, 977–981. [CrossRef] [PubMed]
18. Mukherjee, R.P.; Gottlieb, V.; Hacker, L. Nipple-areolar reconstruction with buried dermal hammock technique. *Ann. Plast. Surg.* **1987**, *19*, 421–423. [CrossRef]
19. Collis, N.; Garrido, A. Maintenance of nipple projection using auricular cartilage. *Plast. Reconstr. Surg.* **2000**, *105*, 2276–2277. [CrossRef]
20. Jang, N.; Kim, J.; Shin, H.W.; Suk, S.W. Long V-Y advancement technique for large nipple reconstruction in Asian women. *Arch. Plast. Surg.* **2021**, *48*, 44–48. [CrossRef]
21. Riccio, C.A.; Zeiderman, M.R.; Chowdhry, S.; Wilhelmi, B.J. Review of nipple reconstruction techniques and introduction of V to Y technique in a bilateral wise pattern mastectomy or reduction mammaplasty. *Eplasty* **2015**, *15*, e11. [PubMed]

https://doi.org/10.3390/life16010088

life

MDPI

Article

Immediate Breast Reconstruction with Fat-Graft-Augmented ICAP Flaps

Francesco Klinger [1], Mattia Federico Cavallero [2,3], Andrea Vittorio Emanuele Lisa [2,4,5,*], Fernando Rosatti [4], Barbara Catania [3], Marco Klinger [2,3], Riccardo Di Giuli [3,6], Simone Furlan [2,3], Roberta Comunian [2,3], Stefano Vaccari [2,3] and Valeriano Vinci [3,6]

1 Department of Health Sciences, Ospedale San Paolo, University of Milan, via Antonio di Rudinì 8, 20142 Milan, Italy
2 Department of Medical Biotechnology and Translational Medicine BIOMETRA, Plastic Reconstructive and Aesthetic Plastic Surgery School, Università degli Studi Di Milano, via Festa del Perdono 7, 20122 Milan, Italy
3 IRCCS Humanitas Research Hospital, via Manzoni 56, Rozzano, 20089 Milan, Italy
4 Department of Plastic and Reconstructive Surgery, European Institute of Oncology, IRCCS, 20139 Milan, Italy
5 Department of Surgical Sciences, University of Rome "Tor Vergata", Cracovia n. 50, 00133 Rome, Italy
6 Department of Biomedical Sciences, Humanitas University, via Rita Levi Montalcini 4, Pieve Emanuele, 20090 Milan, Italy
* Correspondence: andrealisamd@gmail.com

Abstract

Background: Intercostal artery perforator (ICAP) flaps are a reliable option for volume replacement in breast-conserving surgery (BCS), particularly for lower pole defects. However, limited flap volume may reduce their applicability in selected patients. Autologous fat grafting has been proposed to enhance both volume and aesthetic outcomes. **Methods:** This retrospective study evaluated 10 patients undergoing BCS with immediate reconstruction using fat-graft-augmented ICAP flaps. Nine anterior ICAP (AICAP) flaps and one lateral ICAP (LICAP) flap were employed. The outcomes included flap viability, complications, and aesthetic results over a 6-month follow-up. **Results:** Partial flap resorption occurred in 2 patients (20%), both of whom were active smokers. No cases of skin necrosis were observed. Fat grafting volumes ranged from 20 to 60 cc. Aesthetic outcomes were satisfactory, with good restoration of the breast contour and stable integration of the flap and grafted fat. **Conclusions:** The combined use of ICAP flaps and autologous fat grafting is a feasible and effective technique for immediate reconstruction after BCS. It allows volume enhancement, maintains natural contour, and shows low complication rates in properly selected patients. Smoking remains a significant risk factor.

Keywords: breast surgery; breast conservative surgery; immediate breast reconstruction; autologous breast reconstruction; ICAP flap; fat-grafting; fat-graft-augmented ICAP flap

1. Introduction

The integration of oncoplastic techniques into breast-conserving surgery (BCS) has significantly improved aesthetic outcomes in women undergoing partial mastectomy.

In this context, intercostal artery perforator (ICAP) flaps have emerged as a versatile option not only for breast reconstruction but also for breast ptosis correction in post-bariatric patients [1], as well as for reconstructing distant defects of the thorax, sacrum, and axillary region and as free flaps [2].

ICAP flaps are based on perforators from the intercostal vessels, which form an arterial arcade between the aorta and the internal mammary vessels, dividing into vertebral, intercostal, intermuscular, and rectus segments [3].

They are classified into dorsal (DICAP), lateral (LICAP), and anterior (AICAP) types based on the source of the perforators. LICAP flaps are suitable for lateral thoracic defects or can be extended posteriorly for larger reconstructions, while AICAP flaps are ideal for inferomedial breast defects [4].

Microsurgical breast reconstruction can present challenges such as vascular compression by the chest wall [5], impaired skin sensitivity [6], and limited donor-site volume[7]. In this setting, autologous fat grafting serves as an effective adjunct, enhancing outcomes and aesthetic results, as demonstrated in the literature [5–7]. Notably, the flap acts as a vascularized scaffold that supports fat graft integration [8].

This study investigates the combined use of ICAP flaps and autologous fat grafting in breast reconstruction.

By examining clinical outcomes and synthesizing the available evidence, we aim to evaluate the feasibility, benefits, and potential limitations of this innovative approach.

2. Patients and Methods

This retrospective study evaluated 10 patients (10 breasts) who underwent BCS for tumors localized in the inferior breast quadrants, with immediate volume reconstruction using ICAP flaps. Among the 10 ICAP flaps performed, 9 were AICAP flaps and only 1 was a LICAP flap. In two cases, a skin island was preserved in the flap project to reconstruct the skin resection, and in one case, the ICAP flap reconstruction was performed in conjunction with an inverted T mastopexy; autologous fat grafts were used in all cases.

The procedures were performed between 2024 and 2025 at the Humanitas Research Hospital, Milan, Italy.

Patients eligible for this procedure met the following inclusion criteria: (a) having undergone conservative breast surgery, (b) with tumors localized in the inferior breast quadrants, and (c) identification of a well-vascularized perforator artery with an adequate caliber and flow during preoperative color Doppler ultrasound evaluation. Exclusion criteria included a BMI < 20 kg/m^2, diabetes, poor scarring history, and heavy smoking habits.

Preoperative pedicle assessment involved using a Doppler probe (iU22 xMATRIX ultrasound system with a Linear 12–5 MHz transducer, Phillips Electronics, Andover, MA, USA) and direct visualization. Although computed tomography angiography can visualize the pedicle, it was deemed unnecessary for preoperative evaluation.

Intraoperative and postoperative flap viability was assessed through ultrasound evaluation and clinical evaluation conducted, respectively, at 12 and 24 h and at 24–48 h, on postoperative days 7 and 14, and at 1, 3, and 6 months post-surgery.

Data collection included demographic characteristics of the study population, complication rates, and photographic documentation.

3. Operative Technique

The flap is designed along the identified perforator's axis (Figure 1). In this context, preoperative imaging such as Doppler ultrasound is used to locate the dominant perforators.

Anesthesia is performed with particular care to avoid compressive positioning or infiltration near the perforator vessels, which are preoperatively identified and mapped through preoperative imaging using Doppler ultrasound to preserve their integrity throughout the procedure. A total of 10 mL of levobupivacaine 7.5% is selectively infiltrated to facilitate tissue dissection for flap elevation and to reduce postoperative pain, ensuring both surgical precision and patient comfort.

The LICAP or AICAP variant is selected based on the defect's location. The perforator dissection is meticulously performed to preserve vascularity and allow sufficient pedicle length for rotation and inset flexibility (Figures 2 and 3).

In cases where skin island reconstruction is necessary, the flap may be left non-de-epithelialized to maintain a skin paddle for coverage. This approach allows for the preservation of a viable skin component, essential when reconstructing larger skin defects.

Based on aesthetic considerations, the flap axis is designed to align with the inframammary sulcus, ensuring that the resulting scar is positioned discreetly under the brassiere strap.

Adipose tissue is harvested from the donor site in volumes ranging from 20 to 60 cc using a blunt cannula under low-pressure vacuum in order to preserve cell integrity. In our case series, the donor site was consistently located at the hips.

The harvested fat is processed using the Coleman technique [9]. Fat is injected with a blunt cannula to minimize trauma (Figure 4), with fat injection in small aliquots while withdrawing the cannula to create microdroplets and complementing the ICAP flap structure.

To optimize breast reshaping, the ICAP flap can be folded or "rolled" onto itself during the inset. This technique increases the projection and fullness of the reconstructed breast mound.

The inset of the flap is stabilized using Vicryl 2/0 sutures to secure the flap to the pectoralis major fascia, maintaining its intended position, while careful attention must be paid to avoid kinking or compression of the vascular pedicle to maintain adequate perfusion. Following stabilization, Vicryl 2/0 sutures are placed in layers. The final step involves a continuous intradermal closure of the skin and a breast elastic taping with a moderate grade of compression.

Patients are closely monitored during their hospital stay to assess flap viability and detect any early complications. After discharge, weekly outpatient visits are conducted during the first month. Follow-up appointments are then gradually spaced out to ensure long-term outcomes at 1, 3, and 6 months post-surgery.

4. Results

All the patients in the study underwent ICAP flap fat-augmented volume restoration. In one case, a concomitant inverted T mastopexy was performed due to grade III ptosis, according to the Regnault classification [10]. The mean patient age was 55 (range 40–76) years, and the mean BMI was 24.7 (21.6–29.8) Kg/m^2. The mean size of the flap was approximately 13.4 $\times$ 9.6 cm and the average flap thickness was 2.1 (range 1.7–2.4 cm), as visible in Table 1. In all cases where skin reconstruction was performed using a non-de-epithelialized flap, no signs of skin necrosis were observed postoperatively.

Among the 10 patients included in the study, 2 were active smokers and 1 was a former smoker.

All flaps were clinically well-vascularized after complete elevation and were successfully transferred, although partial flap loss was seen in 20% of the cases.

The brassiere cup size remained unchanged in all patients.

Table 1. Demographic and clinical characteristics of the study population. This table summarizes the baseline characteristics of all patients included in the study. Data include patient age (years), type of surgical intervention (AICAP or LICAP flap), body mass index (BMI, kg/m^2), and smoking status (active smoker, former smoker, non-smoker). These parameters provide an overview of the clinical context and potential risk factors influencing surgical outcomes.

AGE (y.o.)	AICAP or LICAP	BMI (kg/m^2)	SMOKING
65	AICAP	29.76	no
65	AICAP	25.64	yes
45	AICAP	18.22	no
39	AICAP	20.81	no
47	AICAP	19.61	yes
76	AICAP	26.84	no
57	AICAP	21.56	no
66	LICAP	24.03	no
40	AICAP	23.05	no
49	AICAP	23.11	no

5. Clinical Examples

Case 1 (Figures 5 and 6): A 55-year-old woman underwent an infero-external and partial infero-internal right breast quadrantectomy followed by plastic reconstruction using an AICAP flap. The flap measured 12 cm × 6 cm and was approximately 2 cm thick. The donor area was directly closed, and the flap was de-epithelialized before being positioned to cover the breast mound. Photographs were taken 60 days postoperatively.

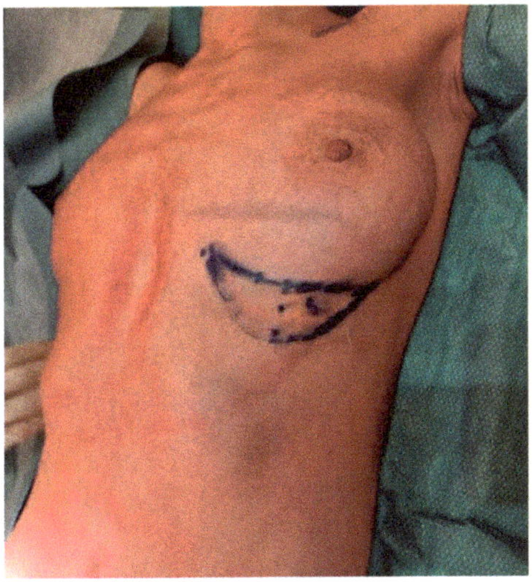

Figure 1. Preoperative marking demonstrating the planned surgical incision, including precise delineation of the flap margins to ensure optimal flap design and accurate tissue mobilization.

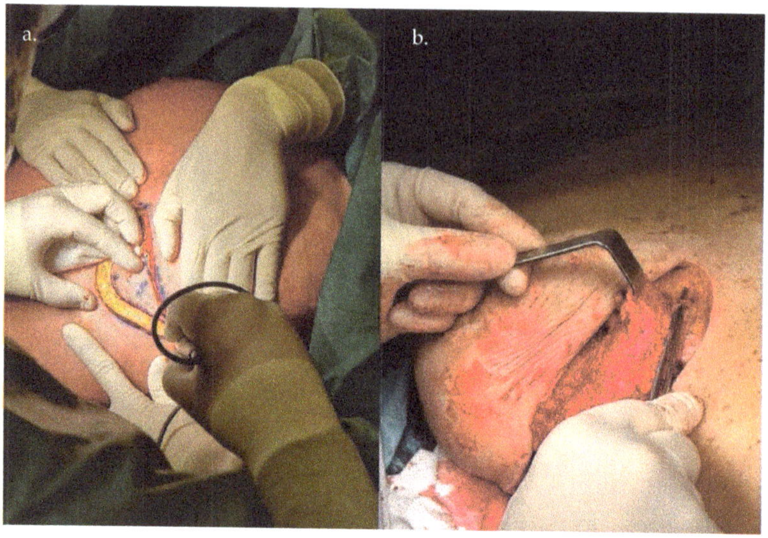

Figure 2. Intraoperative images showing the flap preparation, highlighting the meticulous dissection along the perforator axis (**a**) in order to preserve vascular integrity (**b**).

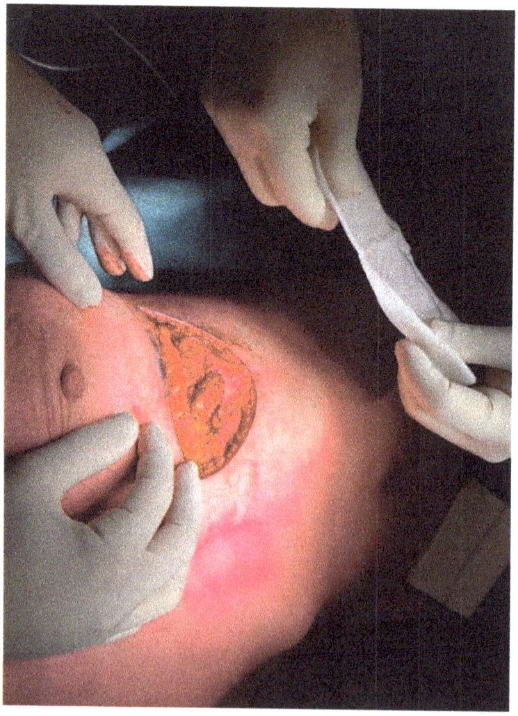

Figure 3. The flap is elevated while maintaining adequate pedicle length for optimal rotation and inset flexibility.

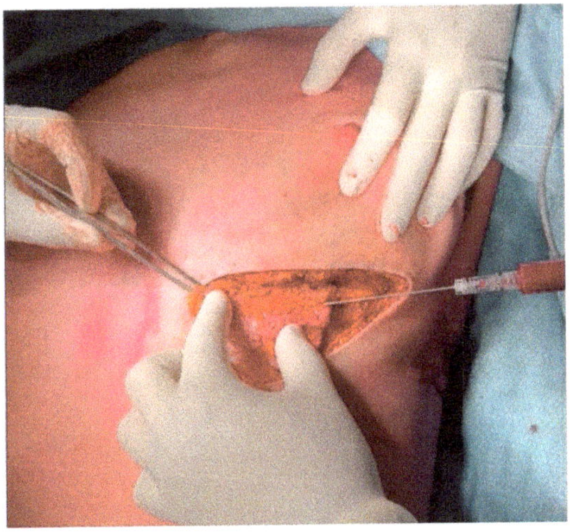

Figure 4. Autologous fat grafting flap augmentation, applying a blunt cannula for precise fat deposition, aiming to enhance volume and contour while minimizing tissue trauma.

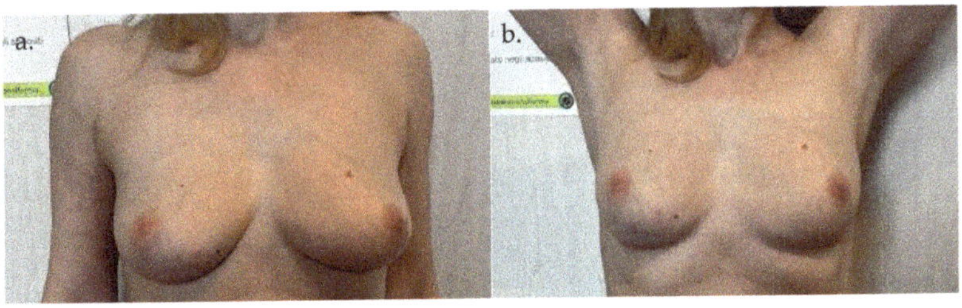

Figure 5. Preoperative images of Case 1. (**a**) Frontal view of the breast region. (**b**) Frontal view including upper arms for assessment of body contour and symmetry.

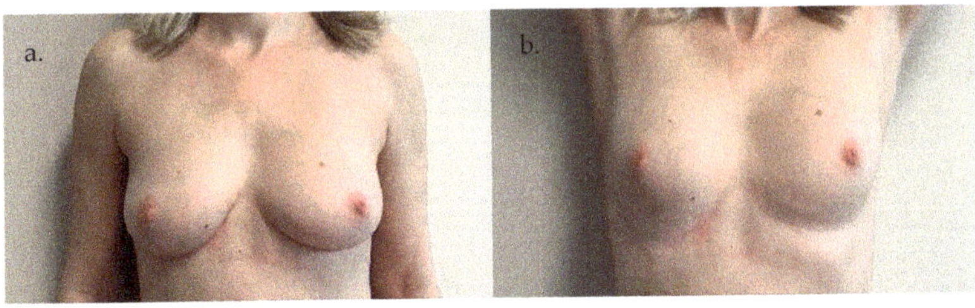

Figure 6. Postoperative images of Case 1, taken 60 days after surgery. (**a**) Frontal view showing breast reconstruction outcome. (**b**) Frontal view with upper arms to illustrate overall aesthetic integration.

Case 2 (Figure 7): A 38-year-old woman underwent an infero-external left breast quadrantectomy breast cancer, followed by immediate volume replacement using a fat-augmented AICAP flap. The flap measured 15 cm × 12 cm and was approximately 2.2 cm thick. After careful elevation and de-epithelialization, the flap was positioned to restore the breast contour, effectively covering the surgical defect. The donor area was directly closed without tension. Postoperative recovery was uneventful, and photographs were taken 120 days after surgery, documenting a satisfactory aesthetic outcome and stable flap integration.

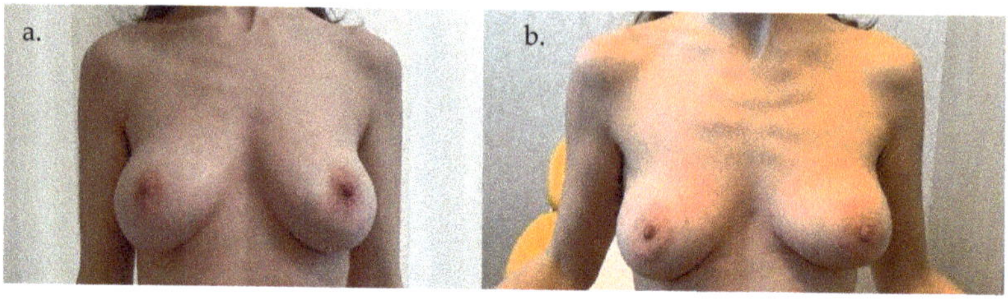

Figure 7. Preoperative (**a**) and postoperative (**b**) images, taken 120 days after surgery, of Case 2. Frontal view illustrating the tumor location and breast morphology prior to quadrantectomy and reconstruction (**a**). Frontal view demonstrating flap integration and left breast contour restoration (**b**).

6. Discussion

Nowadays, numerous oncoplastic techniques have been developed to excise small to medium-sized tumors while ensuring an adequate safety margin and achieving favorable outcomes [11–13]. Although various local techniques are available, selecting the appropriate approach requires careful consideration of multiple factors.

The use of the intercostal neurovascular pedicle to supply skin flaps is a well-established concept in flap surgery, first proposed by Esser et al. [14] in 1931. In 1974, Dibbell et al. [15] reported a clinical case involving a young paraplegic patient with a sacral bed sore. In this case, an intercostal flap incorporating the anterior cutaneous nerve was utilized to provide sensation to the sacral region. Subsequently, Daniel et al. [16] included the lateral cutaneous branch of the intercostal nerve in musculocutaneous flaps. These flaps were harvested as musculocutaneous constructs with random extensions beyond the thoracic cage, often requiring delay procedures to improve their reliability. The extensive anatomical studies conducted by Kerrigan and Daniel [2,17] contributed significantly to refining the clinical indications and surgical techniques for these flaps.

As described by Hamdi et al. [18], these flaps can be based on an intercostal perforator arising from either the costal or muscular segments of the intercostal vessels, enabling the surgeon to construct a reliable skin paddle for chest wall and breast reconstruction. According to these authors [18], these dominant perforators were located on average 3.5 cm from the anterior border of the latissimus dorsi muscle and in the fourth to eighth intercostal spaces, with a higher concentration in the sixth and seventh intercostal spaces. When planned and performed effectively, this anatomical characteristic permits adequate flap vascularization and minimizes vascular pedicle complications [19].

In the present study, ICAP flaps were used to repair lateral and central-inferior skin and breast tissue resected during conservative breast surgery (CBS) in the context of patients with a small to medium breast volume [20]. We observed that using tissue adjacent to the defect provided matching texture and color to the original breast. In addition, it is important to design the skin island flap according to the tumor location to avoid pedicle tension. The utility of the ICAP flap for volume replacement in breast reconstruction has been well documented in the literature [3,21–26]. However, their volume may be insufficient in larger defects, requiring careful patient selection or combination with other flaps [27].

In this context, autologous fat grafting is a safe and effective tool for secondary breast reconstruction, enhancing contour, volume, and overall breast shape and symmetry [28]. This technique is particularly beneficial in breast reshaping, as evaluated by Klinger et al. for composite breast augmentation [29] and for secondary contour deformities, providing a simple and low-morbidity solution [30]. Moreover, autologous fat grafting has been shown to be a relatively safe procedure for refining reconstructed breasts in postmastectomy patients, with low complication rates and no observed cases of locoregional cancer recurrence [31,32]. Additionally, it has demonstrated beneficial effects in alleviating postmastectomy pain syndrome following breast-conserving surgery, likely due to its regenerative and anti-inflammatory properties [33].

To overcome the volume limitations of the ICAP flap in larger defects, the fat-graft-augmented ICAP flap has been introduced, combining perforator flap reliability with autologous fat grafting to enhance volume and contour restoration.

The study by Laporta et al. [7] evaluated the effectiveness of breast reconstruction using DIEP (deep inferior epigastric perforator) flaps augmented with autologous fat grafting in patients with insufficient donor-site volume. The findings indicate that this technique is both feasible and safe, providing satisfactory aesthetic outcomes without increasing the number of secondary procedures or extending the overall treatment duration compared to DIEP flap reconstruction without fat grafting [4].

In this context, as shown in Table 2, the overall complication rate was 20%, with no cases of tissue necrosis. The only complication observed was partial flap resorption, which occurred in two cases, both belonging to the active smoker group. Demiri EC et al. [34] reported an overall complication rate of 38.8%, with only one case of fat necrosis observed in the fat-augmented latissimus dorsi group. The lower flap resorption rate observed in our study can be attributed to the smaller flap volume and the consequently reduced amount of fat augmentation required to achieve the expected surgical result. The absence of tissue necrosis observed in our cohort further supports the stability and reliability of the flap's vascular supply, a concept corroborated by previous studies highlighting the robust perfusion and low complication rates associated with well vascularized perforator flaps [35].

Table 2. Comparison of complication rates between smokers and non-smokers. This table presents the distribution of postoperative complications observed in the study cohort, stratified by smoking status. Data include the total number of patients in each group (active smokers vs. non-smokers/former smokers), the incidence of partial flap resorption, fat necrosis, and other complications. The table highlights the higher rate of partial flap resorption in the smoker group, supporting the negative impact of smoking on microvascular perfusion and tissue healing.

	SMOKERS (Total No. 2)	NON-SMOKERS (Total No. 8)	Total (No. 10)
PARTIAL FLAP LOSS (%)	100	0	20
TISSUE NECROSIS (%)	0	0	0

Obviously, care must be taken in high-risk patients, such as smokers.

As visible in Table 2, the complication rate and the total of cases of partial flap loss were higher in the smoker group, confirming the impact of smoking on postoperative outcomes in breast reconstruction [36].

The limitations of this study include the relatively small number of patients, which may affect the generalizability of the findings. Additionally, the retrospective nature of the study and the lack of an extended follow-up period limit the ability to comprehensively evaluate long-term outcomes and potential late complications. Furthermore, being a single-center study, the findings may reflect institution-specific factors, potentially limiting broader applicability.

7. Conclusions

By combining ICAP flaps with autologous fat grafting, surgeons can achieve natural and harmonious breast reconstructions, leveraging the advantages of both techniques to optimize aesthetic and functional outcomes. The overall complication rate in this study was low, supporting the safety profile of the procedure. Notably, no cases of tissue necrosis were observed, underscoring the reliability of the vascular pedicle and its capacity to sustain adequate perfusion, including support for the survival of the transferred autologous fat. Furthermore, complications were observed exclusively in the smoker group, highlighting a statistically significant correlation between smoking and adverse postoperative outcomes. These findings reinforce the role of ICAP flaps augmented with fat grafting as a reliable and safe option for immediate reconstruction following CBS and emphasize the importance of patient selection and preoperative counseling regarding modifiable risk factors such as smoking.

Author Contributions: Conceptualization, F.K.; Methodology, S.V.; Investigation, S.F.; Data curation, M.F.C. and R.C.; Writing—original draft, M.F.C.; Writing—review & editing, M.F.C., R.D.G. and V.V.; Supervision, A.V.E.L. and F.R.; Project administration, F.K., B.C. and M.K. All authors have read and agreed to the published version of the manuscript.

Funding: This research received no external funding.

Institutional Review Board Statement: The study was conducted in accordance with the Declaration of Helsinki.

Informed Consent Statement: Informed consent was obtained from all subjects involved in this study.

Data Availability Statement: The original contributions presented in this study are included in the article. Further inquiries can be directed to the corresponding author.

Conflicts of Interest: The authors declare no conflicts of interest.

References

1. Kwei, S.; Borud, L.J.; Lee, B.T. Mastopexy with autologous augmentation after massive weight loss: The intercostal artery perforator (ICAP) flap. *Ann. Plast. Surg.* **2006**, *57*, 517–523. [CrossRef] [PubMed]
2. Badran, H.A.; El-Helaly, M.S.; Safe, I. The lateral intercostal neurovascular free flap. *Plast. Reconstr. Surg.* **1984**, *73*, 17–26. [CrossRef] [PubMed]
3. Hamdi, M.; Van Landuyt, K.; de Frene, B.; Roche, N.; Blondeel, P.; Monstrey, S. The versatility of the intercostal artery perforator (ICAP) flaps. *J. Plast. Reconstr. Aesthet. Surg.* **2006**, *59*, 644–652. [CrossRef]
4. Tenna, S.; Brunetti, B.; Morelli Coppola, M.; Persichetti, P. The anterior intercostal artery perforator flap: Clinical applications in partial breast reconstruction. *Plast Reconstr Surg.* **2017**, *140*, 746e–747e. [CrossRef]
5. Bar-Meir, E.D.; Yueh, J.H.; Tobias, A.M.; Lee, B.T. Autologous fat grafting: A technique for stabilization of the microvascular pedicle in DIEP flap reconstruction. *Microsurgery* **2008**, *28*, 495–498. [CrossRef]
6. Heine, N.; Eigenberger, A.; Brebant, V.; Hoesl, V.; Brix, E.; Prantl, L.; Kempa, S. Comparison of skin sensitivity following breast reconstruction with three different techniques: Autologous fat grafting, DIEP flap, and expander/implant. *Clin. Hemorheol. Microcirc.* **2022**, *80*, 389–397. [CrossRef]
7. Laporta, R.; Longo, B.; Sorotos, M.; Pagnoni, M.; Santanelli di Pompeo, F. Breast reconstruction with delayed fat-graft-augmented DIEP flap in patients with insufficient donor-site volume. *Aesthetic Plast. Surg.* **2015**, *39*, 335–345. [CrossRef] [PubMed]
8. Angrigiani, C.; Rancati, A.O.; Masià, J.; Farhadi, J.; Khouri, K.; Acquaviva, J. Modified anterior intercostal artery perforator flap (AICAP) for autologous breast volume restoration after explantation. *J. Plast. Reconstr. Aesthet. Surg.* **2021**, *74*, 2916–2924. [CrossRef]
9. Egro, F.M.; Roy, E.; Rubin, J.P.; Coleman, S.R. Evolution of the Coleman technique. *Plast Reconstr Surg.* **2022**, *150*, 329e–336e. [CrossRef]
10. Regnault, P. Breast ptosis: Definition and treatment. *Clin. Plast. Surg.* **1976**, *3*, 193–203. [CrossRef] [PubMed]
11. Slavin, S.A.; Halperin, T. Reconstruction of the breast conservation deformity. *Semin Plast Surg.* **2004**, *18*, 89–94. [CrossRef]
12. Clough, K.B.; Thomas, S.S.; Fitoussi, A.D.; Couturaud, B.; Reyal, F.; Falcou, M.-C. Reconstruction after conservative treatment for breast cancer: Cosmetic sequelae classification revisited. *Plast Reconstr Surg.* **2004**, *114*, 1743–1749. [CrossRef] [PubMed]

13. Clough, K.B.; Kroll, S.S.; Audretsch, W. An approach to the repair of partial mastectomy defects. *Plast Reconstr Surg.* **1999**, *104*, 409–416. [CrossRef] [PubMed]
14. Esser, J. *Biological or Artery Flaps of the Face*; Institut Esser de Chirurgie Structive: Monaco, 1931.
15. Dibbell, D.G. Use of a long island flap to bring sensation to the sacral area in young paraplegics. *Plast. Reconstr. Surg.* **1974**, *54*, 220–223. [CrossRef]
16. Daniel, R.A. Intercostal neurovascular island skin flap. In *Grabb's Encyclopedia of Flaps*, 2nd ed.; Strauch, B., Vasconz, L., Hall-Findlay, E., Eds.; Lippincott-Raven: Philadelphia, PA, USA, 1998; Chapter 414; Volume III, p. 1632.
17. Daniel, R.K.; Williams, H.B. The free transfer of skin flaps by microvascular anastomoses: An experimental study and a reappraisal. *Plast. Reconstr. Surg.* **1973**, *52*, 16–31. [CrossRef] [PubMed]
18. Hamdi, M.; Spano, A.; Van Landuyt, K.; D'hErde, K.; Blondeel, P.; Monstrey, S. The lateral intercostal artery perforators: Anatomical study and clinical application in breast surgery. *Plast. Reconstr. Surg.* **2008**, *121*, 389–396. [CrossRef]
19. Munhoz, A.M.; Montag, E.; Arruda, E.; Brasil, J.A.; Aldrighi, J.M.; Gemperli, R.; Filassi, J.R.; Ferreira, M.C. Immediate conservative breast surgery reconstruction with perforator flaps: New challenges in the era of partial mastectomy reconstruction? *Breast* **2011**, *20*, 212–217. [CrossRef]
20. Rosatti, F.; Melita, D.; Marchica, P.; Musmarra, I.; Ciancio, F.; DE Lorenzi, F.; Toia, F.; LA Padula, S.; Lombardo, G.A. Local flaps for partial breast reconstruction. *Minerva Surg.* **2025**, *80*, 266–273. [CrossRef] [PubMed]
21. Hakakian, C.S.; Lockhart, R.A.; Kulber, D.A.; Aronowitz, J.A. Lateral intercostal artery perforator flap in breast reconstruction: A simplified pedicle permits an expanded role. *Ann Plast Surg.* **2016**, *76* (Suppl. S3), S184–S190. [CrossRef]
22. Carrasco-López, C.; Julian Ibañez, J.F.; Vilà, J.; Tomás, M.A.L.; López, J.N.; Miguel, I.P.; Fernandez-Llamazares-Rodriguez, J.; Higueras-Suñe, C. Anterior intercostal artery perforator flap in immediate breast reconstruction: Anatomical study and clinical application. *Microsurgery* **2017**, *37*, 603–610. [CrossRef]
23. Mangialardi, M.L.; Baldelli, I.; Salgarello, M.; Raposio, E.M. Breast reconstruction using the lateral thoracic, thoracodorsal, and intercostal arteries perforator flaps. *Plast. Reconstr. Surg. Glob. Open* **2021**, *9*, e3334. [CrossRef] [PubMed]
24. Noguchi, M.; Ohno, Y.; Morioka, E.; Noguchi, M.; Nakano, Y.; Kosaka, T.; Shimada, K. A novel oncoplastic technique for breast cancer localized in the lower pole of the breast. *J. Surg. Oncol.* **2018**, *117*, 137–141. [CrossRef] [PubMed]
25. Schaverien, M.V.; Kuerer, H.M.; Caudle, A.S.; Smith, B.D.; Hwang, R.F.; Robb, G.L. Outcomes of volume replacement oncoplastic breast-conserving surgery using chest wall perforator flaps: Comparison with volume displacement oncoplastic surgery and total breast reconstruction. *Plast. Reconstr. Surg.* **2020**, *146*, 14–27. [CrossRef]
26. Ho, W.; Stallard, S.; Doughty, J.; Mallon, E.; Romics, L. Oncological outcomes and complications after volume replacement oncoplastic breast conservations—The Glasgow experience. *Breast Cancer* **2016**, *10*, 223–228. [CrossRef]
27. Lipman, K.; Graw, G.; Nguyen, D. Lateral intercostal artery perforator (LICAP) flap for breast volume augmentation: Applications for oncoplastic and massive weight loss surgery. *JPRAS Open.* **2021**, *29*, 123–134. [CrossRef]
28. Losken, A.; Pinell, X.A.; Sikoro, K.; Yezhelyev, M.V.; Anderson, E.; Carlson, G.W. Autologous fat grafting in secondary breast reconstruction. *Ann. Plast. Surg.* **2011**, *66*, 518–522. [CrossRef] [PubMed]
29. Maione, L.; Caviggioli, F.; Vinci, V.; Lisa, A.; Barbera, F.; Siliprandi, M.; Battistini, A.; Klinger, F.; Klinger, M. Fat graft in composite breast augmentation with round implants: A new concept for breast reshaping. *Aesthetic Plast. Surg.* **2018**, *42*, 1465–1471. [CrossRef]
30. Kanchwala, S.K.; Glatt, B.S.; Conant, E.F.; Bucky, L.P. Autologous fat grafting to the reconstructed breast: The management of acquired contour deformities. *Plast. Reconstr. Surg.* **2009**, *124*, 409–418. [CrossRef]
31. Kaoutzanis, C.; Xin, M.; Ballard, T.N.S.; Welch, K.B.; Momoh, A.O.; Kozlow, J.H.; Brown, D.L.; Cederna, P.S.; Wilkins, E.G. Autologous fat grafting after breast reconstruction in postmastectomy patients: Complications, biopsy rates, and locoregional cancer recurrence rates. *Ann. Plast. Surg.* **2016**, *76*, 270–275. [CrossRef]
32. Klinger, M.; Losurdo, A.; Lisa, A.V.E.; Morenghi, E.; Vinci, V.; Corsi, F.; Albasini, S.; Leonardi, M.C.; Jereczek-Fossa, B.A.; Veronesi, P.; et al. Safety of autologous fat grafting in breast cancer: A multicenter Italian study among 17 Senonetwork breast units. *Breast Cancer Res. Treat.* **2022**, *191*, 355–363. [CrossRef]
33. Maione, L.; Vinci, V.; Caviggioli, F.; Klinger, F.; Banzatti, B.; Catania, B.; Lisa, A.; Klinger, M. Autologous fat graft in postmastectomy pain syndrome following breast-conservative surgery and radiotherapy. *Aesthetic Plast. Surg.* **2014**, *38*, 528–532. [CrossRef] [PubMed]
34. Demiri, E.C.; Tsimponis, A.; Pagkalos, A.; Georgiadou, E.; Goula, O.C.; Spyropoulou, G.A.; Dionyssiou, D. Fat-augmented latissimus dorsi versus deep inferior epigastric perforator flap: Comparative study in delayed autologous breast reconstruction. *J. Reconstr. Microsurg.* **2021**, *37*, 208–215. [CrossRef] [PubMed]

35. Blondeel, P.N.; Van Landuyt, K.; Monstrey, S.J.; Bozkurt, M.; Dap, L.; Neumann, G.; Mees, R.; Sverdlov, A.; Vandeweyer, E.; Van Laere, L.; et al. The DIEAP flap: A new standard in breast reconstruction? *Br. J. Plast. Surg.* **2003**, *56*, 619–630.
36. Melbourne Breast Cancer Surgery. Smoking and Breast Surgery Complications. Available online: https://www.melbournebreastcancersurgery.com.au/smoking-and-breast-surgery-complications.html (accessed on 20 January 2025).

life

MDPI

Article

Comparative Analyses of Van Nuys Prognostic Index and NCCN Guidelines in Ductal Carcinoma *In Situ* Treatment in a Brazilian Hospital

Marcelo Antonini [1,2,3,*], Raissa Barros Vasconcelos [1], André Mattar [3,4], Mariana Pollone Medeiros [1], Marina Diógenes Teixeira [4], Andressa Gonçalves Amorim [4], Odair Ferraro [1], Larissa Chrispim de Oliveira [4], Marcellus do Nascimento Moreira Ramos [4], Francisco Pimentel Cavalcante [5], Felipe Zerwes [6,7], Marcelo Madeira [8], Eduardo de Camargo Millen [9], Antonio Luiz Frasson [10], Fabricio Palermo Brenelli [11], Gil Facina [12], Henrique Lima Couto [13] and Luiz Henrique Gebrim [14]

1 Department of Breast Surgery, Hospital do Servidor Público Estadual Francisco Morato de Oliveira, Sao Paulo 04039-000, SP, Brazil; draraissa.mastologia@gmail.com (R.B.V.); marianapollone@gmail.com (M.P.M.); odairferraro@hotmail.com (O.F.)
2 Centro de Desenvolvimento de Ensino e Pesquisa do Instituo de Assistência Médica ao Servidor Público Estadual (CEDEP-IAMSPE), Sao Paulo 04039-000, SP, Brazil
3 BBREAST—Brazilian Breast Association Team, Sao Paulo 01258-011, SP, Brazil; mattar.andre@gmail.com
4 Department of Breast Surgery, Centro de Referência da Saúde da Mulher—Hospital da Mulher, Sao Paulo 01206-001, SP, Brazil; mari_diogenes@hotmail.com (M.D.T.); andressaamorim88@hotmail.com (A.G.A.); chrispiml@hotmail.com (L.C.d.O.); marcellusnmr@hotmail.com (M.d.N.M.R.)
5 Department of Breast Surgery, Hospital Geral de Fortaleza, Fortaleza 60150-160, CE, Brazil; fpimentelcavalcante@gmail.com
6 Department of Breast Surgery, Medical School of the Pontifícia Universidade Católica do Rio Grande do Sul (PUCRS), Porto Alegre 90619-900, RS, Brazil; zerwes@hotmail.com
7 Department of Breast Surgery, Oncoclínicas, Porto Alegre 90570-020, RS, Brazil
8 Department of Breast Surgery, Faculdade Israelita de Ciências da Saúde Albert Einstein, Sao Paulo 05653-000, SP, Brazil; marcemadeira@gmail.com
9 Department of Breast Surgery, Americas Oncologia, Rio de Janeiro 22793-080, RJ, Brazil; eduardomillen@gmail.com
10 Department of Breast Surgery, Hospital Albert Einstein, Sao Paulo 05651-901, SP, Brazil; alfrasson.af@gmail.com
11 Department of Breast Surgery, Universidade Estadual de Campinas (UNICAMP), Campinas 13082-859, SP, Brazil; fabriciobrenelli@hotmail.com
12 Department of Gynecology, Universidade Federal de Sao Paulo, Sao Paulo 04023-062, SP, Brazil; facina@unifesp.br
13 Department of Breast Surgery, REDIMAMA-REDIMASTO, Belo Horizonte 30110-022, MG, Brazil; enriquecouto@hotmail.com
14 Department of Breast Surgery, Hospital Beneficiencia Portuguesa, Sao Paulo 01438-000, SP, Brazil; lgebrim1964@gmail.com
* Correspondence: drantonini@uol.com.br; Tel.: +55-11-4573-8105

Abstract: Background: Ductal carcinoma *in situ* (DCIS) is a precursor of invasive breast cancer and its early diagnosis and treatment are essential to prevent progression and recurrences. Risk stratification guidelines, such as the Van Nuys Prognostic Index (VNPI) and those by the National Comprehensive Cancer Network (NCCN), help guide appropriate treatment. This study compares VNPI recommendations for DCIS patients treated at Hospital do Servidor Público Estadual de São Paulo (HSPE) with NCCN guidelines, focusing on treatment conducted and recurrence rates. Methods: This retrospective, cross-sectional study reviewed medical records of 145 patients treated for DCIS at HSPE between January 1996 and June 2022, with a mean follow-up of 60.3 months. Results: Based on VNPI, 38.8% were low risk, 53.2% intermediate risk, and 7.8% high risk. NCCN guidelines classified only 12.9% as low risk and 87.1% as high risk. Treatment included breast-conserving surgery (BCS) with radiotherapy (43.1%), BCS alone (38.8%), and mastectomy (18.1%). There were

18 recurrences (15.5%): 5.2% as DCIS and 10.3% as invasive cancer. Of these recurrences, 5.6% occurred in patients who, according to NCCN, would have received BCS with radiotherapy or mastectomy. Conclusion: By integrating the VNPI with NCCN treatment guidelines, the NCCN's recommendations could potentially reduce local recurrence rates by 5.6%. However, further studies are necessary to evaluate the long-term impact of these guidelines on overall survival outcomes.

Keywords: ductal carcinoma *in situ*; invasive breast cancer; DCIS treatment; Van Nuys Prognostic Index; National Comprehensive Cancer Network

1. Introduction

Breast cancer is the most common malignant neoplasia among women in Brazil and worldwide, excluding non-melanoma skin cancer. An estimated 73,610 new cases are expected in Brazil in 2025, including cases of invasive carcinoma (IBC) and ductal carcinoma *in situ* (DCIS) [1]. DCIS is considered a precursor lesion for IBC, currently accounting for over 20% of new breast cancer diagnoses in the United States, with more than 60,000 cases diagnosed annually [2].

DCIS is a breast neoplasia originating from a terminal duct-lobular unit, characterized by the proliferation of ductal epithelial cells that grow in different architectural patterns [3,4]. This growth is confined to the duct-lobular unit, without extending beyond the myoepithelial layer's basement membrane, and therefore does not lead to stromal invasion [2]. The incidence of DCIS has been rising, primarily due to increased access to mammographic screening [2,5].

DCIS is a heterogeneous entity in terms of presentation, morphology, biomarker expression, genetic alterations, and natural history [6,7]. Most cases are asymptomatic or present with small palpable nodules. Radiologically, 75% of lesions appear as calcifications, and some may go undetected by mammography. Additionally, the area of microcalcifications on imaging may underestimate the true lesion size, which can result in incomplete excision in breast-conserving surgeries [2,8].

Recommended treatment options include mastectomy with sentinel lymph node biopsy, breast-conserving surgery alone or followed by radiotherapy, with the potential addition of hormonal therapy with Tamoxifen or Aromatase Inhibitors in postmenopausal patients for DCIS cases that express hormone receptors [2,9].

Although DCIS is not an invasive disease, its potential for progression underscores the critical importance of timely diagnosis and appropriate treatment [7]. However, many of these lesions have an indolent behavior and may not progress to invasive breast carcinoma (IBC) or may take years to do so [2]. Small retrospective studies have shown that 30% of DCIS cases may progress to IBC over a period of up to 30 years [7]. Thus, predicting the progression of these lesions to invasive disease remains a challenge, as does planning proportionate treatment to avoid overtreatment [5].

Despite its favorable prognosis and lack of impact on life expectancy, a DCIS diagnosis can result in psychological stress [5,6] and its treatment may lead to procedure-related morbidities, work and social activity limitations, as well as changes in self-image [5,10]. To address these concerns, risk calculators and guidelines have been developed as tools for risk estimation and to aid in decision-making following a DCIS diagnosis [11].

In 1996, Silverstein introduced the Van Nuys Prognostic Index (VNPI), which was later updated in 2003, to evaluate prognosis and guide the management of DCIS. The VNPI stratifies the risk of ipsilateral DCIS recurrence by assessing lesion size, margin distance in the surgical specimen, nuclear grade, presence of necrosis, and patient age [2]. Designed to support decision-making, it helps determine the most suitable approach—whether radical surgery, breast-conserving surgery alone, or combined with radiotherapy—while aiming to achieve a local recurrence rate below 20% over 12 years [12,13]. Each variable is assigned a score from 1 to 3. Patients scoring 4–6 are considered low risk, those scoring 7–9 as intermediate risk, and those scoring 10–12 as high risk [12,14].

According to the VNPI study, low-risk patients could undergo wide excision without adjuvant radiotherapy. Intermediate-risk patients would benefit from the inclusion of adjuvant radiotherapy to reduce recurrence rates. Finally, high-risk patients would be candidates for mastectomy, as they exhibited higher recurrence rates even with radiotherapy [2,14].

The National Comprehensive Cancer Network (NCCN) is a nonprofit alliance of U.S. cancer centers that periodically releases guidelines for the diagnosis, treatment, and follow-up of various cancers. NCCN-recommended treatment options for DCIS include mastectomy with sentinel lymph node biopsy, conservative surgery with adjuvant radiotherapy, or isolated conservative surgery with possible addition of Tamoxifen and Aromatase Inhibitors in postmenopausal women with hormone receptor-positive DCIS [2,9].

The NCCN on its 2024 version suggests that only selected patients with low-risk DCIS could be treated with surgery followed by partial breast radiation or even omission of adjuvant radiotherapy, provided they meet the following criteria: detection through screening, low or intermediate nuclear grade, tumor size less than 2.5 cm, and a surgical margin clearance of at least 3 mm [9].

The application of the VNPI and other guidelines varies across different countries and institutions, underscoring the importance of assessing each calculator's effectiveness [11]. Several studies have evaluated the predictive value of VNPI and NCCN criteria in DCIS treatment [11–13]. However, most focus on isolated guideline recommendations rather than a direct comparison between them. Additionally, there are limited data from non-Western populations, such as the Brazilian cohort in our study. This study aims to address this gap by comparing recurrence rates and treatment strategies between VNPI and NCCN classifications.

2. Materials and Methods

2.1. Study Design

We conducted a retrospective observational cross-sectional study by reviewing medical records from the Breast Surgery Department of the Hospital do Servidor Público Estadual (HSPE), Brazil, for patients who underwent surgical treatment for breast cancer (BC) between January 1996 and June 2022.

2.2. Inclusion and Exclusion Criteria

Female patients over 18 years of age, diagnosed with ductal carcinoma in situ DCIS, and treated according to the Van Nuys Prognostic Index (VNPI) were included. We excluded all patients with invasive ductal carcinoma, patients with metastasis, and patients without follow up or incomplete data.

Incomplete data were excluded from the study if critical clinical and pathological variables were missing, including tumor size, nuclear grade, margin status, or recurrence data. However, patients with missing secondary data, such as hormone receptor status or HER-2 expression, were retained in the analysis and the missing values were reported in the results.

2.3. Data Source

The medical records used in this study were retrieved from an institutional database containing all patients who underwent surgical treatment, stored in an Excel spreadsheet. Data extraction was performed manually and cross-checked with individual patient records to ensure accuracy and completeness.

Limitations of retrospective studies, including the possibility of missing data and information bias. To mitigate these issues, we cross-checked data across multiple sources, including electronic and physical medical records, to ensure data accuracy and completeness.

The study included patients diagnosed and treated for ductal carcinoma in situ (DCIS) at the Hospital do Servidor Público Estadual de São Paulo (HSPE) between January 1996 and June 2022. Given the extended time frame of the study, we accounted for changes in clinical practice by stratifying data analysis based on treatment era, ensuring consistency in follow-up and medical record documentation.

2.4. Van Nuys Prognostic Index and Treatment

The VNPI is presented in the Table 1.

Table 1. The Van Nuys Prognostic Index scoring system.

Feature	Score 1	Score 2	Score 3
Size (mm)	≤15	16–40	>40
Margins (mm)	≥10	1–9	<1
Grade Necrosis	Low or intermediate Without necrosis	Intermediate With necrosis	High grade With/without necrosis
Age (years)	>60	40–60	<40

Source: GORRINGE et al. [4].

According to their classification under the VNPI, patients underwent one of three treatments: breast-conserving surgery (BCS) alone, breast-conserving surgery with adjuvant radiotherapy (RT), or mastectomy. Recurrence rates were calculated based on the treatment performed. Next, recurrences were assessed in the low-risk and high-risk groups of patients if classified according to NCCN recommendations.

2.5. Follow-Up Duration

Follow-up duration was standardized across patients, with clinical evaluations every 6 months for the first 5 years and annually thereafter. Imaging follow-up was performed per institutional guidelines. Any variations in post-treatment monitoring were accounted for in the statistical analysis to minimize bias.

2.6. Statistical Analysis

Data recording and statistical analysis were performed using Excel spreadsheets and SPSS 22.0 for Windows. Initially, all variables were analyzed descriptively. For quantitative variables, this analysis involved observing minimum values, and the comparison of means between two groups was carried out using Student's t-test. When the normality assumption was rejected, the non-parametric Mann–Whitney test was used.

To test homogeneity between proportions, the Chi-square test or Fisher's exact test (when expected frequencies were less than 5) was applied. Calculations were conducted using SPSS 22.0 for Windows, with a 5% significance level for the tests.

Multivariate logistic regression analysis was performed to identify independent predictors of recurrence, adjusting for age, nuclear grade, tumor size, margin status, and treatment type. Bonferroni correction was not applied, as multiple hypothesis testing was limited to predefined primary endpoints.

2.7. Ethics

The study was submitted to and approved by the Research Ethics Committee of the Hospital do Servidor Público Estadual-Francisco Morato de Oliveira and registered on the Brazil Platform with approval number CAAE: 60175722.6.0000.5463. The need for informed consent was waived due to the retrospective nature of the study and the confidentiality and integrity of the medical record data were preserved.

3. Results

From January 1996 to June 2022, 3.216 patients were diagnosed with breast cancer at HSPE, of which 394 (12.5%) had metastatic disease, 2.967 (83.9%) had stage I to III breast cancer, and 145 (3.6%) had ductal carcinoma in situ (DCIS). Among the DCIS patients, 29 patients were excluded due to missing critical data. Specifically, 12 cases lacked tumor size information, 8 cases had missing margin status, 5 cases had missing nuclear grade, and 4 cases had no recurrence data available. We can see those data at Figure 1.

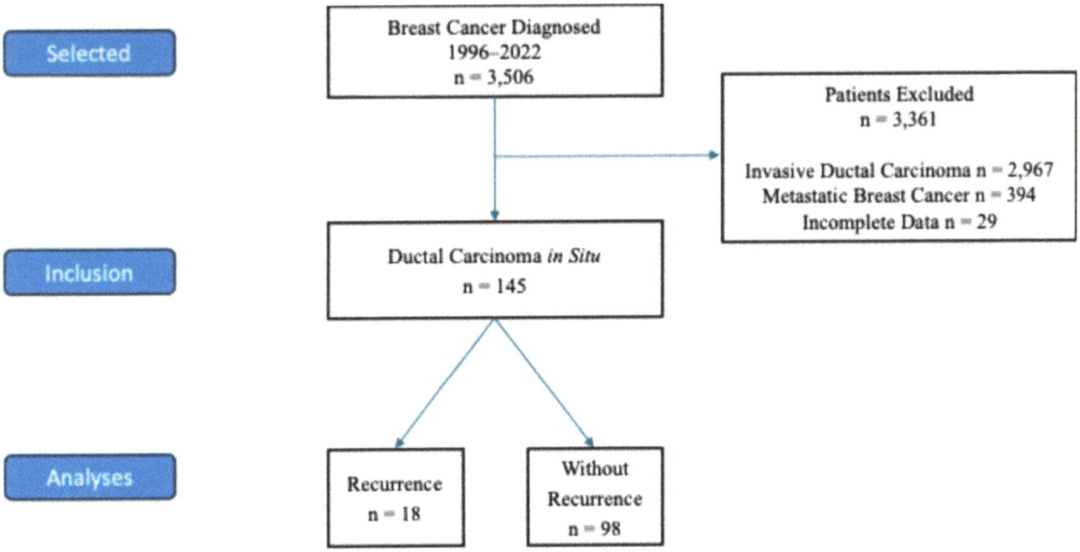

Figure 1. Flowchart of included patients with DCIS.

3.1. Epidemiological Profile

The mean age of the 116 included patients was 60.1 years, with a standard deviation of 11.0 years. Among them, 26 (22.4%) were premenopausal and 90 (77.6%) were post-menopausal ($p < 0.001$). Regarding family history of breast cancer in first-degree relatives, only 16 (13.8%) had a family history, while 100 (86.2%) did not ($p < 0.001$).

The diagnosis was made through mammographic screening in 88 patients (75.9%) and clinically in 28 patients (24.1%) ($p < 0.001$). The histological type was ductal carcinoma in situ, with comedonecrosis present in 70 patients (60.3%) and high-grade lesions in 45 patients (38.8%) ($p < 0.05$).

The size of the lesions found was up to 1.5 cm in 64 patients (39.7%), and margins were larger than 10 mm in 44 patients (41.5%), with lesion size showing a significant difference ($p = 0.006$) and margins showing no significant difference ($p = 0.269$).

Immunohistochemistry was unknown in 64 patients (53.2%). Among those for whom it was known, 44 patients (37.9%) had positive hormone receptors, and 8 (6.9%) had negative hormone receptors, with a significant difference ($p < 0.001$). For HER-2 positivity, eight patients (6.9%) were positive, showing a significant difference ($p = 0.008$).

The type of treatment was breast-conserving surgery in 45 patients (38.8%), breast-conserving surgery with adjuvant radiotherapy in 50 (43.1%), and mastectomy in 21 patients (18.1%), with a significant difference ($p < 0.001$). Due to the extended follow-up period, there was a time when sentinel lymph node biopsy (SLNB) was not performed on DCIS patients undergoing mastectomy. Thus, SLNB was performed in 15 patients: 10 (47.6%) who underwent mastectomy and 5 (5.2%) who underwent breast-conserving surgery, with a significant difference ($p < 0.001$).

The mean follow-up time for the patients was 60.3 months, with a standard deviation of 47.4 months, a maximum of 203 months, and a minimum of 6 months.

Table 2 shows the distributions of the quantitative variables and Table 3 displays the qualitative variables of the epidemiological profile.

Table 2. Descriptive statistics of quantitative variables in the study population.

Variable	Mean	Median	Standard Deviation	CV	Min	Max	n	95% CI
Age (years)	60,1	60	11,0	18%	32	91	116	58.1–62.10
Follow-up (months)	60,3	42	47,4	79%	6	203	116	51.67–68.93
Time to Recurrence (months)	20,9	14	17,6	84%	2	73	18	12.77–29.03

Legend: The confidence interval (CI) values refer to the respective variables of age, follow-up time, and time to recurrence. CV: Coefficient of Variation; Min: minimum/; Max: Maximum; CI: confidence interval.

Table 3. Clinical and pathological characteristics.

Variable		n	%	p-Value
Menopause	No	26	22.4%	<0.001
	Yes	90	77.6%	
Family History	No	100	86.2%	<0.001
	Yes	16	13.8%	
Diagnosis	Clinical	28	24.1%	<0.001
	Screening	88	75.9%	
Nuclear Grade	Low	31	26.7%	0.050
	Intermediate	40	34.5%	
	High	45	38.8%	
Comedonecrosis	No	46	39.7%	0.002
	Yes	70	60.3%	
Size (cm)	<1.5 cm	64	39.7%	0.006
	1.6–4 cm	43	25.9%	
	>4 cm	9	6.9%	
Margins (mm)	<1 mm	36	29.2%	0.269
	1–9 mm	36	29.2%	
	>10 mm	44	41.6%	

Table 3. *Cont.*

Variable		n	%	p-Value
Hormonal Receptor	Unknown	64	55.2%	0.008
	Positive	44	37.9%	
	Negative	8	6.9%	
HER-2 Expression	Unknown	64	55.2%	<0.001
	Positive	43	38.7%	
	Negative	8	6.9%	
Treatment	BCS e RT	50	43.1%	<0.001
	BCS	45	38.8%	
	Mastectomy	21	18.1%	
Sentinel Lymph Node Biopsy	Mastectomy	10	47.6%	<0.001
	BCS	5	5.2%	
Endocrine Therapy	No	104	89.7%	<0.001
	Yes	12	10.3%	
Recurrence	No	98	84.5%	<0.001
	Yes	18	15.5%	
Recurrence-DCIS	No	110	94.8%	<0.001
	Yes	6	5.2%	
Recurrence-IBC	No	104	89.7%	<0.001
	Yes	12	10.3%	

Original database//BCS, breast-conserving surgery; RT, radiotherapy; DCIS, ductal carcinoma in situ; IBC, invasive breast cancer.

3.2. Recurrence

During the follow-up, 18 (15.5%) recurrences were observed, with 6 (5.2%) due to DCIS and 12 (10.3%) due to IBC, showing a significant difference ($p < 0.001$). The mean time to recurrence was 20.9 months, with a standard deviation of 17.6 months, and a maximum time of 73 months and a minimum of 2 months.

A log-rank test was performed to compare recurrence-free survival between invasive and in situ recurrences. The results showed no statistically significant difference between the two survival curves ($\chi^2 = 0.0$, $p = 1.0$). This indicates that the recurrence rates over time for both invasive and in situ groups were similar in this cohort.

A Cox proportional hazards model was applied to compare the recurrence-free survival between invasive and in situ recurrences. The results showed that in situ recurrences had a higher risk over time (HR = 1.87; 95% CI: 0.69–5.12; $p = 0.220$). However, this difference was not statistically significant, indicating that both types of recurrence may follow a similar time-to-event pattern in this cohort. Figures 2 and 3 we demonstrate the in situ and invasive recurrence curves.

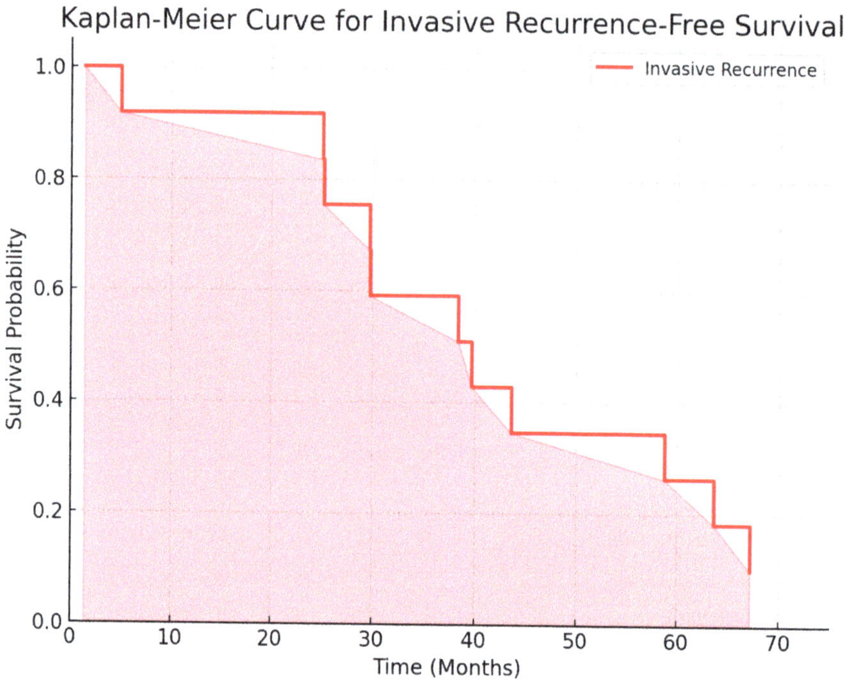

Figure 2. Kaplan–Meier curve for invasive recurrence-free survival.

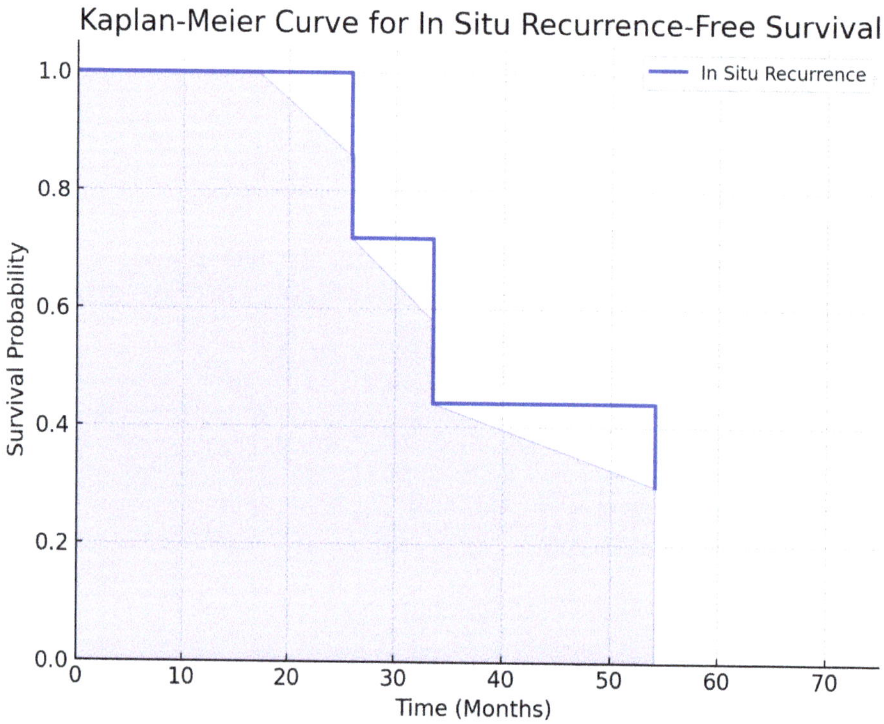

Figure 3. Kaplan–Meier curve for *in situ* recurrence-free survival.

3.3. Evaluation of Recurrences According to IPVN and NCCN Guidelines

The patients classified as low risk according to the IPVN were 45 (38.8%), intermediate risk 62 (53.2%), and high risk 9 (7.8%). The patients classified as low risk according to the NCCN were 15 (12.9%) and high risk 101 (87.1%). In Table 4 we find the distribution according to VNPI classification, treatment, and NCCN classification.

Table 4. Distribution according to VNPI classification, treatment, and NCCN classification.

Classification	Treatment	n	%	*p*-Value
VNPI Low		45	38.8%	<0.001
VNPI Intermediate		62	53.4%	0.294
VNPI High		9	7.8%	<0.001
	BCS e RT	50	43.1%	<0.001
	BCS	45	38.8%	
	Mastectomy	21	18.1%	
NCCN High Risk		101	87.1%	<0.001
NCCN Low Risk		15	12.9%	<0.001

Original database//BCS, breast-conserving surgery; RT, radiotherapy.

The recurrences according to the treatment approach guided by the IPVN were 10 (55.6%) among patients who underwent isolated conservative surgery, 7 (38.9%) among those who underwent surgery followed by radiotherapy, and 1 (5.6%) among those who underwent mastectomy.

The VNPI classified 7.8% of patients as high risk, whereas the NCCN guidelines classified 87.1% in this category. The recurrence rate among patients classified as low or intermediate risk by VNPI was 55.6%, whereas it was 16.7% for patients classified as high risk under NCCN criteria. These findings suggest that VNPI underestimates recurrence risk in some patients, leading to potential undertreatment.

Among recurrences, 17 (94, 4%) patients were treated in accordance with NCCN classification and 1 (5, 6%) was not treated according to NCCN recommendations. In Table 5 we find the recurrence according to NCCN- and VNPI-based treatments.

Table 5. Recurrence according to NCCN- and VNPI-based treatments.

Guideline-Based Treatment		Recurrence		No Recurrence		Total		*p*-Value
		n	%	n	%	n	%	
NCCN	Yes	17	94.4%	84	85.7%	101	87.1%	0.310
	No	1	5.6%	14	14.3%	15	12.9%	
VNPI	BCS	10	55.6%	35	35.7%	45	38.8%	0.175
	Mastectomy	1	5.6%	20	20.4%	21	18.1%	
	BCS and RT	7	38.9%	43	43.9%	50	43.1%	

Original database//BCS-RT, breast-conserving surgery with radiotherapy; MT, mastectomy; BCS, breast-conserving surgery.

Among patients classified as NCCN high risk, 94.4% (17/18) experienced recurrence, compared to 85.7% (84/98) in the NCCN low-risk group (RR = 2.09, 95% CI: 0.81–5.37, $p = 0.156$). The NCCN low-risk subgroup exhibited a recurrence rate of 5.6% (1/18) versus 14.3% (14/98) in the non-recurrent group, with an RR of 0.83 (95% CI: 0.10–6.56, $p = 0.999$).

Regarding VNPI subgroups, the low VNPI group had a recurrence rate of 55.6% (10/18), with a relative risk of 2.76 (95% CI: 1.01–7.51, $p = 0.049$), compared to the intermediate VNPI subgroup, which showed a recurrence rate of 27.7% (5/18) and an RR of 0.36 (95% CI: 0.13–0.99, $p = 0.049$). The high VNPI group demonstrated the highest recurrence rate (16.7%, 3/18) with an RR of 4.13 (95% CI: 1.19–14.41, $p = 0.058$). In Table 6 we find the correlation of recurrence and VNPI and NCC subgroups.

Table 6. Correlation of recurrence and VNPI and NCC subgroups.

Guideline Subgroups		Recurrence		No Recurrence		RR	95% CI	*p*-Value
		n	%	n	%			
NCCN	High	17	94.4%	84	85.7%	2.09	0.81–5.37	0.156
	Low	1	5.6%	14	14.3%	0.83	0.10–6.56	0.999
VNPI	Low	10	55.6%	35	35.7%	2.76	1.01–7.51	0.049
	Intermediate	5	27.7%	57	58.1%	0.36	0.13–0.99	0.049
	High	3	16.7%	6	6.2%	4.13	1.19–14.41	0.058

Original database//RR, relative risk; CI, confidence interval. The reference group for the relative risk (RR) analysis in the VNPI classification is the intermediate VNPI group. The RR values compare the risk of recurrence in the low VNPI and high VNPI groups relative to the intermediate VNPI group.

These findings suggest that higher VNPI and NCCN risk groups are associated with an increased recurrence risk, whereas the intermediate VNPI subgroup presents a lower risk of recurrence. However, statistical significance was observed only for the VNPI low ($p = 0.049$) and intermediate subgroups ($p = 0.049$), indicating potential clinical relevance.

3.4. Multivariate Logistic Regression

A multivariate logistic regression analysis was conducted to identify independent predictors of recurrence in DCIS patients. The results showed that age was significantly associated with an increased risk of recurrence (OR = 1.06; 95% CI: 1.00–1.12; $p = 0.044$), indicating a progressive rise in risk with advancing age. Margin status also showed a trend toward association with recurrence (OR = 1.55; 95% CI: 0.50–4.77), although it did not reach statistical significance ($p = 0.443$).

Nuclear grade (OR = 1.16; 95% CI: 0.55–2.44; $p = 0.701$) and tumor size (OR = 0.78; 95% CI: 0.44–1.39; $p = 0.401$) were not identified as independent risk factors. Treatment type, comparing breast-conserving surgery with mastectomy, showed a trend toward a lower risk with mastectomy (OR = 0.62; 95% CI: 0.18–2.13), but this result was not statistically significant ($p = 0.447$).

These findings on Table 7 suggest that age may be a relevant prognostic factor for recurrence, while the other analyzed variables did not demonstrate a statistically significant association when adjusted for competing factors.

Table 7. Multivariate logistic regression analysis of recurrence predictors in DCIS patients.

Variable	Odds Ratio (OR)	95% CI Lower	95% CI Upper	*p*-Value
Age (years)	1.06	1.0	1.12	0.044
Nuclear Grade	1.16	0.55	2.44	0.701
Tumor Size (cm)	0.78	0.44	1.39	0.401
Margin Status	1.55	0.5	4.77	0.443
Treatment	0.62	0.18	2.13	0.447

Legend: Multivariate logistic regression model evaluating independent predictors of recurrence in patients with ductal carcinoma *in situ* (DCIS). The table presents odds ratios (OR) with 95% confidence intervals (CI) and *p*-values for each variable included in the model. Statistically significant values ($p < 0.05$) are highlighted in bold.

4. Discussions

Several studies have aimed to characterize DCIS and IBC at the molecular level, revealing genetic similarities and likely common origins between the two conditions [2]. Research shows that approximately 30% of untreated DCIS cases will progress to IBC (about 30% to low- and intermediate-grade invasive lesions and around 60% to high-grade lesions within 5–20 years) [15]. However, estimating the progression risk of DCIS remains challenging due to the current standard of care, which generally involves excision with clear margins, radiotherapy, and/or adjuvant endocrine therapy when appropriate [2].

Randomized studies have demonstrated that adding radiotherapy to conservative surgery reduces local recurrence of DCIS, although the impact on overall survival or breast cancer-specific survival remains unclear. The primary role of radiotherapy in this context is to reduce the risk of invasive recurrences, which are associated with increased mortality [2]. While radiotherapy after conservative surgery has been shown to significantly reduce ipsilateral recurrence rates, adjuvant hormonal therapy reduces both ipsilateral and contralateral recurrence rates [16].

DCIS treatment requires a careful balance between preventing progression to invasive disease, avoiding recurrence, and minimizing potential overtreatment effects such as acute or chronic pain, fatigue, sensory disturbances, and other side effects, especially in patients with low-risk lesions [5].

In our study, 38.8% of patients were treated with isolated conservative surgery, and 43.1% underwent surgery combined with radiotherapy. Mastectomy was performed in 18.1% of the sample.

Several phase III trials are ongoing to assess the risk and benefit of active surveillance for low-risk DCIS, with main endpoints including ipsilateral invasive disease-free survival (LORIS), ipsilateral invasive breast cancer-free rate at 2 years (COMET), 5 years (LORETTA), and 10 years (LORD). Due to the lack of consensus on defining low-risk DCIS, these trials use various criteria, including age, tumor grade, lesion size, and other pathological features [15]. On the other hand, it is essential to consider that adopting active surveillance as a DCIS management alternative may lead to anxiety, psychological, and functional disorders in patients, as there is an awareness and fear of disease progression when left untreated [5,17–20].

Multiple factors may contribute to increased recurrence risk in DCIS, with complete lesion excision being the most critical [21]. Clear surgical margins (typically around 2 mm) are associated with reduced recurrence rates, as are low-grade tumors, absence of necrosis, lesion size less than 4 cm, and absence of clinical signs (such as palpable lesions or nipple discharge) [16,21]. These and other factors are used by some scores and guidelines to stratify the post-treatment recurrence risk.

The primary aim of this stratification is to identify low-risk patients who could benefit from less aggressive treatments.

The main guidelines and prognostic indices for DCIS include the Memorial Sloan Kettering Cancer Center Nomogram (MSKCC), University of Southern California Van Nuys Prognostic Index (VNPI), Oncotype DX DCIS, and the National Comprehensive Cancer Network (NCCN) prognostic index, each with limitations and none yet recognized as the optimal tool [11].

The VNPI determines DCIS recurrence risk and classifies patients based on age, tumor size, surgical margins, nuclear grade, and necrosis presence or absence. Low-risk patients are candidates for isolated conservative surgery, intermediate-risk patients for conservative surgery plus radiotherapy, and high-risk patients for mastectomy. The application of the VNPI varies significantly across countries and institutions; for example, over 70% of DCIS cases in Australia and New Zealand are managed according to the VNPI recommendations, compared to less than 16% in the United Kingdom [11]. However, the VNPI has limitations, including criteria changes over the years and the lack of external, independent validation [12,14,21].

VNPI places less emphasis on margin width and nuclear grade than NCCN, which may lead to under-classification of high-risk patients. For example, a patient with small margins (<2 mm) and high nuclear grade might still be categorized as intermediate risk under VNPI, while NCCN would classify them as high risk, leading to different treatment recommendations.

In turn, according to the 2024 NCCN version, whole-breast radiotherapy after conservative surgery reduces tumor recurrence rates by 50–70%. Only selected low-risk patients, particularly those eligible for adjuvant endocrine therapy, may be treated with isolated lesion excision, omitting radiotherapy. For low-risk classification, specific criteria must be met, such as routine screening detection (no clinical signs), low or intermediate nuclear grade, tumor size less than 2.5 cm, and margins greater than 3 mm [9].

A key difference between the VNPI and NCCN guidelines lies in their weighting of prognostic factors. The VNPI assigns numerical scores based on margin size, nuclear grade, tumor size, and patient age, categorizing patients into low, intermediate, or high-risk groups. NCCN guidelines, in contrast, rely more heavily on radiotherapy recommendations and classify nearly all cases as high risk unless strict low-risk criteria are met (e.g., margins >3 mm, tumor <2.5 cm). These differing approaches may contribute to under-classification of high-risk cases by VNPI, potentially leading to increased recurrence rates.

Our study demonstrates a significant difference in risk classification between VNPI and NCCN guidelines. While VNPI classified only 7.8% of patients as high risk, the NCCN guidelines categorized 87.1% in this group. This discrepancy is reflected in recurrence outcomes, where patients classified as low or intermediate risk by VNPI had a recurrence rate of 55.6%, while those classified as high risk under NCCN had a recurrence rate of 16.7%. This highlights the potential for VNPI to underestimate recurrence risk, possibly leading to suboptimal treatment recommendations compared to NCCN guidelines.

In our study sample, patients treated at HSPE were classified according to the VNPI, with 38.8% classified as low risk and thus managed with conservative surgery without adjuvant radiotherapy. In contrast, if classified according to NCCN criteria, only 12.9% would have been considered low risk and treated with surgery alone.

The impact of radiotherapy on DCIS treatment has been assessed in several studies. A systematic review and meta-analysis by Chen Q. et al. compared long-term outcomes of different treatment modalities for low-risk DCIS patients. It was concluded that even in low-risk patients, conservative surgery with adjuvant radiotherapy was associated with a lower disease progression risk and survival benefit compared to surgery alone [15].

In the present study, 5.6% of recurrences occurred among patients who underwent mastectomy, 38.8% among those treated with conservative surgery followed by radiotherapy, and 55.5% among those treated with isolated surgery, demonstrating worse outcomes in this low-risk VNPI classified group.

A systematic review by Goodwin et al. evaluated 3925 women with DCIS and observed ipsilateral and contralateral recurrence rates of 1–2%, lower than in our population, but similarly demonstrated worse outcomes for those treated with isolated conservative surgery. Patients who underwent mastectomy had recurrence rates of 0.5–1% per year in the contralateral breast, while those with conservative surgery had 5-year recurrence rates of 16%, reduced to 8% with radiotherapy addition [21].

The present data indicate that the VNPI was not an adequate stratification tool for our population, potentially underestimating risk for many patients. If classified by NCCN criteria, most would have been classified as high risk and managed with adjuvant radiotherapy or mastectomy. Among recurrences, 5.6% patients were under classified by VNPI and were not treated according to NCCN recommendations.

Although the NCCN guidelines' emphasis on radiotherapy may reduce recurrence rates, it is important to consider potential long-term treatment-related morbidity, including radiation-induced fibrosis, lymphedema, and secondary malignancies. Future studies should evaluate whether the benefits of reducing recurrence outweigh these risks, particularly in low-risk patients.

In accordance, a retrospective study by Kunkiel and Niwinska evaluated VNPI's prognostic value in 525 DCIS patients initially treated with lesion excision in a public hospital in Poland. Based on histopathological features of the specimen and VNPI risk, patients were triaged for observation, adjuvant radiotherapy, or mastectomy. After evaluating 5-, 10-, and 15-year recurrence-free survival rates, it was observed that individual risk groups did not differ significantly and VNPI was not an optimal tool for DCIS patients [11].

Throughout the study period (1996–2022), significant advancements in breast cancer management, particularly in ductal carcinoma in situ (DCIS) treatment, have been implemented. These include refinements in surgical techniques, radiation protocols, and systemic therapies, as well as improved imaging modalities and risk stratification tools. Given these changes, maintaining consistency in follow-up and outcome assessment presented a methodological challenge.

To address this, we stratified our analysis based on treatment era, ensuring that variations in clinical practice were accounted for. This stratification allowed us to compare recurrence rates within distinct periods, reducing bias introduced by evolving treatment standards. Additionally, we considered the impact of changes in diagnostic and therapeutic approaches on recurrence rates, acknowledging that newer strategies may lead to different clinical outcomes.

Despite these precautions, we recognize that our findings are influenced by institutional treatment guidelines and resource availability over time. Therefore, the conclusions drawn from this study should be interpreted in the context of our specific patient population and may not be directly generalizable to other healthcare settings.

Our study has some limitations inherent to observational designs, including potential inaccuracies in registries and inference limitations. Additionally, missing data due to loss to follow-up may bias results, particularly if related to the outcome.

Future perspectives in the treatment of DCIS increasingly focus on the role of genomic assays to refine patient stratification and personalize therapy. Tools such as the Oncotype DX DCIS Score [22] and the DCISionRT Decision Score [23] offer promising avenues to identify patients at higher or lower risk of local recurrence after breast-conserving surgery, thus guiding decisions on adjuvant radiotherapy and endocrine therapy. While these assays have shown potential in improving risk stratification beyond traditional clinicopathological factors, cost and accessibility remain significant barriers to widespread adoption, particularly in resource-limited settings. Furthermore, the predictive accuracy of these assays requires ongoing external validation in diverse patient populations to enhance clinical utility. Notably, prospective trials like the ELISA study are incorporating the Oncotype DX DCIS Score to define low-risk populations for whom radiotherapy may be safely omitted, marking a critical step forward in de-escalation strategies [24].

5. Conclusions

Our study provides an important comparative analysis of the VNPI and NCCN guidelines for the treatment of DCIS in a Brazilian cohort. The findings suggest that NCCN's treatment recommendations may be associated with lower local recurrence rates when compared to VNPI-based decisions. However, our findings also highlight potential over-treatment concerns, given NCCN's greater emphasis on radiotherapy. Kaplan–Meier survival analysis further supports these findings, showing no statistically significant difference in recurrence-free survival between invasive and in situ recurrences. Future prospective studies should further explore optimal risk stratification and treatment de-escalation strategies.

However, we recognize the limitations inherent to our retrospective study, particularly regarding the extended follow-up period and potential variations in clinical practices over time.

We acknowledge that our results are based on a single-institution database and may not be directly generalizable to other healthcare settings. Additionally, changes in breast cancer management from 1996 to 2022 necessitate a cautious interpretation of our findings. While our data provide valuable insights into the applicability of these guidelines, future prospective studies are needed to validate these results in broader, more diverse populations.

Author Contributions: Conceptualization, M.A. and A.M.; methodology, M.A. and R.B.V.; software, M.d.N.M.R.; validation, M.A., A.M. and L.H.G.; formal analysis, M.A. and M.P.M.; investigation, R.B.V., M.D.T., A.G.A. and L.C.d.O.; resources, O.F. and F.P.C.; data curation, M.A. and R.B.V.; writing—original draft preparation, M.A. and R.B.V.; writing—review and editing, M.A., R.B.V., A.M., M.P.M., M.D.T., A.G.A., O.F., L.C.d.O., M.d.N.M.R., F.P.C., F.Z., M.M., E.d.C.M., A.L.F., F.P.B., G.F., H.L.C. and L.H.G.; visualization, E.d.C.M. and M.M.; supervision, A.L.F. and G.F.; project administration, M.A. and H.L.C. All authors have read and agreed to the published version of the manuscript.

Funding: This research received no external funding.

Institutional Review Board Statement: The study was conducted in accordance with the Declaration of Helsinki and approved by the Institutional Review Board (or Ethics Committee) of the Hospital do Servidor Público Estadual-Francisco Morato de Oliveira (protocol code CAAE: 60175722.6.0000.5463, approved on 20 July 2022.

Informed Consent Statement: The need for informed consent was waived due to the retrospective nature of the study and the confidentiality and integrity of the medical record data were preserved.

Data Availability Statement: The data are fully available from the author upon request.

Acknowledgments: We would like to express our sincere gratitude to all the professionals involved in this research, such as the nameless patients who indirectly contribute to medical advances. This work was elaborated with support of Hospital do Servidor Público Estadual de São Paulo, which provided access to medical records, essential equipment, and facilities. Finally, we extend our heartfelt appreciation to our families for their continuous support and motivation.

Conflicts of Interest: None of the authors have conflicts of interest to declare.

References

1. Instituto Nacional do Câncer (INCA). Incidência do Câncer de Mama no Brasil, Regiões e Estados. 2022. Available online: https://www.gov.br/inca/pt-br/assuntos/gestor-e-profissional-de-saude/controle-do-cancer-de-mama/dados-e-numeros/incidencia (accessed on 12 November 2024).
2. Kalwaniya, D.S.; Gairola, M.; Gupta, S.; Pawan, G. Ductal Carcinoma in Situ: A Detailed Review of Current Practices. *Cureus* **2023**, *15*, e37932. [CrossRef]
3. Salvatorelli, L.; Puzzo, L.; Vecchio, G.M.; Caltabiano, R.; Virzì, V.; Magro, G. Ductal Carcinoma In Situ of the Breast: An Update with Emphasis on Radiological and Morphological Features as Predictive Prognostic Factors. *Cancers* **2020**, *12*, 609. [CrossRef] [PubMed]
4. Gorringe, K.L.; Fox, S.B. Ductal Carcinoma In Situ Biology, Biomarkers, and Diagnosis. *Front. Oncol.* **2017**, *7*, 248. [CrossRef] [PubMed]
5. Rajeswaran, T.; Gojsevic, M.; Chan, A.W.; Wong, H.C.; Lee, S.F.; Bernard, R.; Marta, G.N.; Pogoda, K.; Kwan, J.Y.; Kuszaj, O.; et al. Quality of life issues in patients with ductal carcinoma in situ: A systematic review. *Support. Care Cancer* **2024**, *32*, 695. [CrossRef] [PubMed]

6. Bane, A. Ductal Carcinoma In Situ: What the Pathologist Needs to Know and Why. *Int. J. Breast Cancer* **2013**, *2013*, 1–7. [CrossRef]
7. Partridge, A.; Adloff, K.; Blood, E.; Dees, E.C.; Kaelin, C.; Golshan, M.; Ligibel, J.; De Moor, J.S.; Weeks, J.; Emmons, K.; et al. Risk Perceptions and Psychosocial Outcomes of Women With Ductal Carcinoma In Situ: Longitudinal Results From a Cohort Study. *JNCI J. Natl. Cancer Inst.* **2008**, *100*, 243–251. [CrossRef]
8. Barreau, B.; de Mascarel, I.; Feuga, C.; MacGrogan, G.; Dilhuydy, M.H.; Picot, V.; Dilhuydy, J.M.; de Lara, C.T.; Bussières, E.; Schreer, I. Mammography of ductal carcinoma in situ of the breast: Review of 909 cases with radiographic–pathologic correlations. *Eur. J. Radiol.* **2005**, *54*, 55–61. [CrossRef]
9. Gradishar, W.J.; Moran, M.S.; Abraham, J.; Abramson, V.; Aft, R.; Agnese, D.; Allison, K.H.; Anderson, B.; Bailey, J.; Burstein, H.J.; et al. Breast Cancer, Version 3.2024, NCCN Clinical Practice Guidelines in Oncology. *J. Natl. Compr. Canc. Netw.* **2024**, *22*, 331–357. [CrossRef] [PubMed]
10. Ganz, P.A. Quality-of-Life Issues in Patients With Ductal Carcinoma In Situ. *JNCI Monogr.* **2010**, *2010*, 218–222. [CrossRef]
11. Kunkiel, M.; Niwińska, A. Assessment of the usefulness of prognostic Van Nuys Prognostic Index in the treatment in ductal carcinoma in situ in 15-year observation. *Sci. Rep.* **2021**, *11*, 22645. [CrossRef]
12. Silverstein, M.J.; Lagios, M.D.; Craig, P.H.; Waisman, J.R.; Lewinsky, B.S.; Colburn, W.J.; Poller, D.N. A prognostic index for ductal carcinoma in situ of the breast. *Cancer* **1996**, *77*, 2267–2274. [CrossRef]
13. Lei, R.Y.; Carter, D.L.; Antell, A.G.; Nowels, M.A.; Tole, S.P.; Bennett, J.P.; Turner, M.; Baehner, F.L.; Leonard, C.E. A Comparison of Predicted Ipsilateral Tumor Recurrence Risks in Patients With Ductal Carcinoma in Situ of the Breast After Breast-Conserving Surgery by Breast Radiation Oncologists, the Van Nuys Prognostic Index, the Memorial Sloan Kettering Cancer Center DCIS Nomogram, and the 12-Gene DCIS Score Assay. *Adv. Radiat. Oncol.* **2021**, *6*, 100607. [CrossRef]
14. Di Saverio, S.; Catena, F.; Santini, D.; Ansaloni, L.; Fogacci, T.; Mignani, S.; Leone, A.; Gazzotti, F.; Gagliardi, S.; De Cataldis, A.; et al. 259 Patients with DCIS of the breast applying USC/Van Nuys prognostic index: A retrospective review with long term follow up. *Breast Cancer Res. Treat.* **2008**, *109*, 405–416. [CrossRef] [PubMed]
15. Chen, Q.; Campbell, I.; Elwood, M.; Cavadino, A.; Aye, P.S.; Tin Tin, S. Outcomes from low-risk ductal carcinoma in situ: A systematic review and meta-analysis. *Breast Cancer Res. Treat.* **2024**, *208*, 237–251. [CrossRef]
16. Solin, L.J. Management of Ductal Carcinoma In Situ (DCIS) of the Breast: Present Approaches and Future Directions. *Curr. Oncol. Rep.* **2019**, *21*, 33. [CrossRef]
17. Francis, A.; Thomas, J.; Fallowfield, L.; Wallis, M.; Bartlett, J.M.; Brookes, C.; Roberts, T.; Pirrie, S.; Gaunt, C.; Young, J.; et al. Addressing overtreatment of screen detected DCIS; the LORIS trial. *Eur. J. Cancer* **2015**, *51*, 2296–2303. [CrossRef] [PubMed]
18. Hwang, E.S.; Hyslop, T.; Lynch, T.; Frank, E.; Pinto, D.; Basila, D.; Collyar, D.; Bennett, A.; Kaplan, C.; Rosenberg, S.; et al. The COMET (Comparison of Operative versus Monitoring and Endocrine Therapy) trial: A phase III randomised controlled clinical trial for low-risk ductal carcinoma in situ (DCIS). *BMJ Open* **2019**, *9*, e026797. [CrossRef]
19. Nakayama, S.; Masuda, H.; Miura, S.; Kuwayama, T.; Hashimoto, R.; Taruno, K.; Sawada, T.; Akashi-Tanaka, S.; Nakamura, S. Identifying ductal carcinoma in situ cases not requiring surgery to exclude postoperative upgrade to invasive ductal carcinoma. *Breast Cancer* **2022**, *29*, 610–617. [CrossRef]
20. Elshof, L.E.; Tryfonidis, K.; Slaets, L.; van Leeuwen-Stok, A.E.; Skinner, V.P.; Dif, N.; Pijnappel, R.M.; Bijker, N.; Emiel, J.; Wesseling, J. Feasibility of a prospective, randomised, open-label, international multicentre, phase III, non-inferiority trial to assess the safety of active surveillance for low risk ductal carcinoma in situ—The LORD study. *Eur. J. Cancer* **2015**, *51*, 1497–1510. [CrossRef]
21. Goodwin, A.; Parker, S.; Ghersi, D.; Wilcken, N. Post-operative radiotherapy for ductal carcinoma in situ of the breast. *Cochrane Database Syst. Rev.* **2013**. [CrossRef]
22. Rakovitch, E.; Nofech-Mozes, S.; Hanna, W.; Baehner, F.L.; Saskin, R.; Butler, S.M.; Tuck, A.; Sengupta, S.; Elavathil, L.; Jani, P.A.; et al. A population-based validation study of the DCIS Score predicting recurrence risk in individuals treated by breast-conserving surgery alone. *Breast Cancer Res. Treat.* **2015**, *152*, 389–398. [CrossRef] [PubMed] [PubMed Central]
23. Shah, C.; Bremer, T.; Cox, C.; Whitworth, P.; Patel, R.; Patel, A.; Brown, E.; Gold, L.; Rock, D.; Riley, L.; et al. The Clinical Utility of DCISionRT® on Radiation Therapy Decision Making in Patients with Ductal Carcinoma In Situ Following Breast-Conserving Surgery. *Ann. Surg. Oncol.* **2021**, *28*, 5974–5984, Erratum in: *Ann. Surg. Oncol.* **2021**, *28* (Suppl. S3), 878. [CrossRef] [PubMed] [PubMed Central]
24. Hahn, E.; Rodin, D.; Sutradhar, R.; Nofech-Mozes, S.; Trebinjac, S.; Paszat, L.F.; Rakovitch, E. Can Molecular Biomarkers Help Reduce the Overtreatment of DCIS? *Curr. Oncol.* **2023**, *30*, 5795–5806. [CrossRef] [PubMed] [PubMed Central]

life

MDPI

Article

Extracting Knowledge from Machine Learning Models to Diagnose Breast Cancer

José Manuel Martínez-Ramírez [1], Cristobal Carmona [1,2,3], María Jesús Ramírez-Expósito [4] and José Manuel Martínez-Martos [4,*]

[1] Department of Computer Science, University of Jaén, E-23071 Jaén, Spain; jmmr0049@red.ujaen.es (J.M.M.-R.); ccarmona@ujaen.es (C.C.)
[2] Andalusian Research Institute in Data Science and Computational Intelligence, DASCI, University of Jaén, E-23071 Jaén, Spain
[3] Leicester School of Pharmacy, DeMontfort University, Leicester LE1 7RH, UK
[4] Experimental and Clinical Physiopathology Research Group CVI-1039, Department of Health Sciences, University of Jaén, E-23071 Jaén, Spain; mramirez@ujaen.es
* Correspondence: jmmartos@ujaen.es; Tel.: +34-953-212-600

Abstract: This study explored the application of explainable machine learning models to enhance breast cancer diagnosis using serum biomarkers, contrary to many studies that focus on medical images and demographic data. The primary objective was to develop models that are not only accurate but also provide insights into the factors driving predictions, addressing the need for trustworthy AI in healthcare. Several classification models were evaluated, including OneR, JRIP, the FURIA, J48, the ADTree, and the Random Forest, all of which are known for their explainability. The dataset included a variety of biomarkers, such as electrolytes, metal ions, marker proteins, enzymes, lipid profiles, peptide hormones, steroid hormones, and hormone receptors. The Random Forest model achieved the highest accuracy at 99.401%, followed closely by JRIP, the FURIA, and the ADTree at 98.802%. OneR and J48 achieved 98.204% accuracy. Notably, the models identified oxytocin as a key predictive biomarker, with most models featuring it in their rules. Other significant parameters included GnRH, β-endorphin, vasopressin, IRAP, and APB, as well as factors like iron, cholinesterase, the total protein, progesterone, 5-nucleotidase, and the BMI, which are considered clinically relevant to breast cancer pathogenesis. This study discusses the roles of the identified parameters in cancer development, thus underscoring the potential of explainable machine learning models for enhancing early breast cancer diagnosis by focusing on explainability and the use of serum biomarkers.The combination of both can lead to improved early detection and personalized treatments, emphasizing the potential of these methods in clinical settings. The identified markers also provide additional research and therapeutic targets for breast cancer pathogenesis and a deep understanding of their interactions, advancing personalized approaches to breast cancer management.

Keywords: breast cancer; serum biomarkers; explainable AI; oxytocin; early diagnosis; peptide hormones; IRAP; progesterone

1. Introduction

Breast cancer is a disease in which breast cells grow abnormally and uncontrollably [1–3]. These cells can form a tumor, which may become palpable as a lump [2]. Invasive ductal carcinoma is the most common type of breast cancer. It originates in the cells lining the ducts of the breast and spreads outside the duct into the surrounding breast tissue [1]. Approximately 80% of breast cancers are ductal carcinomas. Ductal carcinoma in situ (DCIS) is a non-invasive cancer that originated in the cells lining the ducts of the breast and has not spread outside the duct into the surrounding breast tissue [4]. Invasive lobular carcinoma originates in the lobules of the breast, where milk is produced, and spreads into the surrounding breast tissue [2,3].

The risk of breast cancer increases with age. Approximately 66.6% of breast cancer patients are older than 55 years [1,5–8]. Having a close relative with breast cancer, especially a first-degree relative, is also a risk factor [6,9]. Certain genetic mutations, such as mutations in the BRCA1 and BRCA2 genes, significantly increase the risk of breast cancer [8]. Prolonged exposure to estrogen, such as the early onset of menstruation, late menopause, or the use of hormone therapy, also increases the risk [6,8]. With regard to lifestyle, alcohol consumption, obesity, and a lack of physical activity are associated with an increased risk of the disease [2,6].

Prevention currently includes breast self-examination to detect any changes, such as lumps, skin changes, or nipple discharge. Screening and early detection methods include mammography [9–11], an X-ray of the breast used to detect breast cancer in the early stages, even before a lump can be felt, breast ultrasounds [8,11–13], which use sound waves to determine whether a lump in the breast is solid or a fluid-filled cyst, and breast MRI [8,13,14], an imaging test that uses a magnetic field and radio waves to create more detailed images of breast tissue and which is used to detect cancer in women at high risk or to better evaluate a finding from mammography or a physical examination. For definitive diagnosis, a biopsy is performed [5,10]. Ultimately, the treatment of breast cancer depends on several factors, such as the stage of the cancer, the type of cancer, and the age and general health of the patient. The treatment options include surgery, which may involve the removal of the tumor, part of the breast, or the whole breast (mastectomy) [5,8,10]; radiotherapy, which uses high-energy X-rays to destroy cancer cells [10,11]; chemotherapy, which uses drugs to destroy cancer cells [8,10,13,15]; hormone therapy, which is used to block the effects of hormones on cancer cells; and targeted therapy, which uses drugs that target specific characteristics of cancer cells [8,10,15].

Breast cancer is the most common cancer among women in the United States. It is estimated that there were about 2.9 million new cases in 2022 [7,16,17]. Globally, breast cancer accounts for 11.7% of cancer diagnoses and contributes significantly to mortality, especially among women over the age of 50 [1,8,18]. In 2020, there were 2.2 million new cases and 684,996 deaths from breast cancer [18]. The breast cancer incidence in 2020 was highest in Asia (45.4%) and Europe (23.5%), followed by North America (12.5%) and Latin America and the Caribbean (9.29%) [18]. Africa and Oceania reported lower incidence rates of 8.25% and 1.14%, respectively [7,18]. Asia also led in terms of breast cancer mortality with 50.5% of deaths, while Europe's mortality rate stood at 20.7%, probably reflecting better healthcare outcomes [18]. Africa's considerable mortality rate of 12.5% highlights critical challenges in medical care. In Spain, breast cancer causes 35,000 new cases per year and is the leading cause of cancer-related mortality in women, with 6651 deaths per year [6,18].

An early diagnosis of breast cancer is directly associated with a higher chance of survival [5,10–12]. This is because early detection allows for treatment to be initiated in a timely manner, which increases the chances of success and reduces the risk of the disease spreading to other parts of the body (metastasis) [1,5,11,12]. The treatment options are also broader and less aggressive [6,11]. Late diagnosis may imply the need for more invasive treatments, such as a mastectomy instead of breast-conserving surgery [5,9,10]. In addition, chemotherapy and radiotherapy, which can have considerable side effects, may be necessary at later stages [5,10,15]. In general, breast cancer treatment is most effective when initiated early in the disease because cancer cells are more sensitive when the tumor is small and has not spread [5,10–12]. The early diagnosis of breast cancer therefore allows patients to preserve a better quality of life as treatments are less invasive and have fewer side effects, allowing patients to maintain a more active and healthy life [11]. Finally, the early detection of breast cancer can reduce long-term healthcare costs because early-stage treatments are generally less expensive than treatments for advanced breast cancer [6,11].

In any case, it is important to emphasize the importance of early diagnosis in breast cancer management because it is crucial for improving survival rates and preventing breast cancer recurrence [5,6,10–12]. In fact, technology-based diagnostic tools are emerging. For example, a liquid biopsy is a non-invasive method for breast cancer screening that utilizes cancer-related biomolecules in the blood. Surface-enhanced Raman spectroscopy (SERS) combined with machine learning algorithms has shown great potential for rapid and accurate breast cancer detection. This technology enhances the Raman spectroscopy signal of serum using a specialized substrate, allowing for the discovery of new biomarkers. SERS combined with machine learning could achieve 100% accuracy, sensitivity, and specificity in distinguishing breast cancer patients from controls [19,20]. Similarly, Raman spectroscopy has been used with deep learning algorithms for screening triple-negative or HER-2-positive breast cancers [21].

Most data mining and machine learning studies mainly focus on the use of medical images and demographic data in studies of breast cancer diagnosis and prognosis [1,5,8,9,12–14,16,22–24]. However, serum biochemical parameters have been much less studied. Nevertheless, it is important to note that the integration of serum biomarkers can improve the accuracy of machine learning models for breast cancer risk prediction, early diagnosis, treatment response prediction, or recurrence detection [8,13,16,17]. Furthermore, the use of serum biomarkers offers a less invasive and more efficient approach for the early detection and monitoring of disease, also allowing for a binary classification of whether a patient has breast cancer or not. Unlike imaging techniques, which may have limitations in terms of their sensitivity and specificity, serum biomarkers can reflect early molecular changes associated with cancer development. In the context of the medical workflow, the extraction of serum biomarkers is performed through a blood test, a routine and minimally invasive procedure. This sample can be obtained at various stages of the healthcare process, from initial screening to monitoring the response to treatment. While conventional screening focuses primarily on mammography and other imaging techniques such as ultrasounds and MRI, it is important to note that these techniques have certain limitations. For example, mammography may not be as effective in women with dense breast tissue, and both ultrasounds and MRI can be expensive and are not available everywhere. In addition, these imaging techniques often detect cancer when a palpable tumor has already formed. Therefore, several artificial intelligence (AI) tools could be applied to a dataset of serum biomarkers with information on both patients and healthy individuals. By defining a classification problem [25], models could be trained in order to extract information from the previously mentioned dataset to diagnose whether a patient had breast cancer based exclusively on these biomarkers, which could prove useful to diagnose breast cancer when used along with other detection methods.

Though these tools often achieve highly accurate predictions, given the context of the problem, it is of the utmost importance that the results are not only precise but also explainable. The implemented systems should be able to explicitly define the reasons why their predictions occur, which in turn will allow the specialists who make use of them to discover new patterns in the data that may aid further research. For this purpose, black-box models, such as deep learning models, were ignored, and only models that offer rules or trees that grant information on specific markers that are relevant for the diagnosis of breast cancer were used in an effort to achieve trustworhy AI (TAI) [26].

2. Materials and Methods

2.1. Subjects

A total of 198 women were recruited in the Unit of Breast Pathology at the University Hospital of Jaén, and 78 volunteer women without breast cancer were also included as a control group. This study was approved by the Ethical Committee of the University Hospital of Jaén, and all subjects signed a declaration of free, informed consent. All women with breast cancer were diagnosed with ductal infiltrating carcinoma. A total of 83 of these women (39 premenopausal and 44 postmenopausal) did not receive neoadjuvant chemotherapy, whereas 115 of them (63 premenopausal and 52 postmenopausal) received neoadjuvant chemotherapy before surgery. The clinicopathological characteristics of the studied patients have been previously described [27–31]. The patient characterization included the age at diagnosis, tumor size, tumor histology, pathologic T classification, Scarff–Bloom–Richardson grade, hormonal and HER-2/neu status, and molecular subtype. The control group consisted of 78 women, aged 28 to 69 years old (premenopausal women with regular menstrual periods, n = 38; postmenopausal women with spontaneous menopause for at least one year, n = 40), with no previous history of any type of cancer, chemotherapy, hormonal or antioxidant therapy, or chronic diseases. Women were excluded if they were current smokers, regular alcohol consumers, antioxidant supplement users, pregnant or lactating, presented hepatic, cardiac, or renal dysfunction, had undergone hormonal therapy, used drugs, or had hypertension, diabetes, and other eventual chronic conditions. Samples from patients treated with NACT were obtained after the completion of chemotherapy treatment and in parallel to samples from patients not treated with NACT and control volunteers in order to be processed under the same conditions. Blood samples were obtained after an overnight fast by venous arm puncture in tubes without anticoagulants. Blood specimens were allowed to clot and centrifuged at $3000\times g$ for 10 min at 4 °C to obtain the serum. Serum samples were collected, rapidly frozen in liquid nitrogen, and kept at −80 °C until usage for assays. All procedures performed in studies involving human participants were in accordance with the ethical standards of the institutional and/or national research committee and with the 1964 Helsinki declaration and its later amendments or comparable ethical standards.

2.2. Datasets and Pre-Processing

The study subjects were all women of varying ages, as described in Section 2.1. This study aimed to make use of explainable models to accurately predict if a patient had breast cancer.

The dataset included data previously described in [27–35] and additional unpublished results. The dataset included electrolytes (sodium, potassium, chloride, calcium, phosphorus, and magnesium), metal ions (iron), albumin, the total protein, and marker proteins (Cerb2, also known as HER2/neu, P53, Ki67, and Bcl-2). It also included enzymes such as adenosine deaminase, CKMB (creatine kinase muscle–brain isoform), amylase, gamma-glutamyl-transferase (GGT), ALAT (alanine aminotransferase) and ASAT (aspartate aminotransferase), acid and alkaline phosphatases, CPK (creatine phosphokinase), and LDH (lactate dehydrogenase). Additionally, it included glucose; non-protein nitrogenous compounds like creatinine, urea, and uric acid; and a lipid profile with the total cholesterol, HDL (high-density lipoprotein), LDL (low-density lipoprotein), and triglycerides. Furthermore, it contained renin–angiotensin system-regulating aminopeptidases (APN, aminopeptidase N; APB, aminopeptidase B; ASAP, aspartyl aminopeptidase; APA, aminopeptidase A; and IRAP, insulin-regulated aminopeptidase). The dataset also contained apolipoproteins A1 and B; hormones including GnRH (gonadotropin-releasing hormone), TSH (thyroid-stimulating hormone), FSH (follicle-stimulating hormone), LH (luteinizing hormone), GH (growth hormone), thyroxine, cortisol, angiotensin, vasopressin, oxytocin, β-endorphin, estradiol, progesterone, and prolactin, along with oxytocinase and pyrrolidone carboxypeptidase hormone-regulatory enzymes; and hormone receptors (ERs, estrogen receptors; PRs, progesterone receptors).

The dataset had several issues, and as such, several pre-processing techniques were applied. The first issue that the dataset had was that some of the attributes in it were specific for patients that had breast cancer and were not applicable to people that belonged to the control class. Consequently, all those attributes were ignored, since they would negatively impact the results of the model due to a bias towards one class if a patient presented any value whatsoever for those attributes. However, no examples were removed from the dataset, as it contained no errors and all data were relevant to the study. As such, it was not necessary to filter the dataset beyond removing those values.

Moreover, many of the attributes had several missing values. In order to fix this issue, the average of the existing values was calculated for each individual class (control or breast cancer), and the result was used to get rid of missing values.

Additionally, since the dataset was quite unbalanced, including roughly twice as many examples of breast cancer patients as healthy people, SMOTE [36] was used as an oversampling technique on the minority class, increasing the amount of examples it included by 50%. There were still more examples of the majority class, but it made the difference less notable. Though more examples could have been added, it would have caused the accuracy of the results to decrease due to an increase in noise in the dataset.

Lastly, the strategy used for the training of the models was a leave-one-out strategy, in which each model was trained with all examples in the dataset but one, which was used to test its accuracy. This process was repeated a number of times equal to the amount of examples in the dataset, so that every individual example could be used as a test once, as seen in Figure 1.

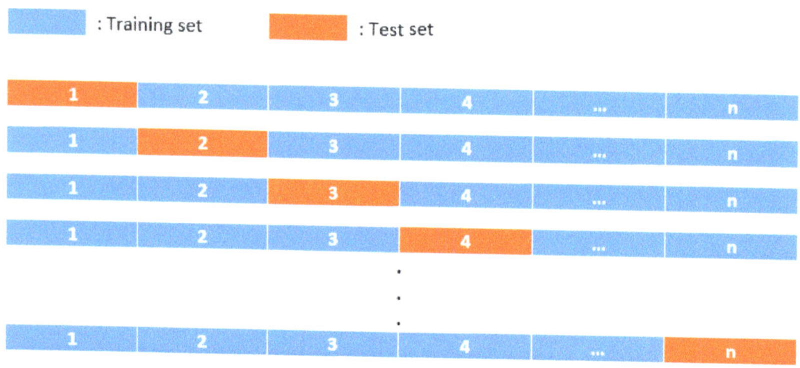

Figure 1. Behavior of the leave-one-out strategy.

2.3. Classification Models

In order to solve the classification problem, several models were used in order to compare the results and discover which was most appropriate for the problem at hand. This research focused on models that are both easy to understand and offer highly accurate results.

In order to determine which explainable models to use, the main criterion that was considered for the present study was the output of the model, so that all models provided rules or trees that justified their results and that could be directly linked to the biomarkers that influenced them, thus offering evidence of what biomarkers can provide relevant information that can be easily understood by doctors, potentially aiding them when diagnosing patients. Within these two categories, models that presented highly accurate results (as described in Section 3.1) were selected.

This section includes an explanation of all the models used, as well as the relevant hyperparameters for each model. These hyperparameters were selected due to them providing the most accurate results for each model and were obtained by fine-tuning the models on the training and test datasets.

2.3.1. OneR

One Rule (OneR) [37] is a very simple model that attempts to predict the label of a given dataset by generating one singular rule. It is the natural evolution of the ZeroR model, which simply predicts the majority label. While ZeroR offers no valuable information for predictions, it can be used as a baseline for all the other models, which should achieve higher results.

OneR generates an output, that is, a rule, in two steps. At first, it analyzes all the attributes of the dataset and creates a rule for each of them. Afterwards, it tests every single generated rule to determine which one achieves the highest accuracy. That rule is the one that the model will use for all future predictions.

In spite of its simplicity, OneR can achieve remarkable results on some datasets, particularly those that contain classes that heavily depend on a given attribute or that are relatively simple.

For this specific research, OneR was tested out with a minimum bucket size of 6 for discretizing numerical values, and 4 decimal places were considered.

2.3.2. JRIP

JRIP is the java implementation of the RIPPER [38] algorithm, which generates a set of rules that are then used to predict the labels of later examples. The algorithm creates the ruleset in two stages:

- Building stage: This stage generates rules until the error rate increases beyond a given threshold or no examples remain. A rule is generated by using one attribute, and antecedents are added to it by following a greedy approach [39] until the rule is perfect. Afterwards, rules are incrementally pruned to prevent overfitting.
- Optimization stage: after generating the original ruleset, several variants of each rule are generated and pruned by following the same process as the building stage. One variant is generated by adding more antecedents, and the other is generated from an empty rule. The minimal description length for each original rule and its variants is computed, and the rule that has the lowest description length is selected and added into the final ruleset.

The result is a series of rules that are executed in order and that can accurately describe the behavior of a dataset, which allows for the prediction of items but most importantly allows experts to learn and understand how and why the algorithm labels each example.

In this research, the error threshold was 50%, and two variant rules were generated during the optimization stage.

2.3.3. FURIA

The Fuzzy Unordered Rule Induction Algorithm [40], or FURIA for short, is an algorithm for generating a ruleset that allows for the prediction of labels. It was designed as an extension of JRIP.

The main difference between both algorithms is that the FURIA makes use of fuzzy logic. This implies that rules do not have specific values that need to be met for the antecedent to become relevant; instead, a rule can be triggered with values that are in a certain range with a given confidence factor. It also includes an efficient rule-stretching method to deal with uncovered examples and unordered rulesets.

All of these changes allow the FURIA to improve the results of JRIP in regard to certain problems, particularly for those examples with a large number of attributes or that would require a large ruleset in order to achieve accurate predictions.

As the FURIA is based on JRIP, the hyperparameters used for the FURIA were akin to those used in JRIP. Furthermore, the fuzzy AND operator was defined as the Product T-Norm.

2.3.4. J48

J48 [41] is a decision tree that originated from C4.5 [42]. A J48 tree consists of several nodes which can lead to two or more nodes based on the value of certain attributes of the data. The lowest level nodes, also called leaf nodes, return a given label as the output. Whenever a new example is to be labeled, one must simply follow the nodes of the tree by assessing the values of the attributes of the example.

The tree is built through several iterations of the same algorithm, which takes place in three main stages:

- Stage 1: If all or almost all the elements of a node are of the same class or label, such a node is considered a leaf node, and the algorithm stops for that node.
- Stage 2: Otherwise, for each element in the node, all attributes are considered and tested. The attribute that presents the largest gain in data is selected, and the node is split into several child nodes. Each child node is defined by the threshold values of the selected attribute, so that the items of the parent node are distributed accordingly between all child nodes.
- Stage 3: Once the entire tree has been built by repeating stages 1 and 2, it can be pruned [43] by removing some nodes, thus preventing overfitting and sometimes increasing the accuracy.

J48 has been thoroughly used in many applications and has proven to be able to deal with particular characteristics, lost or missing attribute estimations of the data, and varying attribute costs [44].

The J48 tree used in this research had a confidence factor of 0.25 and a minimum of 2 objects per leaf node. It was also designed not to use the actual value of nodes when it came to splitting them, instead using the value that achieved the highest accuracy. Lastly, during the pruning phase, subtree raising was considered so as to not fully remove them from the resulting tree.

2.3.5. ADTree

The Alternating Decision Tree (ADTree) [45] is a tree-based algorithm. However, unlike decision trees such as J48, the ADTree has two different types of nodes: decision nodes and prediction nodes. Decision nodes work much like nodes in a decision tree would, specifying a singular predicate condition. Prediction nodes, however, have a singular value associated with them.

The ADTree represents each possible label as a numerical value. This way, when an example is examined, it follows the path of the tree, using the decision nodes to decide which prediction nodes are taken into consideration. Once this process is over, the numbers associated with each decision node are added up. The result of this operation is compared to the value of each label. The label with the closest value will be the one predicted. Moreover, this result can double as a confidence factor: if it is very close to the value of the label, it is more likely that the prediction is accurate.

The ADTree has proven to be as robust as boosted decision trees. Although it tends to be more complex than other algorithms that can achieve similar results, several optimizations have been developed to overcome this issue [46].

All of the previously discussed improvements were taken into consideration when designing the ADTree model for this research. Furthermore, the tree was built in 10 boosting iterations, each of which was designed to attempt to expand all possible paths.

2.3.6. Random Forest

The Random Forest model [47] is an ensemble model, and as such, it combines several simpler tree-based models, such as J48, to compute a final result.

What differentiates the Random Forest from other ensemble algorithms is the fact that, for every tree, instead of the full set of attributes, only a fraction of them is considered. Afterwards, the results of each individual tree are combined by using a weighted sum or a majority vote in order to predict the label for a given example.

Since each tree considers different attributes and the information is combined later on, the algorithm becomes more robust, particularly when compared with the individual trees that make up the model. Furthermore, this allows for the partial mitigation of overfitting, as well as any other issues that may arise as a consequence of a high number of attributes.

The Random Forest was generated by using J48 trees, and all previously applicable hyperparameters were still used here. Furthermore, the model was tested by using several different numbers of trees. The best results were achieved by using 20 trees, which were the ones that were considered.

3. Results

3.1. Metrics

As the accuracy may not be sufficient to determine the quality of a model that deals with the diagnosis of cancer, several metrics were used to compare the models to one another. To understand these metrics, a series of basic definitions upon which they are based is presented:

- True Positive (TP): The amount of patients with breast cancer that were labeled as such.
- True Negative (TN): The amount of patients without breast cancer that were labeled as such.
- False Positive (FP): The amount of patients without breast cancer that were labeled as having breast cancer.
- False Negative (FN): The amount of patients with breast cancer that were labeled as healthy.

With these definitions, we can define the following metrics, which were used to measure the performance of our models:

- Accuracy: $(TP + TN)/(TP + FP + TN + FN)$
- Precision: $TP/(TP + FP)$
- Recall: $TP/(TP + FN)$
- F1 Score: $2 \times (Precision \times Recall)/(Precision + Recall)$

3.2. General Results

This section includes the results for all models when attempting to detect if a patient had cancer, which can be found in Table 1. Further subsections show the specific results of each model.

Table 1. Results for each model when diagnosing a patient.

	Accuracy (%)	Precision	Recall	F1
OneR	98.204 ± 0.134	0.991	0.982	0.987
JRIP	98.802 ± 0.110	0.991	0.991	0.991
FURIA	98.802 ± 0.111	0.991	0.991	0.991
J48	98.204± 0.134	0.991	0.982	0.987
ADTree	98.802 ± 0.110	0.991	0.991	0.991
Random Forest	99.401 ± 0.103	0.991	1.000	0.996

3.2.1. OneR

OneR achieved an accuracy of 98.204% by using one singular rule, which implies that the attributes in it could be extremely relevant when attempting to diagnose breast cancer. The rule it generated can be found in Table 2.

Table 2. Rule generated by OneR.

				Precision	Recall
IF	Oxytocin < 35.75	THEN	diagnosis = BreastCancer	0.991	0.982

3.2.2. JRIP

JRIP generated a ruleset that achieved an accuracy of 98.802% when diagnosing breast cancer. This ruleset can be found in Table 3.

It should be noted that JRIP only generated one rule, which was the same as the one generated by OneR except for some tweaking of specific values, further proving that oxytocin is a key value to consider when diagnosing breast cancer.

Table 3. Rule generated by JRIP.

				Precision	Recall
IF	Oxytocin < 36.7	THEN	diagnosis = BreastCancer	0.991	0.991

3.2.3. FURIA

Unlike the previous algorithms, the FURIA applies fuzzy logic and a confidence factor (CF) to determine how likely a predicted diagnosis is based on the specific values of the data it uses. For each attribute, the entire range of values was divided evenly across five labels: very low, low, medium, high, and very high. Each label was defined by four values; if the example value was between the first and second values or between the third and fourth values, the chosen label was not certain, and the confidence factor was considered. If no other labels were more likely, that label was used. If it was between the second and third values, the label was certain, acting much like those of the previous algorithms. These rules, which achieved an accuracy of 98.802%, are described in Table 4.

The relevant values for the labels are presented hereunder:

- Oxytocin: Very low oxytocin levels were defined by [-Inf, -Inf, 34.8, 36.7]. Very high oxytocin levels were defined by [31.6, 36.7, inf, inf].
- IRAP: Very low levels were defined by [-inf, -inf, 360, 363.859944].

Table 4. Rules generated by FURIA.

				Precision	Recall
IF	Oxytocin IS very low	THEN	diagnosis = BreastCancer (CF = 0.99)	0.991	0.991
IF	Oxytocin IS very high AND IRAP IS very low	THEN	diagnosis = Control (CF = 0.98)	0.982	0.982

3.2.4. J48

Unlike the previous algorithms, J48 generates a tree which must be followed in order to determine the diagnosis of a given patient. However, such a tree can be easily transformed into an algorithmic set of rules, which can be found in Table 5.

It should be noted that, once more, J48 generated one single rule while achieving an accuracy of 98.204%. However, this time a different attribute was used.

Table 5. Rules generated by J48.

				Precision	Recall
IF	GnRH < 17.5	THEN	diagnosis = BreastCancer	0.991	0.982

3.2.5. ADTree

The ADTree, much like the previous algorithm, also generates a tree. However, in this case there are two kinds of nodes: decision nodes, to determine which branch of the tree is followed, and prediction nodes, which modify the value of the diagnosis score. Based on the final score, a final diagnosis is provided. The ADTree achieved an accuracy of 98.802%. The results can be represented as a ruleset, which is shown in Table 6. It should be noted that a base value for the score was provided, so that $defaultScore = -0.545$.

Table 6. Rules generated by ADTree.

IF	oxytocin < 35.75	THEN	score = score − 2.258
IF	oxytocin >= 35.75	THEN	score = score + 2.012
IF	oxytocin < 85.9	THEN	score = score − 1.151
IF	oxytocin >= 85.9	THEN	score = score + 1.185
IF	APB < 133.396	THEN	score = score + 0.684
IF	APB >= 133.396	THEN	score = score − 0.711
IF	β-Endorphin < 133.396	THEN	score = score − 0.742
IF	β-Endorphin >= 133.396	THEN	score = score + 0.858
IF	GnRH < 20.6	THEN	score = score − 0.303
IF	GnRH >= 20.6	THEN	score = score + 0.378
IF	5′-Nucleotidase < 1.45	THEN	score + 0.26
IF	5′-Nucleotidase >= 1.45	THEN	score = score − 0.224
IF	APB < 114.584	THEN	score = score − 0.179
IF	APB >= 114.584	THEN	score = score − 0.177
IF	score < 0	THEN	diagnosis = BreastCancer
ELSE	diagnosis = Control		

3.2.6. Random Forest

The Random Forest created a total of 25 J48 trees in order to analyze the data and diagnose breast cancer in patients. By using these 25 trees, it achieved an accuracy of

99.401%. By using only the five most relevant trees, the accuracy was over 98%. Those five trees can be found in Tables 7, 8, 9, 10, and 11, respectively.

It should be noted that the model correctly diagnosed all cancer patients, as shown by the perfect score for its recall.

Table 7. First tree generated by Random Forest.

				Precision	Recall
IF	Vasopressin < 2.68 AND β-Endorphin < 23.3 AND cholinesterase < 7186	THEN	diagnosis = BreastCancer	1.000	0.223
IF	Vasopressin < 2.68 AND β-Endorphin < 23.3 AND cholinesterase >= 7186	THEN	diagnosis = control	1.000	0.055
IF	Vasopressin < 2.68 AND β-Endorphin >= 23.3	THEN	diagnosis = control	0.975	0.710
IF	Vasopressin >= 2.68 AND GnRH < 22.48	THEN	diagnosis = BreastCancer	0.989	0.830
IF	Vasopressin >= 2.68 AND GnRH >= 22.48	THEN	diagnosis = Control	1.000	0.127

Table 8. Second tree generated by Random Forest.

				Precision	Recall
IF	GnRH < 20.6 AND Cholinesterase < 8342.5	THEN	diagnosis = BreastCancer	0.991	1.000
IF	GnRH < 20.6 AND Cholinesterase >= 8342.5	THEN	diagnosis = Control	1.000	0.019
IF	GnRH >= 20.6	THEN	diagnosis = Control	0.981	0.964

Table 9. Third tree generated by Random Forest.

				Precision	Recall
IF	β-Endorphin < 32.55 AND Protein < 6.79	THEN	diagnosis = BreastCancer	0.984	0.580
IF	β-Endorphin < 32.55 AND Protein >= 6.79 AND Progesterone < 2.15 AND BMI < 28.71 AND oxytocin < 104.24	THEN	diagnosis = BreastCancer	1.000	0.054
IF	β-Endorphin < 32.55 AND Protein >= 6.79 AND Progesterone < 2.15 AND BMI < 28.71 AND oxytocin >= 104.24	THEN	diagnosis = Control	1.000	0.109
IF	β-Endorphin < 32.55 AND Protein >= 6.79 AND Progesterone < 2.15 AND BMI >= 28.71	THEN	diagnosis = BreastCancer	1.000	0.348
IF	β-Endorphin < 32.55 AND Protein >= 6.79 AND Progesterone >= 2.15	THEN	diagnosis = Control	1.000	0.145
IF	β-Endorphin >= 32.55	THEN	diagnosis = Control	0.980	0.890

Table 10. Fourth tree generated by Random Forest.

				Precision	Recall
IF	Cholinesterase < 5174	THEN	diagnosis = BreastCancer	0.987	0.688
IF	Cholinesterase >= 5174 AND oxytocin < 35.75	THEN	diagnosis = BreastCancer	1.000	0.277
IF	Cholinesterase >= 5174 AND oxytocin >= 35.75	THEN	diagnosis = control	0.982	0.982

Table 11. Fifth tree generated by Random Forest.

				Precision	Recall
IF	β-Endorphin < 28.2 AND Iron < 78.87	THEN	diagnosis = BreastCancer	0.987	0.679
IF	β-Endorphin < 28.2 AND Iron >= 78.87 AND IRAP < 224.96	THEN	diagnosis = BreastCancer	1.000	0.036
IF	β-Endorphin < 28.2 AND Iron >= 78.87 AND 224.96 < IRAP < 264.45	THEN	diagnosis = Control	1.000	0.109
IF	β-Endorphin < 28.2 AND Iron >= 78.87 AND IRAP >= 264.45	THEN	diagnosis = BreastCancer	1.000	0.303
IF	β-Endorphin >= 28.2	THEN	diagnosis = Control	0.980	0.909

4. Discussion

Though deep learning and AI techniques have been used for the detection of breast cancer in several recent articles in the literature [23,24,48], most of them are fully focused on the prediction of the cancer in itself and less so on the information that can be extracted during the process.

As such, many of the models in the literature oftentimes make use of structures such as autoencoders (AEs) for anomaly detection in images or convolutional neural networks (CNNs) for the analysis of screening mammograms. In [49,50], AEs were used to great effect, achieving an accuracy of roughly 98.5%. In [51], CNNs were used to predict breast cancer using ultrasound images, achieving an accuracy of up to 94%. Similarly, in [52], a CNN model achieved 95.7% accuracy on mammography images. Lastly, in [53], a Quantum Hybrid CNN achieved results of 97.661%.

Though these models can achieve highly accurate results on their own, they lack explainability, since they are black-box models that offer no reasons as to why they predict a given label. Though in many applications of these models this is not an issue, when it comes to healthcare, it is of the utmost importance that experts make informed decisions based on data that back them up. As such, using explainable and trustworthy methods, even if their results are slightly less accurate than those provided by more advanced models, may be preferable. Several studies have made use of machine learning models, such as the Random Forest, to predict breast cancer: in [54], a perfect result was achieved using the Random Forest to diagnose breast cancer according to cell attributes extracted from images, while in [55], a result of 93% was achieved when predicting whether a patient was at risk of developing breast cancer, but the results were based on data gathered from images as well.

Though many other articles have previously used machine learning techniques to detect and prevent breast cancer in the literature [56–60], their results are based on the analysis of images or the exclusive use of medical parameters such as symptoms, weight, and height, among others. Furthermore, whilst several of the proposed models make use of machine learning to predict breast cancer or its recurrence risk, none of them make use of biomarkers and, while the most relevant factors for predictions are sometimes mentioned, the relation and interactions between these factors and breast cancer or its recurrence are hardly explained. This manuscript not only uses biomarkers as a key indicator to predict breast cancer, but also properly defines the given interactions between different biomarkers and the specific threshold values that could be relevant to researchers attempting to diagnose breast cancer.

As can be seen, most of the literature revolves around the use of mammographies and symptom data and extracting information from them, while clinical biomarkers are seldom used. Regardless, the accuracy provided by all the models in this contribution, particularly the Random Forest, which achieved an accuracy of 99.401%, rivals that of some of the state-of-the-art literature while additionally providing valuable information on biomarkers that can guide researchers when it comes to predicting breast cancer.

The used models also provide another advantage: they are easy to understand once generated, and as long as the data that were used to train them are not biased, their results will also be trustworthy. Once a model has generated a ruleset, all that is needed is to check all the generated rules one by one to determine which one is to be used whilst considering biomarker data from a new patient who is to be diagnosed. This prevents biases and limitations of the models when interpreting the results of the models if they have been properly trained.

Nevertheless, these models are not without their own limitations: they require large amounts of quality information in order to properly define the interactions between biomarkers and their potential effect on breast cancer. If data for a specific patient lack values for biomarkers, the models may offer the wrong predictions. Furthermore, model training also has to be carefully revised to avoid biases and overfitting, which may cause models to perfectly predict the training set but offer predictions that are lacking when used outside of it. However, as long as these limitations are taken into consideration and fixed by using pre-processing techniques such as balancing the datasets and leave-one-out validation, the models are able to provide valuable information and accurate results.

In this regard, several important parameters related to breast cancer pathogenesis appear to be involved in several models described here. These parameters can be seen in Table 12 and include peptide hormones, their regulatory proteolytic enzymes, steroid hormones, enzyme activities, and the BMI, among others, being the basis for the explainability of the different models. Of these, oxytocin appeared in all the proposed models except J48.

The following sections provide a thorough explanation of these parameters and their relevance when predicting breast cancer according to the literature.

Table 12. Parameters deemed relevant for the prediction of breast cancer for each tested model.

Parameter	OneR	JRIP	FURIA	J48	ADTree	Random Forest
Hormones	Oxytocin	Oxytocin	Oxytocin	GnRH	Oxytocin, GnRH, β-endorphin	Oxytocin, GnRH, β-endorphin, vasopressin, progesterone
Hormone-regulating proteolytic enzymes			IRAP		APB	IRAP
Other enzyme activities					5′-Nucleotidase	Cholinesterase
Metal ions						Iron
Other						Total protein, BMI

IRAP: insulin-regulated aminopeptidase; GnRH: gonadotrophin-releasing hormone; APB: aminopeptidase B; BMI: body mass index.

4.1. Peptide Hormones

Oxytocin, a neuropeptide hormone known for its roles in reproduction and social behavior, has emerged as a potentially influential agent in breast cancer development and progression. Its role in this disease, however, is characterized by remarkable complexity, manifesting both proliferative and antiproliferative effects depending on an intricate network of factors including the cell type, hormone concentration, duration of treatment, and interactions with the tumor microenvironment [61–67]. In fact, oxytocin can inhibit breast cancer cell proliferation through the activation of its receptor (OTR), a member of the G protein-coupled receptor family, triggering an intracellular signaling cascade that can lead to apoptosis, cell cycle arrest, and the suppression of pro-tumor pathways such as PI3K/Akt and ERK [68–73]. In vivo studies using animal models of breast cancer have corroborated these findings, demonstrating a reduction in the tumor growth and volume following oxytocin administration [63,64,68,73]. On the other hand, the ability of oxytocin to modulate estrogen receptor (ER) expression and function adds another layer of complexity to its interaction with breast cancer [74,75]. ERs, which play a crucial role in breast cell growth and development, are also important therapeutic targets in breast cancer treatment. Thus, oxytocin, through the modulation of ER expression and activity, could influence the response of tumor cells to hormonal therapies [74,76]. The impact of oxytocin on the expression of microRNAs (miRNAs) has also been the subject of research. miRNAs, small non-coding RNA molecules that regulate gene expression, have emerged as important regulators in cancer. Studies have suggested that oxytocin may influence the expression of specific miRNAs, such as miR-195, which are involved in suppressing tumor growth [73,77]. Despite evidence pointing towards a potential anti-tumor effect of oxytocin, several studies have suggested that this peptide hormone, under certain conditions, may promote breast cancer cell proliferation. The stimulation of angiogenesis, the formation of new blood vessels that nourish the tumor, is one of the mechanisms proposed to explain this effect [78]. In the same way, the overexpression of OTRs, observed in some studies, has been associated with an increase in the metastatic potential of triple-negative breast cancer cells [64,79]. The discrepancy in the results obtained in different studies highlights the importance of context in assessing the effects of OT in breast cancer and could also explain the influence of oxytocin levels in the rules generated in the different models.

GnRH also appeared as an explainable hormonal factor in the J48, ADtree, and Random Forest models. This peptide hormone, produced in the hypothalamus, stimulates the release of luteinizing hormone (LH) and follicle-stimulating hormone (FSH) by the pituitary gland, which in turn induces steroid production in the gonads, playing a key role in the regulation of the reproductive system but also enhancing the growth and progression of estrogen-dependent tumors, as occurs with breast cancer [80,81]. In fact, GnRH analogs are used to suppress ovarian function based on the premise that reducing the circulating estrogen levels can inhibit hormone-dependent tumor growth [82]. It has been recently described that, in premenopausal women with hormone receptor-positive (HR+) breast cancer, ovarian function suppression using GnRH analogs such as goserelin and leuprolide has become an important therapeutic strategy [82]. A meta-analysis of 25 randomized clinical trials, involving nearly 15,000 premenopausal women with HR+ breast cancer, also demonstrated that ovarian ablation or ovarian function suppression using GnRH significantly reduced the 15-year risk of recurrence and mortality [82]. Finally, in addition to the role of GnRH in suppressing ovarian function, there has been research on the use of GnRH analogs as vectors for the targeted delivery of chemotherapeutic drugs to tumor cells. Preclinical studies have shown that the conjugation of GnRH analogs to cytotoxic agents, such as doxorubicin, can increase the treatment efficacy and reduce systemic toxicity [83,84].

The ADtree and Random Forest models also included β-endorphin as an explainable parameter. Neurobehavioural stress can influence the growth and progression of various types of cancer, including breast cancer [35,85]. This is due to the ability of psychological factors to alter the functioning of the immune and endocrine systems, which can affect the tumor growth and spread. Indeed, stress relief in cancer patients promotes faster recovery and a sense of well-being in those receiving radiotherapy or chemotherapy [35,85]. Within this context, opioid peptides, and in particular β-endorphin, have emerged as possible mediators of immune alterations that may contribute to tumor development [35]. β-endorphin plays a crucial role in the bidirectional connection between the immune and neuroendocrine systems, which could explain the effects of stress on the immune capacity against cancer [35,86]. Several studies have shown that healthy women have elevated levels of β-endorphin, with these being even higher in postmenopausal women. However, in women with breast cancer, significantly lower levels are observed, with no differences between premenopausal and postmenopausal women [35]. These findings suggest that decreased levels of β-endorphin may be associated with the development of breast cancer. Neoadjuvant chemotherapy, a common treatment for breast cancer, only improves β-endorphin levels in postmenopausal women, without reaching the levels of healthy women. Furthermore, both premenopausal and postmenopausal women treated with neoadjuvant chemotherapy maintain elevated cortisol levels, indicating a persistent stressful situation [35]. Our findings underscore the significance of our explainable artificial intelligence models in elucidating the critical role of β-endorphin modulation in breast cancer patients to the medical and research community. They could not only facilitate the potential therapeutic implications of β-endorphin regulation, thereby informing clinicians and researchers about novel avenues for intervention in breast cancer management, but could also enhance the understanding of the neuroendocrine basis of breast cancer, potentially guiding future research directions and clinical strategies. In fact, Sánchez et al. (2023) have pointed out that the regulation of stress levels by modulation with β-endorphin could be an alternative pharmacological therapy against tumor growth and development.

Vasopressin was other explainable parameter found in the Random Forest model. Also known as antidiuretic hormone, it has been studied in breast cancer since the presence of vasopressin receptors (V1a, V1b, and V2) in breast cancer cell lines, including MCF-7, were demonstrated [87,88]. Although it has traditionally been attributed a proliferative effect, recent research has suggested that its influence in breast cancer may be more complex and dependent on factors such as the dose, receptor subtype, and cellular context. Thus, a recent study by Alkafaas et al. [89] explored the effects of vasopressin on luminal A breast cancer cells, using a concentration of 100 nM. Their results suggested an anti-proliferative and anti-metastatic effect of vasopressin, with increased apoptosis, the increased expression of apoptosis markers such as Bax and Caspase-3, and decreased cell invasion. In addition, the increased expression of the autophagy marker LC3II and decreased Akt protein activation were found, suggesting the involvement of these pathways in the response to AVP. Garona et al. [90,91] investigated the desmopressin analog [V4Q5]dDAVP, a selective V2 receptor agonist, in preclinical models of breast cancer. They found that this analog inhibited tumor growth, angiogenesis, and metastasis in in vitro and in vivo models. Their findings in MDA-MB-231 xenografts showed reduced tumor growth and angiogenesis with its administration, while in immunocompetent mice with F3II mammary tumors, the complete inhibition of metastatic progression was observed.

4.2. Peptide Hormone-Regulating Proteolytic Enzymes

Apart from the previously described peptide hormones, our FURIA, ADTree, and Random Forest models also included the peptide hormone-regulating proteolytic enzymes insulin-regulated aminopeptidase (IRAP) and aminopeptidase B (APB). IRAP, also known as oxytocinase, is a zinc metallopeptidase with important regulatory functions. IRAP is the only known enzyme that cleaves oxytocin and vasopressin in vivo [92,93], two important peptide hormones also included in our proposed models. In addition to its role in hormone regulation, IRAP is also the high-affinity binding site for angiotensin IV (AngIV) type 4 (AT4) receptor ligands [94,95] and is linked to insulin-dependent glucose transporters via the translocation of glucose transporter type 4 (GLUT4) [96]. Previous studies have shown an association between the IRAP activity and the number and size of mammary tumors in an animal model of breast cancer [97,98]. A highly significant increase in IRAP activity has also been found in breast cancer tissue from female patients [99]. Another study by Ramírez-Expósito et al. [31] analyzed the IRAP activity in premenopausal and postmenopausal women with breast cancer, whether or not they had been treated with neoadjuvant chemotherapy. The results showed significant changes in the specific activity of circulating IRAP, with variations dependent on the hormonal status and chemotherapy administration. Thus, IRAP is involved in mechanisms related not only to the regulation of oxytocin and/or vasopressin, but also to the local mammary renin–angiotensin system (RAS) through AngIV and its role in glucose transport through the IRAP/GLUT4 system. The local RAS has been implicated in several of the hallmarks of cancer, including the processes of tumor cell survival, proliferation, and dissemination. The best known RAS signaling is mediated by angiotensin II (AngII) through its type 1 receptor (AT1), which increases cell proliferation and modulates vascular cell growth during angiogenesis [100]. The study by Ramírez-Expósito et al. [31,101] suggested an increase in the circulating levels of AngIV, a substrate of IRAP. AT4 receptor ligands dose-dependently inhibit the catalytic activity of IRAP [102]. Therefore, the changes in the IRAP-specific activity observed in women with breast cancer may be related to local RAS dysregulation. Additionally, mammalian cells use glucose as a major source of energy production. The presence of GLUT4 has been demonstrated in insulin-independent tissues, such as the mammary gland, and also in breast cancer cells, where it plays a role in cell cycle progression [103,104].

IRAP and GLUT4 colocalize in specialized vesicles that translocate to the cell membrane in response to stimuli such as insulin. GLUT4 inhibition in breast cancer cells has been shown to inhibit cell proliferation and decrease cell viability [105]. The changes in IRAP activity observed in [31] could affect the availability of IRAP and, consequently, the translocation of GLUT4 to the cell membrane, which would influence glucose metabolism in tumor cells.

Aminopeptidase B (APB) has also been studied in relation to breast cancer because of its role in the RAS [106]. Thus, APB activity has shown significant patterns of alteration [107]. Martinez et al. [99] found that the membrane-bound APB activity was significantly elevated in neoplastic tissue compared to unaffected tissue, suggesting that APB may play an active role in the tumor microenvironment. Furthermore, in a study by Carrera et al. [108], a decrease in soluble APB was observed in mammary tissue from rats with N-methyl-nitrosourea (NMU)-induced tumors, suggesting that the subcellular location of this enzyme could be an important factor in its influence on a tumor. APB is an enzyme involved in the conversion of angiotensin III (AngIII) to AngIV [108]. Despite the important role of Ang II in cell proliferation, angiogenesis, and tumor growth [106], the metabolism of Ang II, via RAS regulatory enzymes such as APB, is thought to affect tumor progression. Ruíz-Sanjuan et al. [106] suggested that elevated levels of APB may indicate an increased conversion of AngIII to Ang IV, a peptide that binds to AT4 receptors and is also associated with IRAP activity, as previously described. In a two-year follow-up study, Ramirez-Exposito et al. [28] found that APB levels remained significantly unchanged in women with breast cancer. However, a small increase in the APB levels was observed in premenopausal women treated with chemotherapy but was not sustained afterwards. These data suggest that the initial changes in APB activity, associated with tumor development or the effect of chemotherapy, are largely stable over time. Therefore, APB may influence breast cancer development through its role in angiotensin regulation, which may alter cell proliferation and angiogenesis in the tumor microenvironment.

4.3. Progesterone

The steroid hormone progesterone was another explainable variable proposed by the Random Forest model. The relationship between progesterone and breast cancer is complex and has been the subject of numerous studies for decades, generating controversy as to whether progesterone is a stimulating or inhibitory factor in the development of this disease [109,110]. Initially, the presence of progesterone receptors (PRs) was recognized as a marker of a response to endocrine therapies and as an indicator of the presence of functional estrogen receptors [109,110]. However, subsequent research has revealed that the action of progesterone can vary depending on factors such as the dose, the type of progestin used, the presence or absence of estrogens, and the clinical context [110,111]. Progesterone exerts its effects through its progesterone receptors (PRs), which are transcription factors that regulate various physiological processes in mammary cells [110]. PRs are found in two isoforms, A and B, which differ in their structure and function [112,113]. These isoforms can operate through different signaling pathways in target cells [112,113]. The activity of PRs is also modulated by the binding of cofactors and by post-translational modifications, such as phosphorylation [113,114]. Progesterone plays a crucial role in the regulation of the cell cycle and proliferation in mammary tissue [111]. It has been observed that progesterone can induce cell proliferation through the activation of the RANKL and WNT4 pathways [110,111,115–119]. In addition, it has been proposed that progesterone can influence the expansion of cancer stem cells, which could contribute to tumor progression [110,115,118,119]. However, there is also evidence that progesterone can have anti-proliferative effects in certain circumstances [120]. The interaction between progesterone and estrogen is fundamental in the context of breast cancer. PRs are target genes regulated by estrogen receptors (ERs) [110,116]. PRs can modulate ER activity, which can have both agonist and antagonist effects [121]. Some studies have suggested that the presence of PRs in ER+ breast tumors is associated with a better response to endocrine therapy [122]. However, the loss of PRs in these tumors has been linked to resistance to treatments such as tamoxifen [123–125]. The relationship between endogenous progesterone and the breast cancer risk is not entirely clear [111]. Some studies suggest that elevated levels of progesterone during the luteal phase of the menstrual cycle may increase the risk of breast cancer, especially in women carrying BRCA mutations [111]. It has also been proposed that cumulative exposure to progesterone throughout a woman's reproductive life could influence the risk [111]. Therefore, progesterone exerts a complex role in the development of breast cancer, with actions that can be both promoting and inhibitory, depending on the cellular and molecular context [110,111].

4.4. Other Enzyme Activities

The ADTree model included the 5′-nucleotidase activity as an explainable variable, whereas the Random Forest included the cholinesterase activity as an explainable variable. Ecto-5′-nucleotidase (CD73) is emerging as an enzyme of great interest due to its involvement in the regulation of the tumor microenvironment and its influence on disease progression [126–129]. This enzyme, located on the cell surface, catalyzes the hydrolysis of 5′-AMP to adenosine and inorganic phosphate [128,130]. Adenosine, the product of this reaction, acts as a potent immunosuppressor in the tumor microenvironment, promoting immune evasion and tumor growth [127,129,131,132]. Its presence in breast cancer is correlated with a worse prognosis, a higher risk of metastasis, and resistance to chemotherapy [127,133]. In particular, in TNBC, a high expression of CD73 has been observed, contributing to treatment resistance and tumor progression [127,134]. Also, CD73 participates in the promotion of tumor angiogenesis, tumor growth, and metastasis [135,136]. On the other hand, some studies have suggested the involvement of cholinesterases in mammary tumorigenesis. In fact, Ruiz-Espejo et al. [137] found that acetylcholinesterase activity doubled in breast tumor tissue compared to in normal tissue. Additionally, the glycosylation of acetylcholinesterase was affected in breast cancer [137]. Bernardi et al. [138] observed a high frequency of amplifications in the gene encoding this enzyme in sporadic breast tumors, which could be responsible for its increased expression. Furthermore, the promoter of the acetylcholinesterase gene contains binding sites for estrogen receptors [139], and the abundance of estrogen receptors in mammary tissue also suggests that these could be involved in regulating acetylcholinesterase expression. Other factors, such as the stimulation of muscarinic receptors, can induce the synthesis of transcription factors that activate the transcription of the acetylcholinesterase gene [140]. Additionally, stimulation by cytokines like interleukin-1β (IL-1β), which is actively synthesized in breast cancer [141], could increase the expression of acetylcholinesterase mRNA and, consequently, its enzymatic activity [142]. Therefore, tumor-specific genetic and biological alterations could influence acetylcholinesterase activity, or its alteration may be a consequence of neoplastic transformations that contribute to the maintenance of the tumor process [138].

4.5. Iron

The Random Forest model also proposed serum-circulating iron as an explainable variable for breast cancer diagnosis. Iron is an essential micronutrient for the human body, playing a crucial role in various physiological processes such as oxygen transport, immune function, and energy production [143]. However, excess iron can generate oxidative stress and promote the activation of oncogenes, making it a potentially implicated factor in cancer development [144,145]. Previous epidemiological studies have yielded conflicting results on the association between the serum iron status and breast cancer risk [146]. Some studies have reported a positive association, while others have found no significant relationship [147–152]. This inconsistency in results may be attributed to limitations inherent to observational studies, such as the presence of confounding factors and the possibility of reverse causality [153]. To address these limitations, Hou et al. [146] conducted a two-sample Mendelian randomization study to explore the causal relationship between the serum iron status and breast cancer risk. This method used genetic variants as instruments to estimate the causal effect of exposure on an outcome, minimizing bias from confounding factors [154,155]. The results of this study suggested that serum transferrin levels might be positively associated with the risk of estrogen receptor-negative (ER-negative) breast cancer, although no significant effect was observed on the risk of breast cancer in general or its ER-positive subtypes [146]. In another study, Hua et al. [156] retrospectively analyzed the prognostic value of the serum iron level in women with early-stage triple-negative breast cancer (TNBC). They divided patients into two groups based on a cut-off point for the serum iron level and found that those with lower iron levels had significantly longer disease-free survival and overall survival compared to patients with higher levels. These findings suggest that the serum iron level could be a potential prognostic biomarker in patients with early-stage TNBC. It is important to note that most studies assess the iron status in the body by measuring the circulating levels of iron-binding proteins, such as transferrin and ferritin [157,158]. However, the direct measurement of serum iron might provide a more accurate assessment of the iron status in the body [150–152,159].

4.6. Body Mass Index

The BMI was another of the explainable variables proposed by the Random Forest model. One of the primary mechanisms by which the BMI influences breast cancer outcomes is through its impact on the mammographic imaging quality. Increased adipose tissue in the breast can significantly affect the sensitivity and specificity of mammography, potentially leading to delayed or missed diagnoses. Calvo et al. [160] have described significant differences between the BMI and mammographic density, with a higher BMI in patients with type A density ("fatty breast") and a lower BMI in patients with type D density ("dense breast"). These differences were more relevant in premenopausal women than in postmenopausal women, although no association was found between the mammographic density and the molecular subtype of cancer [161]. On the other hand, Cao et al. [162] found a positive association between the BMI and breast cancer risk. Women with obesity (BMI >= 28 kg/m^2) had a 2.09-fold higher risk of developing breast cancer compared to women with a normal weight (BMI of 18.5–23.9 kg/m^2). A family history of cancer also increased the risk of breast cancer, and an additive interaction was observed between the BMI and family history. In other study, Jiang et al. [163] analyzed the association between the BMI and breast cancer risk in a large cohort of Chinese women, confirming the association with an increased risk of breast cancer, especially in postmenopausal women. Wada et al. [164] also found that in postmenopausal women, a high BMI was significantly associated with an increased risk of breast cancer, but no significant association was observed in premenopausal women. Cubelos-Fernand et al. [165] used the so-called "Clínica Universidad de Navarra Body Fatness Estimator (CUN-BAE)" to assess the relationship between body fat and postmenopausal breast cancer. An increased risk of breast cancer was also observed in women with a higher BMI and CUN-BAE score, with covariates such as the age at recruitment, age at menarche, number of children, and breastfeeding duration also appearing in the multivariate model. Holm et al. [166] investigated the relationship between C-reactive protein (CRP) quartiles and the disease-free survival and overall survival in patients with stage I-III breast cancer. A high CRP level was found to be associated with worse survival in all BMI groups, but the association was more pronounced in overweight and obese women. Jiang et al. [163] focused their study on the relationship between the BMI and the incidence of chronic arm lymphedema (BCRL) after breast cancer surgery. Patients with a BMI >= 30 kg/m^2 had a significantly higher rate of BCRL at 1, 2, and 3 years compared with patients with a BMI < 25 kg/m^2. Ogiya et al. [167] also investigated the incidence of lymphedema after breast cancer surgery and its relationship with the BMI. A BMI >= 25 kg/m^2 was found to be significantly associated with an increased risk of lymphedema at 1 year after surgery. Quartuccio et al. [168] found that the BMI appears to correlate with an increased recurrence rate, especially in the breast, and increased glucose uptake in postmenopausal patients with recurrent breast cancer. Therefore, a complex relationship also exists between the BMI and breast cancer, with a high BMI, especially in postmenopausal women, increasing the risk of breast cancer development, an important variable that must also be taken into account [169].

4.7. Total Protein

Very few studies have investigated the relationship between the total serum protein levels and breast cancer. One study revealed that breast cancer patients had significantly higher total serum protein levels compared to the control group [170]. This finding is in agreement with previous studies that have explored the assessment of the total serum protein in patients with brain tumors [171] and with significant increases in the serum protein levels in patients with lung cancer [172]. The increase in the total serum protein concentration may be due to the fact that the total serum protein is composed of albumin and other proteins, collectively called globulins. It is known that the serum albumin concentration can change under oxidative stress, such as that associated with cancer [173]. Furthermore, as plasma circulates through tissues, it picks up proteins released due to physiological events such as tissue remodeling, trauma, and cell death, leading to an increase in the total serum protein.

5. Conclusions

This research provides a solid foundation for future clinical research and practical applications, directly challenging conventional approaches to breast cancer diagnosis. The use of XAI models and the identification of relevant serum biomarkers offer a new perspective for early detection, understanding pathogenesis, and developing more effective and personalized therapies. The results not only enrich the scientific knowledge but also open a wide array of possibilities for transforming breast cancer management. They include new approaches to early detection, proposing an alternative to traditional imaging techniques, such as mammography, by focusing on the analysis of serum biomarkers and providing new targets for research and the development of more precise and less invasive diagnostic tools.

Secondly, they provide a deeper understanding of pathogenesis due to the ability of XAI models to generate rules and decision trees that allow researchers to better understand the underlying biological mechanisms in breast cancer development. For example, the identification of oxytocin as a primary predictive factor across several models suggests that this hormone plays a more complex and multifaceted role in pathogenesis than previously thought, which could impact the development of targeted therapies. The detection of complex interactions among hormonal, enzymatic, and metabolic factors opens new avenues for understanding the molecular alterations that drive the disease.

Thirdly, this study lays the groundwork for the design of more individualized treatments by providing information on the specific biomarkers relevant to a patient's breast cancer risk, an approach that could be especially valuable in managing breast cancers with high heterogeneity and resistance to conventional therapies. Ultimately, this research proposes a transformation in breast cancer management by promoting a more integrated and personalized approach, based on a deeper understanding of the disease and the use of more precise diagnostic tools. The adoption of XAI models could lead to earlier detection, which translates to better survival rates and a reduction in the use of more invasive and costly treatments. Moreover, the models' capacity to generate explainable information promotes greater confidence in the diagnostic process.

Lastly, the practical challenges of implementing these models in real-world clinical settings should be addressed. All of the used models are highly efficient and simple to implement and can be run even on low-end computers. For these models to be used in a real-world clinical setting, all that would be required would be to store the trained models on a computer, gather all relevant biomarker data from a given patient who has to be diagnosed, and follow the rules generated by a given model. Since several models have been proposed, the data could be run through any of them. This additionally prevent biases and limitations: if all models predict that the patient is likely to develop breast cancer, the chance of a misdiagnosis severely decreases.

Author Contributions: Conceptualization, J.M.M.-R., C.C., M.J.R.-E. and J.M.M.-M.; methodology, J.M.M.-R. and C.C.; software, J.M.M.-R. and C.C.; validation, M.J.R.-E. and J.M.M.-M.; formal analysis, J.M.M.-R.; investigation, J.M.M.-R. and J.M.M.-M.; resources, C.C.; data curation, M.J.R.-E. and J.M.M.-M.; writing—original draft preparation, J.M.M.-R. and J.M.M.-M.; writing—review and editing, J.M.M.-R., C.C., M.J.R.-E. and J.M.M.-M.; supervision, J.M.M.-R., C.C., M.J.R.-E. and J.M.M.-M.; project administration, C.C.; funding acquisition, C.C. All authors have read and agreed to the published version of the manuscript.

Funding: This work was financed by the Ministry of Science, Innovation and Universities, Spain, with the code PID2023-149511OB-I00.

Institutional Review Board Statement: The study was conducted in accordance with the Declaration of Helsinki and approved by the local Clinical Research Ethical Committee at the University Hospital of Jaén (protocol code 0756-N-18), as stated in the corresponding references.

Informed Consent Statement: Informed consent was obtained from all subjects involved in the study.

Data Availability Statement: The data are unavailable due to privacy or ethical restrictions.

Conflicts of Interest: The authors declare no conflicts of interest. The funders had no role in the design of the study; in the collection, analyses, or interpretation of data; in the writing of the manuscript; or in the decision to publish the results.

Abbreviations

The following abbreviations are used in this manuscript:

ADTree	Alternating Decision Tree
AI	Artificial intelligence
ALAT	Alanine aminotransferase
AngII	Angiotensin II
AngIII	Angiotensin III
AngIV	Angiotensin IV
APA	Aminopeptidase A
APB	Aminopeptidase B
APN	Aminopeptidase N
ASAP	Aspartyl aminopeptidase
ASAT	Aspartate aminotransferase
AT1	Angiotensin II type 1 receptor

BCRL	Breast cancer-related lymphedema
CF	Confidence factor
CKMB	Creatine kinase muscle–brain isoform
CNN	Convolutional neural network
CPK	Creatine phosphokinase
CRP	C-reactive protein
CUN-BAE	Clínica Universidad de Navarra Body Fatness Estimator
DCIS	Ductal carcinoma in situ
ER	Estrogen receptor
FN	False Negative
FP	False Positive
FSH	Follicle-stimulating hormone
FURIA	Fuzzy Unordered Rule Induction Algorithm
GGT	Gamma-glutamyl-transferase
GH	Growth hormone
GLUT4	Glucose transporter type 4
GnRH	Gonadotropin-releasing hormone
HDL	High-density lipoprotein
HR+	Hormone receptor-positive
IRAP	Insulin-regulated aminopeptidase
JRIP	Java implementation of the RIPPER algorithm
LDH	Lactate dehydrogenase
LDL	Low-density lipoprotein
LH	Luteinizing hormone
miRNAs	MicroRNAs
MRI	Magnetic resonance imaging
NMU	N-methyl-nitrosourea
OneR	One Rule
OTR	Oxytocin receptor
PR	Progesterone receptor
RANKL	Receptor activator of nuclear factor kappa-B ligand
RAS	Renin–angiotensin system
RNN	Recurrent neural network
TAI	Trustworthy artificial intelligence
TN	True Negative
TNBC	Triple-negative breast cancer
TP	True Positive
TSH	Thyroid-stimulating hormone
WNT4	Wingless-related integration site 4

References

1. Alsheikhy, A.A.; Said, Y.; Shawly, T.; Alzahrani, A.K.; Lahza, H. Biomedical Diagnosis of Breast Cancer Using Deep Learning and Multiple Classifiers. *Diagnostics* **2022**, *12*, 2863. [CrossRef] [PubMed]
2. Shankar, G.S.; Onyema, E.M.; Kavin, B.P.; Gude, V.; Prasad, B.S. Breast Cancer Diagnosis Using Virtualization and Extreme Learning Algorithm Based on Deep Feed Forward Networks. *Biomed. Eng. Comput. Biol.* **2024**, *15*, 11795972241278907. [CrossRef]
3. Teoh, J.R.; Hasikin, K.; Lai, K.W.; Wu, X.; Li, C. Enhancing early breast cancer diagnosis through automated microcalcification detection using an optimized ensemble deep learning framework. *PeerJ Comput. Sci.* **2024**, *10*, e2082. [CrossRef]
4. Kanbayashi, C.; Iwata, H. Update on the management of ductal carcinoma in situ of the breast: Current approach and future perspectives. *Jpn. J. Clin. Oncol.* **2024**, *55*, 4–11. [CrossRef]
5. Dewangan, K.K.; Dewangan, D.K.; Sahu, S.P.; Janghel, R. Breast cancer diagnosis in an early stage using novel deep learning with hybrid optimization technique. *Multimed. Tools Appl.* **2022**, *81*, 13935–13960. [CrossRef]
6. Sanchez Galindo, B.; Lopez-Torres Hidalgo, J.; Lopez Gonzalez, A.; Sanchez Martinez, C.M.; Castano Diaz, M.; Rabanales Sotos, J. Time intervals for diagnosis and treatment in breast cancer patients. *Aten. Primaria* **2024**, *57*, 103153. [CrossRef]
7. Sung, H.; Ferlay, J.; Siegel, R.L.; Laversanne, M.; Soerjomataram, I.; Jemal, A.; Bray, F. Global Cancer Statistics 2020: GLOBOCAN Estimates of Incidence and Mortality Worldwide for 36 Cancers in 185 Countries. *CA Cancer J. Clin.* **2021**, *71*, 209–249. [CrossRef]

8. Uchikov, P.; Khalid, U.; Dedaj-Salad, G.H.; Ghale, D.; Rajadurai, H.; Kraeva, M.; Kraev, K.; Hristov, B.; Doykov, M.; Mitova, V.; et al. Artificial Intelligence in Breast Cancer Diagnosis and Treatment: Advances in Imaging, Pathology, and Personalized Care. *Life* **2024**, *14*, 1451. [CrossRef]
9. Kavitha, T.; Mathai, P.P.; Karthikeyan, C.; Ashok, M.; Kohar, R.; Avanija, J.; Neelakandan, S. Deep Learning Based Capsule Neural Network Model for Breast Cancer Diagnosis Using Mammogram Images. *Interdiscip. Sci.* **2022**, *14*, 113–129. [CrossRef] [PubMed]
10. Aljuaid, H.; Alturki, N.; Alsubaie, N.; Cavallaro, L.; Liotta, A. Computer-aided diagnosis for breast cancer classification using deep neural networks and transfer learning. *Comput. Methods Programs Biomed.* **2022**, *223*, 106951. [CrossRef] [PubMed]
11. Devi, S.; Kaul Ghanekar, R.; Pande, J.A.; Dumbre, D.; Chavan, R.; Gupta, H. Prediction and Diagnosis of Breast Cancer Using Machine and Modern Deep Learning Models. *Asian Pac. J. Cancer Prev.* **2024**, *25*, 1077–1085. [CrossRef]
12. Alam, T.; Shia, W.C.; Hsu, F.R.; Hassan, T. Improving Breast Cancer Detection and Diagnosis through Semantic Segmentation Using the Unet3+ Deep Learning Framework. *Biomedicines* **2023**, *11*, 1536. [CrossRef] [PubMed]
13. Hu, Q.; Whitney, H.M.; Giger, M.L. A deep learning methodology for improved breast cancer diagnosis using multiparametric MRI. *Sci. Rep.* **2020**, *10*, 10536. [CrossRef] [PubMed]
14. Kaur, J.; Kaur, P. A systematic literature analysis of multi-organ cancer diagnosis using deep learning techniques. *Comput. Biol. Med.* **2024**, *179*, 108910. [CrossRef] [PubMed]
15. Schettini, F.; Braso-Maristany, F.; Pascual, T.; Lorman-Carbo, N.; Nucera, S.; Bergamino, M.; Galvan, P.; Conte, B.; Segui, E.; Garcia Fructuoso, I.; et al. Identifying predictors of treatment response and molecular changes induced by neoadjuvant chemotherapy and endocrine therapy in hormone receptor-positive/HER2-negative breast cancer: The NEOENDO translational study. *ESMO Open* **2024**, *9*, 103989. [CrossRef] [PubMed]
16. Zhang, M.; Li, S.; Xue, M.; Zhu, Q. Two-stage classification strategy for breast cancer diagnosis using ultrasound-guided diffuse optical tomography and deep learning. *J. Biomed. Opt.* **2023**, *28*, 086002. [CrossRef] [PubMed]
17. Zhang, Y.; Liu, Y.L.; Nie, K.; Zhou, J.; Chen, Z.; Chen, J.H.; Wang, X.; Kim, B.; Parajuli, R.; Mehta, R.S.; et al. Deep Learning-based Automatic Diagnosis of Breast Cancer on MRI Using Mask R-CNN for Detection Followed by ResNet50 for Classification. *Acad. Radiol.* **2023**, *30* (Suppl. 2), S161–S171. [CrossRef] [PubMed]
18. Ejiyi, C.J.; Qin, Z.; Agbesi, V.K.; Yi, D.; Atwereboannah, A.A.; Chikwendu, I.A.; Bamisile, O.F.; Bakanina Kissanga, G.M.; Bamisile, O.O. Advancing cancer diagnosis and prognostication through deep learning mastery in breast, colon, and lung histopathology with ResoMergeNet. *Comput. Biol. Med.* **2024**, *185*, 109494. [CrossRef]
19. Cheng, Z.; Li, H.; Chen, C.; Lv, X.; Zuo, E.; Xie, X.; Li, Z.; Liu, P.; Li, H.; Chen, C. Application of serum SERS technology based on thermally annealed silver nanoparticle composite substrate in breast cancer. *Photodiagnosis Photodyn. Ther.* **2023**, *41*, 103284. [CrossRef]
20. Ma, X.; Cheng, H.; Hou, J.; Jia, Z.; Wu, G.; Lü, X.; Li, H.; Zheng, X.; Chen, C. Detection of breast cancer based on novel porous silicon Bragg reflector surface-enhanced Raman spectroscopy-active structure. *Chin. Opt. Lett.* **2020**, *18*, 051701. [CrossRef]
21. Zeng, Q.; Chen, C.; Chen, C.; Song, H.; Li, M.; Yan, J.; Lv, X. Serum Raman spectroscopy combined with convolutional neural network for rapid diagnosis of HER2-positive and triple-negative breast cancer. *Spectrochim. Acta A Mol. Biomol. Spectrosc.* **2023**, *286*, 122000. [CrossRef]
22. Samala, R.K.; Heang-Ping, C.; Hadjiiski, L.; Helvie, M.A.; Richter, C.D.; Cha, K.H. Breast Cancer Diagnosis in Digital Breast Tomosynthesis: Effects of Training Sample Size on Multi-Stage Transfer Learning Using Deep Neural Nets. *IEEE Trans. Med. Imaging* **2019**, *38*, 686–696. [CrossRef] [PubMed]
23. Yang, J.; Hussein Kadir, D. Data mining techniques in breast cancer diagnosis at the cellular-molecular level. *J. Cancer Res. Clin. Oncol.* **2023**, *149*, 12605–12620. [CrossRef] [PubMed]
24. Zhang, S.; Yang, B.; Yang, H.; Zhao, J.; Zhang, Y.; Gao, Y.; Monteiro, O.; Zhang, K.; Liu, B.; Wang, S. Potential rapid intraoperative cancer diagnosis using dynamic full-field optical coherence tomography and deep learning: A prospective cohort study in breast cancer patients. *Sci. Bull.* **2024**, *69*, 1748–1756. [CrossRef]
25. Hastie, T.; Tibshirani, R.; Friedman, J. *The Elements of Statistical Learning: Data Mining, Inference and Prediction*, 2nd ed.; Springer: Berlin/Heidelberg, Germany, 2009.
26. Cuadros, M. The EU Trustworthy Artificial Intelligence: How the EU Wants to Regulate the Artificial Intelligence Practices. In *Actas del Congreso Internacional de Ingeniería de Sistemas*; University of Lima: Lima, Peru, 2021; pp. 173–181. [CrossRef]
27. Ramirez-Exposito, M.J.; Carrera-Gonzalez Mdel, P.; Mayas, M.D.; Duenas, B.; Martinez-Ferrol, J.; Martinez-Martos, J.M. Neoadjuvant chemotherapy modifies serum angiotensinase activities in women with breast cancer. *Maturitas* **2012**, *72*, 79–83. [CrossRef] [PubMed]
28. Ramirez-Exposito, M.J.; Duenas-Rodriguez, B.; Martinez-Martos, J.M. Circulating renin-angiotensin system-regulating specific aminopeptidase activities in pre- and post- menopausal women with breast cancer treated or not with neoadyuvant chemotherapy. A two years follow up study. *Breast* **2019**, *43*, 28–30. [CrossRef]

29. Ramirez-Exposito, M.J.; Sanchez-Lopez, E.; Cueto-Urena, C.; Duenas, B.; Carrera-Gonzalez, P.; Navarro-Cecilia, J.; Mayas, M.D.; Arias de Saavedra, J.M.; Sanchez-Agesta, R.; Martinez-Martos, J.M. Circulating oxidative stress parameters in pre- and post-menopausal healthy women and in women suffering from breast cancer treated or not with neoadjuvant chemotherapy. *Exp. Gerontol.* **2014**, *58*, 34–42. [CrossRef]
30. Ramirez-Exposito, M.J.; Urbano-Polo, N.; Duenas, B.; Navarro-Cecilia, J.; Ramirez-Tortosa, C.; Martin-Salvago, M.D.; Martinez-Martos, J.M. Redox status in the sentinel lymph node of women with breast cancer. *Upsala J. Med. Sci.* **2017**, *122*, 207–216. [CrossRef] [PubMed]
31. Ramirez-Exposito, M.J.; Martinez-Martos, J.M.; Duenas-Rodriguez, B.; Navarro-Cecilia, J.; Carrera-Gonzalez, M.P. Neoadjuvant chemotherapy modifies serum pyrrolidone carboxypeptidase specific activity in women with breast cancer and influences circulating levels of GnRH and gonadotropins. *Breast Cancer Res. Treat.* **2020**, *182*, 751–760. [CrossRef]
32. Carrera-Gonzalez Mdel, P.; Ramirez-Exposito, M.J.; Duenas, B.; Martinez-Ferrol, J.; Mayas, M.D.; Martinez-Martos, J.M. Putative relationship between hormonal status and serum pyrrolidone carboxypeptidase activity in pre- and post- menopausal women with breast cancer. *Breast* **2012**, *21*, 751–754. [CrossRef]
33. Martinez-Martos, J.M.; del Pilar Carrera-Gonzalez, M.; Duenas, B.; Mayas, M.D.; Garcia, M.J.; Ramirez-Exposito, M.J. Renin angiotensin system-regulating aminopeptidase activities in serum of pre- and postmenopausal women with breast cancer. *Breast* **2011**, *20*, 444–447. [CrossRef] [PubMed]
34. Navarro-Cecilia, J.; Duenas-Rodriguez, B.; Luque-Lopez, C.; Ramirez-Exposito, M.J.; Martinez-Ferrol, J.; Ruiz-Mateas, A.; Urena, C.; Carrera-Gonzalez, M.P.; Mayas, M.D.; Martinez-Martos, J.M. Intraoperative sentinel node biopsy by one-step nucleic acid amplification (OSNA) avoids axillary lymphadenectomy in women with breast cancer treated with neoadjuvant chemotherapy. *Eur. J. Surg. Oncol.* **2013**, *39*, 873–879. [CrossRef] [PubMed]
35. Ramirez-Exposito, M.J.; Duenas-Rodriguez, B.; Carrera-Gonzalez, M.P.; Navarro-Cecilia, J.; Martinez-Martos, J.M. Circulating levels of beta-endorphin and cortisol in breast cancer. *Compr. Psychoneuroendocrinol.* **2021**, *5*, 100028. [CrossRef] [PubMed]
36. Chawla, N.V.; Bowyer, K.W.; Hall, L.O.; Kegelmeyer, W.P. SMOTE: Synthetic Minority Over-sampling Technique. *J. Artif. Intell. Res.* **2002**, *16*, 321–357. [CrossRef]
37. Holte, R.C. Very Simple Classification Rules Perform Well on Most Commonly Used Datasets. *Mach. Learn.* **1993**, *11*, 63–90. [CrossRef]
38. Cohen, W.W. Fast Effective Rule Induction. In *Machine Learning Proceedings 1995, Proceedings of the Twelfth International Conference on Machine Learning, Tahoe City, CA, USA, 9–12 July 1995*; Morgan Kaufmann: Burlington, MA, USA, 1995; pp. 115–123.
39. Wang, Y. Review on greedy algorithm. *Theor. Nat. Sci.* **2023**, *14*, 233–239. [CrossRef]
40. Hühn, J.; Hüllermeier, E. FURIA: An algorithm for unordered fuzzy rule induction. *Data Min. Knowl. Discov.* **2009**, *19*, 293–319. [CrossRef]
41. Ihya, R.; Namir, A.; Filali, S.; Aitdaoud, M.; Guerss, f.Z. J48 algorithms of machine learning for predicting user's the acceptance of an E-orientation systems. In Proceedings of the SCA '19: The 4th International Conference on Smart City Applications, Casablanca, Morocco, 2–4 October 2019; pp. 1–8. [CrossRef]
42. Quinlan, J.R. *C4.5: Programs for Machine Learning*; Morgan Kaufmann Publishers Inc.: San Francisco, CA, USA, 1993.
43. Venkatesan, E.; Thambusamy, V. Performance Analysis of Decision Tree Algorithms for Breast Cancer Classification. *Indian J. Sci. Technol.* **2015**, *8*, IPL0625. [CrossRef]
44. Saravanan, N.; Gayathri, V. Performance and Classification Evaluation of J48 Algorithm and Kendall's Based J48 Algorithm (KNJ48). *Int. J. Comput. Trends Technol.* **2018**, *59*, 73–80. [CrossRef]
45. Freund, Y.; Mason, L. The Alternating Decision Tree Learning Algorithm. In Proceedings of the Sixteenth International Conference on Machine Learning, Bled, Slovenia, 27–30 June 1999; ICML '99; pp. 124–133.
46. Pfahringer, B.; Holmes, G.; Kirkby, R. Optimizing the Induction of Alternating Decision Trees. In Proceedings of the 5th Pacific-Asia Conference on Knowledge Discovery and Data Mining, Hong Kong, China, 16–18 April 2001; PAKDD '01; pp. 477–487.
47. Breiman, L. Random Forests. *Mach. Learn.* **2001**, *45*, 5–32. [CrossRef]
48. Nasser, M.; Yusof, U.K. Deep learning based methods for breast cancer diagnosis: A systematic review and future direction. *Diagnostics* **2023**, *13*, 161. [CrossRef] [PubMed]
49. Manimurugan, S.; Karthikeyan, P.; Aborokbah, M.; Narmatha, C.; Ganesan, S. Breast cancer diagnosis model using stacked autoencoder with particle swarm optimization. *Ain Shams Eng. J.* **2024**, *15*, 102734. [CrossRef]
50. Singh, S.J.; Rajaraman, R.; Verlekar, T.T. Breast Cancer Prediction Using Auto-Encoders. In Proceedings of the Data Management, Analytics and Innovation, Pune, India, 14–16 January 2022; Goswami, S., Barara, I.S., Goje, A., Mohan, C., Bruckstein, A.M., Eds.; Springer: Singapore, 2023; pp. 121–132.
51. Alotaibi, M.; Aljouie, A.; Alluhaidan, N.; Qureshi, W.; Almatar, H.; Alduhayan, R.; Alsomaie, B.; Almazroa, A. Breast cancer classification based on convolutional neural network and image fusion approaches using ultrasound images. *Heliyon* **2023**, *9*, e22406. [CrossRef] [PubMed]

52. Das, H.S.; Das, A.; Neog, A.; Mallik, S.; Bora, K.; Zhao, Z. Breast cancer detection: Shallow convolutional neural network against deep convolutional neural networks based approach. *Front. Genet.* **2022**, *13*, 1097207. [CrossRef]
53. Xiang, Q.; Li, D.; Hu, Z.; Yuan, Y.; Sun, Y.; Zhu, Y.; Fu, Y.; Jiang, Y.; Hua, X. Quantum classical hybrid convolutional neural networks for breast cancer diagnosis. *Sci. Rep.* **2024**, *14*, 24699. [CrossRef] [PubMed]
54. Minnoor, M.; Baths, V. Diagnosis of Breast Cancer Using Random Forests. *Procedia Comput. Sci.* **2023**, *218*, 429–437. [CrossRef]
55. Whig, P.; Kouser, S.; Bhatia, A.B.; Nadikattu, R.R. Chapter 3—Prediction of breast cancer diagnosis using random forest classifier. In *Computational Intelligence and Modelling Techniques for Disease Detection in Mammogram Images*; Hemanth, D.J., Ed.; Academic Press: Cambridge, MA, USA, 2024; pp. 55–73. [CrossRef]
56. Clift, A.K.; Dodwell, D.; Lord, S.; Petrou, S.; Brady, M.; Collins, G.S.; Hippisley-Cox, J. Development and internal-external validation of statistical and machine learning models for breast cancer prognostication: Cohort study. *BMJ* **2023**, *381*, e073800. [CrossRef] [PubMed]
57. Khalid, A.; Mehmood, A.; Alabrah, A.; Alkhamees, B.F.; Amin, F.; AlSalman, H.; Choi, G.S. Breast Cancer Detection and Prevention Using Machine Learning. *Diagnostics* **2023**, *13*, 3113. [CrossRef] [PubMed]
58. Sammut, S.J.; Crispin-Ortuzar, M.; Chin, S.F.; Provenzano, E.; Bardwell, H.A.; Ma, W.; Cope, W.; Dariush, A.; Dawson, S.J.; Abraham, J.E.; et al. Multi-omic machine learning predictor of breast cancer therapy response. *Nature* **2022**, *601*, 623–629. [CrossRef]
59. Stark, G.F.; Hart, G.R.; Nartowt, B.J.; Deng, J. Predicting breast cancer risk using personal health data and machine learning models. *PLoS ONE* **2019**, *14*, e0226765. [CrossRef] [PubMed]
60. Zuo, D.; Yang, L.; Jin, Y.; Qi, H.; Liu, Y.; Ren, L. Machine learning-based models for the prediction of breast cancer recurrence risk. *BMC Med. Inf. Decis. Mak.* **2023**, *23*, 276. [CrossRef]
61. Ito, Y.; Kobayashi, T.; Kimura, T.; Matsuura, N.; Wakasugi, E.; Takeda, T.; Shimano, T.; Kubota, Y.; Nobunaga, T.; Makino, Y.; et al. Investigation of the oxytocin receptor expression in human breast cancer tissue using newly established monoclonal antibodies. *Endocrinology* **1996**, *137*, 773–779. [CrossRef] [PubMed]
62. Lerman, B.; Harricharran, T.; Ogunwobi, O.O. Oxytocin and cancer: An emerging link. *World J. Clin. Oncol.* **2018**, *9*, 74–82. [CrossRef]
63. Liu, H.; Gruber, C.W.; Alewood, P.F.; Moller, A.; Muttenthaler, M. The oxytocin receptor signalling system and breast cancer: A critical review. *Oncogene* **2020**, *39*, 5917–5932. [CrossRef] [PubMed]
64. Liu, S.; Huang, R.; Li, A.; Yu, S.; Yao, S.; Xu, J.; Tang, L.; Li, W.; Gan, C.; Cheng, H. The role of the oxytocin system in the resilience of patients with breast cancer. *Front. Oncol.* **2023**, *13*, 1187477. [CrossRef]
65. Miranda-Olvera, A.D.; Ferro-Flores, G.; Pedraza-Lopez, M.; de Murphy, C.A.; De Leon-Rodriguez, L.M. Synthesis of oxytocin HYNIC derivatives as potential diagnostic agents for breast cancer. *Bioconjug. Chem.* **2007**, *18*, 1560–1567. [CrossRef] [PubMed]
66. Murrell, T.G. The potential for oxytocin (OT) to prevent breast cancer: A hypothesis. *Breast Cancer Res. Treat.* **1995**, *35*, 225–229. [CrossRef]
67. Suijkerbuijk, K.P.; van der Wall, E.; Meijrink, H.; Pan, X.; Borel Rinkes, I.H.; Ausems, M.G.; van Diest, P.J. Successful oxytocin-assisted nipple aspiration in women at increased risk for breast cancer. *Fam. Cancer* **2010**, *9*, 321–325. [CrossRef] [PubMed]
68. Alizadeh, A.M.; Heydari, Z.; Rahimi, M.; Bazgir, B.; Shirvani, H.; Alipour, S.; Heidarian, Y.; Khalighfard, S.; Isanejad, A. Oxytocin mediates the beneficial effects of the exercise training on breast cancer. *Exp. Physiol.* **2018**, *103*, 222–235. [CrossRef] [PubMed]
69. Ariana, M.; Pornour, M.; Mehr, S.S.; Vaseghi, H.; Ganji, S.M.; Alivand, M.R.; Salari, M.; Akbari, M.E. Preventive effects of oxytocin and oxytocin receptor in breast cancer pathogenesis. *Per. Med.* **2019**, *16*, 25–34. [CrossRef] [PubMed]
70. Cassoni, P.; Sapino, A.; Negro, F.; Bussolati, G. Oxytocin inhibits proliferation of human breast cancer cell lines. *Virchows Arch.* **1994**, *425*, 467–472. [CrossRef]
71. Cassoni, P.; Sapino, A.; Fortunati, N.; Munaron, L.; Chini, B.; Bussolati, G. Oxytocin inhibits the proliferation of MDA-MB231 human breast-cancer cells via cyclic adenosine monophosphate and protein kinase A. *Int. J. Cancer* **1997**, *72*, 340–344. [CrossRef]
72. Kalaba, P.; Sanchez de la Rosa, C.; Moller, A.; Alewood, P.F.; Muttenthaler, M. Targeting the Oxytocin Receptor for Breast Cancer Management: A Niche for Peptide Tracers. *J. Med. Chem.* **2024**, *67*, 1625–1640. [CrossRef] [PubMed]
73. Khori, V.; Alizadeh, A.M.; Khalighfard, S.; Heidarian, Y.; Khodayari, H. Oxytocin effects on the inhibition of the NF-kappaB/miR195 pathway in mice breast cancer. *Peptides* **2018**, *107*, 54–60. [CrossRef]
74. Amico, J.A.; Rauk, P.N.; Cai, H.M. Estradiol and progesterone regulate oxytocin receptor binding and expression in human breast cancer cell lines. *Endocrine* **2002**, *18*, 79–84. [CrossRef]
75. Kalinina, T.S.; Kononchuk, V.V.; Sidorov, S.V.; Obukhova, D.A.; Abdullin, G.R.; Gulyaeva, L.F. Oxytocin receptor expression is associated with estrogen receptor status in breast tumors. *Biomed. Khim.* **2021**, *67*, 360–365. [CrossRef] [PubMed]
76. Cassoni, P.; Catalano, M.G.; Sapino, A.; Marrocco, T.; Fazzari, A.; Bussolati, G.; Fortunati, N. Oxytocin modulates estrogen receptor alpha expression and function in MCF7 human breast cancer cells. *Int. J. Oncol.* **2002**, *21*, 375–378. [PubMed]
77. Behtaji, S.; Ghafouri-Fard, S.; Sayad, A.; Sattari, A.; Rederstorff, M.; Taheri, M. Identification of oxytocin-related lncRNAs and assessment of their expression in breast cancer. *Sci. Rep.* **2021**, *11*, 6471. [CrossRef]

78. Cassoni, P.; Marrocco, T.; Deaglio, S.; Sapino, A.; Bussolati, G. Biological relevance of oxytocin and oxytocin receptors in cancer cells and primary tumors. *Ann. Oncol.* **2001**, *12* (Suppl. 2), S37–S39. [CrossRef] [PubMed]
79. Liu, H.; Muttenthaler, M. High Oxytocin Receptor Expression Linked to Increased Cell Migration and Reduced Survival in Patients with Triple-Negative Breast Cancer. *Biomedicines* **2022**, *10*, 1595. [CrossRef] [PubMed]
80. Alfaris, I.; Asselah, J.; Aziz, H.; Bouganim, N.; Mousavi, N. The Cardiovascular Risks Associated with Aromatase Inhibitors, Tamoxifen, and GnRH Agonists in Women with Breast Cancer. *Curr. Atheroscler. Rep.* **2023**, *25*, 145–154. [CrossRef] [PubMed]
81. Sukumar, J.S.; Quiroga, D.; Kassem, M.; Grimm, M.; Shinde, N.V.; Appiah, L.; Palettas, M.; Stephens, J.; Gatti-Mays, M.E.; Pariser, A.; et al. Patient preferences and adherence to adjuvant GnRH analogs among premenopausal women with hormone receptor positive breast cancer. *Breast Cancer Res. Treat.* **2021**, *190*, 183–188. [CrossRef]
82. LeVasseur, N.; Manna, M.; Jerzak, K.J. An Overview of Long-Acting GnRH Agonists in Premenopausal Breast Cancer Patients: Survivorship Challenges and Management. *Curr. Oncol.* **2024**, *31*, 4209–4224. [CrossRef]
83. Bergami, M.; Manfrini, O.; Cenko, E.; Bugiardini, R. Combined Therapy with Anthracyclines and GnRH Analogues for Breast Cancer: Impact on Ischemic Heart Disease. *J. Clin. Med.* **2023**, *12*, 6791. [CrossRef] [PubMed]
84. Ma, L.; Yang, B.; Wu, J. Revisiting ovarian function suppression with GnRH agonists for premenopausal women with breast cancer: Who should use and the impact on survival outcomes. *Cancer Treat. Rev.* **2024**, *129*, 102770. [CrossRef] [PubMed]
85. Sarkar, D.K.; Zhang, C. Beta-endorphin neuron regulates stress response and innate immunity to prevent breast cancer growth and progression. *Vitam. Horm.* **2013**, *93*, 263–276. [CrossRef]
86. Sanchez, M.L.; Rodriguez, F.D.; Covenas, R. Involvement of the Opioid Peptide Family in Cancer Progression. *Biomedicines* **2023**, *11*, 1993. [CrossRef] [PubMed]
87. Iannucci, N.B.; Ripoll, G.V.; Garona, J.; Cascone, O.; Ciccia, G.N.; Gomez, D.E.; Alonso, D.F. Antiproliferative effect of 1-deamino-8-D-arginine vasopressin analogs on human breast cancer cells. *Future Med. Chem.* **2011**, *3*, 1987–1993. [CrossRef]
88. North, W.G.; Fay, M.J.; Du, J. MCF-7 breast cancer cells express normal forms of all vasopressin receptors plus an abnormal V2R. *Peptides* **1999**, *20*, 837–842. [CrossRef]
89. Alkafaas, S.S.; Loutfy, S.A.; Diab, T.; Hessien, M. Vasopressin induces apoptosis but does not enhance the antiproliferative effect of dynamin 2 or PI3K/Akt inhibition in luminal A breast cancer cells. *Med. Oncol.* **2022**, *40*, 35. [CrossRef] [PubMed]
90. Garona, J.; Pifano, M.; Orlando, U.D.; Pastrian, M.B.; Iannucci, N.B.; Ortega, H.H.; Podesta, E.J.; Gomez, D.E.; Ripoll, G.V.; Alonso, D.F. The novel desmopressin analogue [V4Q5]dDAVP inhibits angiogenesis, tumour growth and metastases in vasopressin type 2 receptor-expressing breast cancer models. *Int. J. Oncol.* **2015**, *46*, 2335–2345. [CrossRef] [PubMed]
91. Garona, J.; Alonso, D.F. Perioperative biology in primary breast cancer: Selective targeting of vasopressin type 2 receptor using desmopressin as a novel therapeutic approach. *Breast Cancer Res. Treat.* **2016**, *158*, 597–599. [CrossRef]
92. Keller, S.R. The insulin-regulated aminopeptidase: A companion and regulator of GLUT4. *Front. Biosci.* **2003**, *8*, s410–s420. [CrossRef]
93. Wallis, M.G.; Lankford, M.F.; Keller, S.R. Vasopressin is a physiological substrate for the insulin-regulated aminopeptidase IRAP. *Am. J. Physiol. Endocrinol. Metab.* **2007**, *293*, E1092–E1102. [CrossRef] [PubMed]
94. Albiston, A.L.; McDowall, S.G.; Matsacos, D.; Sim, P.; Clune, E.; Mustafa, T.; Lee, J.; Mendelsohn, F.A.; Simpson, R.J.; Connolly, L.M.; et al. Evidence that the angiotensin IV (AT(4)) receptor is the enzyme insulin-regulated aminopeptidase. *J. Biol. Chem.* **2001**, *276*, 48623–48626. [CrossRef]
95. Albiston, A.L.; Mustafa, T.; McDowall, S.G.; Mendelsohn, F.A.; Lee, J.; Chai, S.Y. AT4 receptor is insulin-regulated membrane aminopeptidase: Potential mechanisms of memory enhancement. *Trends Endocrinol. Metab.* **2003**, *14*, 72–77. [CrossRef] [PubMed]
96. Keller, S.R. Role of the insulin-regulated aminopeptidase IRAP in insulin action and diabetes. *Biol. Pharm. Bull.* **2004**, *27*, 761–764. [CrossRef] [PubMed]
97. Carrera, M.P.; Ramirez-Exposito, M.J.; Valenzuela, M.T.; Duenas, B.; Garcia, M.J.; Mayas, M.D.; Martinez-Martos, J.M. Renin-angiotensin system-regulating aminopeptidase activities are modified in the pineal gland of rats with breast cancer induced by N-methyl-nitrosourea. *Cancer Investig.* **2006**, *24*, 149–153. [CrossRef]
98. Carrera-Gonzalez, M.P.; Ramirez-Exposito, M.J.; de Saavedra, J.M.; Sanchez-Agesta, R.; Mayas, M.D.; Martinez-Martos, J.M. Hypothalamus-pituitary-thyroid axis disruption in rats with breast cancer is related to an altered endogenous oxytocin/insulin-regulated aminopeptidase (IRAP) system. *Tumour Biol.* **2011**, *32*, 543–549. [CrossRef] [PubMed]
99. Martinez, J.M.; Prieto, I.; Ramirez, M.J.; Cueva, C.; Alba, F.; Ramirez, M. Aminopeptidase activities in breast cancer tissue. *Clin. Chem.* **1999**, *45*, 1797–1802. [CrossRef] [PubMed]
100. Arrieta, O.; Villarreal-Garza, C.; Vizcaino, G.; Pineda, B.; Hernandez-Pedro, N.; Guevara-Salazar, P.; Wegman-Ostrosky, T.; Villanueva-Rodriguez, G.; Gamboa-Dominguez, A. Association between AT1 and AT2 angiotensin II receptor expression with cell proliferation and angiogenesis in operable breast cancer. *Tumour Biol.* **2015**, *36*, 5627–5634. [CrossRef] [PubMed]
101. Ramirez-Exposito, M.J.; Duenas-Rodriguez, B.; Carrera-Gonzalez, M.P.; Navarro-Cecilia, J.; Martinez-Martos, J.M. Insulin-Regulated Aminopeptidase in Women with Breast Cancer: A Role beyond the Regulation of Oxytocin and Vasopressin. *Cancers* **2020**, *12*, 3252. [CrossRef] [PubMed]

102. Lew, R.A.; Mustafa, T.; Ye, S.; McDowall, S.G.; Chai, S.Y.; Albiston, A.L. Angiotensin AT4 ligands are potent, competitive inhibitors of insulin regulated aminopeptidase (IRAP). *J. Neurochem.* **2003**, *86*, 344–350. [CrossRef] [PubMed]
103. Medina, R.A.; Meneses, A.M.; Vera, J.C.; Guzman, C.; Nualart, F.; Astuya, A.; Garcia, M.A.; Kato, S.; Carvajal, A.; Pinto, M.; et al. Estrogen and progesterone up-regulate glucose transporter expression in ZR-75-1 human breast cancer cells. *Endocrinology* **2003**, *144*, 4527–4535. [CrossRef]
104. Harmon, A.W.; Patel, Y.M. Naringenin inhibits glucose uptake in MCF-7 breast cancer cells: A mechanism for impaired cellular proliferation. *Breast Cancer Res. Treat.* **2004**, *85*, 103–110. [CrossRef] [PubMed]
105. Garrido, P.; Osorio, F.G.; Moran, J.; Cabello, E.; Alonso, A.; Freije, J.M.; Gonzalez, C. Loss of GLUT4 induces metabolic reprogramming and impairs viability of breast cancer cells. *J. Cell Physiol.* **2015**, *230*, 191–198. [CrossRef] [PubMed]
106. Ruiz-Sanjuan, M.D.; Martinez-Martos, J.M.; Carrera-Gonzalez, M.P.; Mayas, M.D.; Garcia, M.J.; Arrazola, M.; Ramirez-Exposito, M.J. Normolipidic dietary fat modifies circulating Renin-Angiotensin system-regulating aminopeptidase activities in rat with breast cancer. *Integr. Cancer Ther.* **2015**, *14*, 149–155. [CrossRef]
107. Ma, Y.; Yang, X.; Pan, P.; Yang, J.; Wu, X.; Wang, D.; Gao, H. Bestatin attenuates breast cancer stemness by targeting puromycin-sensitive aminopeptidase. *Discov. Oncol.* **2024**, *15*, 197. [CrossRef] [PubMed]
108. Carrera, M.P.; Ramirez-Exposito, M.J.; Mayas, M.D.; Garcia, M.J.; Martinez-Martos, J.M. Mammary renin-angiotensin system-regulating aminopeptidase activities are modified in rats with breast cancer. *Tumour Biol.* **2010**, *31*, 583–588. [CrossRef] [PubMed]
109. Dwyer, A.R.; Truong, T.H.; Ostrander, J.H.; Lange, C.A. 90 YEARS OF PROGESTERONE: Steroid receptors as MAPK signaling sensors in breast cancer: Let the fates decide. *J. Mol. Endocrinol.* **2020**, *65*, T35–T48. [CrossRef]
110. Horwitz, K.B.; Sartorius, C.A. 90 YEARS OF PROGESTERONE: Progesterone and progesterone receptors in breast cancer: Past, present, future. *J. Mol. Endocrinol.* **2020**, *65*, T49–T63. [CrossRef] [PubMed]
111. Trabert, B.; Sherman, M.E.; Kannan, N.; Stanczyk, F.Z. Progesterone and Breast Cancer. *Endocr. Rev.* **2020**, *41*, 320–344. [CrossRef]
112. Kastner, P.; Krust, A.; Turcotte, B.; Stropp, U.; Tora, L.; Gronemeyer, H.; Chambon, P. Two distinct estrogen-regulated promoters generate transcripts encoding the two functionally different human progesterone receptor forms A and B. *EMBO J.* **1990**, *9*, 1603–1614. [CrossRef] [PubMed]
113. Li, Z.; Wei, H.; Li, S.; Wu, P.; Mao, X. The Role of Progesterone Receptors in Breast Cancer. *Drug Des. Dev. Ther.* **2022**, *16*, 305–314. [CrossRef] [PubMed]
114. Beato, M.; Wright, R.H.G.; Dily, F.L. 90 YEARS OF PROGESTERONE: Molecular mechanisms of progesterone receptor action on the breast cancer genome. *J. Mol. Endocrinol.* **2020**, *65*, T65–T79. [CrossRef] [PubMed]
115. Bighetti-Trevisan, R.L.; Sousa, L.O.; Castilho, R.M.; Almeida, L.O. Cancer Stem Cells: Powerful Targets to Improve Current Anticancer Therapeutics. *Stem Cells Int.* **2019**, *2019*, 9618065. [CrossRef] [PubMed]
116. Brisken, C.; Scabia, V. 90 YEARS OF PROGESTERONE: Progesterone receptor signaling in the normal breast and its implications for cancer. *J. Mol. Endocrinol.* **2020**, *65*, T81–T94. [CrossRef]
117. Tanos, T.; Sflomos, G.; Echeverria, P.C.; Ayyanan, A.; Gutierrez, M.; Delaloye, J.F.; Raffoul, W.; Fiche, M.; Dougall, W.; Schneider, P.; et al. Progesterone/RANKL is a major regulatory axis in the human breast. *Sci. Transl. Med.* **2013**, *5*, 182ra55. [CrossRef] [PubMed]
118. Truong, T.H.; Lange, C.A.; Ostrander, J.H. Targeting steroid receptor co-activators to inhibit breast cancer stem cells. *Oncoscience* **2018**, *5*, 281–282. [CrossRef] [PubMed]
119. Joshi, P.A.; Jackson, H.W.; Beristain, A.G.; Di Grappa, M.A.; Mote, P.A.; Clarke, C.L.; Stingl, J.; Waterhouse, P.D.; Khokha, R. Progesterone induces adult mammary stem cell expansion. *Nature* **2010**, *465*, 803–807. [CrossRef]
120. Diep, C.H.; Charles, N.J.; Gilks, C.B.; Kalloger, S.E.; Argenta, P.A.; Lange, C.A. Progesterone receptors induce FOXO1-dependent senescence in ovarian cancer cells. *Cell Cycle* **2013**, *12*, 1433–1449. [CrossRef] [PubMed]
121. Singhal, H.; Greene, M.E.; Tarulli, G.; Zarnke, A.L.; Bourgo, R.J.; Laine, M.; Chang, Y.F.; Ma, S.; Dembo, A.G.; Raj, G.V.; et al. Genomic agonism and phenotypic antagonism between estrogen and progesterone receptors in breast cancer. *Sci. Adv.* **2016**, *2*, e1501924. [CrossRef] [PubMed]
122. Bardou, V.J.; Arpino, G.; Elledge, R.M.; Osborne, C.K.; Clark, G.M. Progesterone receptor status significantly improves outcome prediction over estrogen receptor status alone for adjuvant endocrine therapy in two large breast cancer databases. *J. Clin. Oncol.* **2003**, *21*, 1973–1979. [CrossRef]
123. Arpino, G.; Weiss, H.; Lee, A.V.; Schiff, R.; De Placido, S.; Osborne, C.K.; Elledge, R.M. Estrogen receptor-positive, progesterone receptor-negative breast cancer: Association with growth factor receptor expression and tamoxifen resistance. *J. Natl. Cancer Inst.* **2005**, *97*, 1254–1261. [CrossRef] [PubMed]
124. Cui, X.; Schiff, R.; Arpino, G.; Osborne, C.K.; Lee, A.V. Biology of progesterone receptor loss in breast cancer and its implications for endocrine therapy. *J. Clin. Oncol.* **2005**, *23*, 7721–7735. [CrossRef] [PubMed]
125. Piva, M.; Domenici, G.; Iriondo, O.; Rabano, M.; Simoes, B.M.; Comaills, V.; Barredo, I.; Lopez-Ruiz, J.A.; Zabalza, I.; Kypta, R.; et al. Sox2 promotes tamoxifen resistance in breast cancer cells. *EMBO Mol. Med.* **2014**, *6*, 66–79. [CrossRef]

126. Bricard, G.; Cadassou, O.; Cassagnes, L.E.; Cros-Perrial, E.; Payen-Gay, L.; Puy, J.Y.; Lefebvre-Tournier, I.; Tozzi, M.G.; Dumontet, C.; Jordheim, L.P. The cytosolic 5′-nucleotidase cN-II lowers the adaptability to glucose deprivation in human breast cancer cells. *Oncotarget* **2017**, *8*, 67380–67393. [CrossRef] [PubMed]
127. da Silva, F.C.; de Melo Neto, A.B.; Martins, C.A.; de Sousa Cardoso, T.C.; de Souza Gomes, M.; de Araujo, T.G.; Furstenau, C.R. Is the regulation by miRNAs of NTPDase1 and ecto-5′-nucleotidase genes involved with the different profiles of breast cancer subtypes? *Purinergic Signal.* **2022**, *18*, 123–133. [CrossRef]
128. Jedrzejewska, A.; Kutryb-Zajac, B.; Krol, O.; Harasim, G.; Franczak, M.; Jablonska, P.; Slominska, E.; Smolenski, R.T. The decreased serum activity of cytosolic 5′-nucleotidase IA as a potential marker of breast cancer-associated muscle inflammation. *Nucleosides Nucleotides Nucleic Acids* **2022**, *41*, 273–284. [CrossRef]
129. Yu, J.; Wang, X.; Lu, Q.; Wang, J.; Li, L.; Liao, X.; Zhu, W.; Lv, L.; Zhi, X.; Yu, J.; et al. Extracellular 5′-nucleotidase (CD73) promotes human breast cancer cells growth through AKT/GSK-3beta/beta-catenin/cyclinD1 signaling pathway. *Int. J. Cancer* **2018**, *142*, 959–967. [CrossRef] [PubMed]
130. Rocha-Vieira, T.C.; Lacerda-Abreu, M.A.; Carvalho-Kelly, L.F.; Santos-Araujo, S.; Gondim, K.C.; Meyer-Fernandes, J.R. Comparative characterisation of an ecto-5′-nucleotidase (CD73) in non-tumoral MCF10-A breast cells and triple-negative MDA-MB-231 breast cancer cells. *Cell Biol. Int.* **2024**, *48*, 1354–1363. [CrossRef]
131. Stagg, J.; Smyth, M.J. Extracellular adenosine triphosphate and adenosine in cancer. *Oncogene* **2010**, *29*, 5346–5358. [CrossRef] [PubMed]
132. Stagg, J.; Divisekera, U.; McLaughlin, N.; Sharkey, J.; Pommey, S.; Denoyer, D.; Dwyer, K.M.; Smyth, M.J. Anti-CD73 antibody therapy inhibits breast tumor growth and metastasis. *Proc. Natl. Acad. Sci. USA* **2010**, *107*, 1547–1552. [CrossRef] [PubMed]
133. Wang, L.; Zhou, X.; Zhou, T.; Ma, D.; Chen, S.; Zhi, X.; Yin, L.; Shao, Z.; Ou, Z.; Zhou, P. Ecto-5′-nucleotidase promotes invasion, migration and adhesion of human breast cancer cells. *J. Cancer Res. Clin. Oncol.* **2008**, *134*, 365–372. [CrossRef]
134. Qi, Y.; Li, M.; Li, S.; Zeng, D.; Xiao, Y.; Li, J.; Ye, Q.; Bremer, E.; Zhang, G.J. Notch1 promotes resistance to cisplatin by up-regulating Ecto-5′-nucleotidase (CD73) in triple-negative breast cancer cells. *Cell Death Discov.* **2023**, *9*, 204. [CrossRef] [PubMed]
135. Wang, L.; Tang, S.; Wang, Y.; Xu, S.; Yu, J.; Zhi, X.; Ou, Z.; Yang, J.; Zhou, P.; Shao, Z. Ecto-5′-nucleotidase (CD73) promotes tumor angiogenesis. *Clin. Exp. Metastasis* **2013**, *30*, 671–680. [CrossRef] [PubMed]
136. Yu, J.; Liao, X.; Li, L.; Lv, L.; Zhi, X.; Yu, J.; Zhou, P. A preliminary study of the role of extracellular-5′- nucleotidase in breast cancer stem cells and epithelial-mesenchymal transition. *Vitr. Cell Dev. Biol. Anim.* **2017**, *53*, 132–140. [CrossRef] [PubMed]
137. Ruiz-Espejo, F.; Cabezas-Herrera, J.; Illana, J.; Campoy, F.J.; Vidal, C.J. Cholinesterase activity and acetylcholinesterase glycosylation are altered in human breast cancer. *Breast Cancer Res. Treat.* **2002**, *72*, 11–22. [CrossRef] [PubMed]
138. Bernardi, C.C.; Ribeiro Ede, S.; Cavalli, I.J.; Chautard-Freire-Maia, E.A.; Souza, R.L. Amplification and deletion of the ACHE and BCHE cholinesterase genes in sporadic breast cancer. *Cancer Genet. Cytogenet.* **2010**, *197*, 158–165. [CrossRef] [PubMed]
139. Grisaru, D.; Sternfeld, M.; Eldor, A.; Glick, D.; Soreq, H. Structural roles of acetylcholinesterase variants in biology and pathology. *Eur. J. Biochem.* **1999**, *264*, 672–686. [CrossRef]
140. Nitsch, R.M.; Rossner, S.; Albrecht, C.; Mayhaus, M.; Enderich, J.; Schliebs, R.; Wegner, M.; Arendt, T.; von der Kammer, H. Muscarinic acetylcholine receptors activate the acetylcholinesterase gene promoter. *J. Physiol. Paris* **1998**, *92*, 257–264. [CrossRef] [PubMed]
141. Speirs, V.; Kerin, M.J.; Newton, C.J.; Walton, D.S.; Green, A.R.; Desai, S.B.; Atkin, S.L. Evidence for transcriptional activation of ERalpha by IL-1beta in breast cancer cells. *Int. J. Oncol.* **1999**, *15*, 1251–1254. [CrossRef]
142. Li, Y.; Liu, L.; Kang, J.; Sheng, J.G.; Barger, S.W.; Mrak, R.E.; Griffin, W.S. Neuronal-glial interactions mediated by interleukin-1 enhance neuronal acetylcholinesterase activity and mRNA expression. *J. Neurosci.* **2000**, *20*, 149–155. [CrossRef]
143. Ponka, P.; Beaumont, C.; Richardson, D.R. Function and regulation of transferrin and ferritin. *Semin. Hematol.* **1998**, *35*, 35–54. [PubMed]
144. Toyokuni, S. Iron-induced carcinogenesis: The role of redox regulation. *Free Radic. Biol. Med.* **1996**, *20*, 553–566. [CrossRef] [PubMed]
145. Huang, X. Iron overload and its association with cancer risk in humans: Evidence for iron as a carcinogenic metal. *Mutat. Res.* **2003**, *533*, 153–171. [CrossRef] [PubMed]
146. Hou, C.; Hou, Q.; Xie, X.; Wang, H.; Chen, Y.; Lu, T.; Wu, Q.; Liang, Y.; Hu, Y.; Mao, Y. Serum iron status and the risk of breast cancer in the European population: A two-sample Mendelian randomisation study. *Genes Nutr.* **2021**, *16*, 9. [CrossRef] [PubMed]
147. Cui, Y.; Vogt, S.; Olson, N.; Glass, A.G.; Rohan, T.E. Levels of zinc, selenium, calcium, and iron in benign breast tissue and risk of subsequent breast cancer. *Cancer Epidemiol. Biomark. Prev.* **2007**, *16*, 1682–1685. [CrossRef] [PubMed]
148. Gaur, A.; Collins, H.; Wulaningsih, W.; Holmberg, L.; Garmo, H.; Hammar, N.; Walldius, G.; Jungner, I.; Van Hemelrijck, M. Iron metabolism and risk of cancer in the Swedish AMORIS study. *Cancer Causes Control* **2013**, *24*, 1393–1402. [CrossRef] [PubMed]
149. Quintana Pacheco, D.A.; Sookthai, D.; Graf, M.E.; Schubel, R.; Johnson, T.; Katzke, V.A.; Kaaks, R.; Kuhn, T. Iron status in relation to cancer risk and mortality: Findings from a population-based prospective study. *Int. J. Cancer* **2018**, *143*, 561–569. [CrossRef]

150. Sanagoo, A.; Kiani, F.; Saei Gharenaz, M.; Sayehmiri, F.; Koohi, F.; Jouybari, L.; Dousti, M. A systematic review and meta-analysis on the association of serum and tumor tissue iron and risk of breast cancer. *Casp. J. Intern. Med.* **2020**, *11*, 1–11. [CrossRef]
151. Von Holle, A.; O'Brien, K.M.; Sandler, D.P.; Janicek, R.; Weinberg, C.R. Association Between Serum Iron Biomarkers and Breast Cancer. *Cancer Epidemiol. Biomark. Prev.* **2021**, *30*, 422–425. [CrossRef]
152. Von Holle, A.; Thompson, R.L.; O'Brien, K.M.; Sandler, D.P.; Weinberg, C.R. Associations Between Serum Iron Biomarkers and Breast Cancer Tumor Size. *Cancer Res. Commun.* **2024**, *4*, 182–185. [CrossRef] [PubMed]
153. Kabat, G.C.; Rohan, T.E. Does excess iron play a role in breast carcinogenesis? An unresolved hypothesis. *Cancer Causes Control* **2007**, *18*, 1047–1053. [CrossRef]
154. Ebrahim, S.; Davey Smith, G. Mendelian randomization: Can genetic epidemiology help redress the failures of observational epidemiology? *Hum. Genet.* **2008**, *123*, 15–33. [CrossRef] [PubMed]
155. Smith, G.D.; Ebrahim, S. 'Mendelian randomization': Can genetic epidemiology contribute to understanding environmental determinants of disease? *Int. J. Epidemiol.* **2003**, *32*, 1–22. [CrossRef] [PubMed]
156. Hua, X.; Duan, F.; Huang, J.; Bi, X.; Xia, W.; Song, C.; Wang, L.; Jiang, C.; Yuan, Z. A Novel Prognostic Model Based on the Serum Iron Level for Patients With Early-Stage Triple-Negative Breast Cancer. *Front. Cell Dev. Biol.* **2021**, *9*, 777215. [CrossRef] [PubMed]
157. Hambidge, M. Biomarkers of trace mineral intake and status. *J. Nutr.* **2003**, *133*, 948S–955S. [CrossRef] [PubMed]
158. Fonseca-Nunes, A.; Jakszyn, P.; Agudo, A. Iron and cancer risk–a systematic review and meta-analysis of the epidemiological evidence. *Cancer Epidemiol. Biomark. Prev.* **2014**, *23*, 12–31. [CrossRef] [PubMed]
159. Torti, S.V.; Manz, D.H.; Paul, B.T.; Blanchette-Farra, N.; Torti, F.M. Iron and Cancer. *Annu. Rev. Nutr.* **2018**, *38*, 97–125. [CrossRef]
160. Calvo, I.; Gonzalez-Rodriguez, M.; Neria, F.; Gallegos, I.; Garcia-Sanchez, L.; Sanchez-Gomez, R.; Perez, S.; Arenas, M.F.; Estevez, L.G. An analysis of the association between breast density and body mass index with breast cancer molecular subtypes in early breast cancer: Data from a Spanish population. *Clin. Transl. Oncol.* **2024**, *26*, 2541–2548. [CrossRef] [PubMed]
161. Johnson, K.S.; Conant, E.F.; Soo, M.S. Molecular Subtypes of Breast Cancer: A Review for Breast Radiologists. *J. Breast Imaging* **2021**, *3*, 12–24. [CrossRef] [PubMed]
162. Cao, J.; Li, J.; Zhang, Z.; Qin, G.; Pang, Y.; Wu, M.; Gu, K.; Xu, H. Interaction between body mass index and family history of cancer on the risk of female breast cancer. *Sci. Rep.* **2024**, *14*, 4927. [CrossRef]
163. Jiang, Q.; Hu, H.; Liao, J.; Duan, P.; Li, Z.; Tan, J. Body Mass Index and Breast Cancer-Related Lymphedema: A Retrospective Cohort Study. *J. Surg. Oncol.* **2024**. [CrossRef] [PubMed]
164. Wada, K.; Kuboyama, K.; Abe, S.K.; Rahman, M.S.; Islam, M.R.; Saito, E.; Nagata, C.; Sawada, N.; Tamakoshi, A.; Shu, X.O.; et al. Body mass index and breast cancer risk in premenopausal and postmenopausal East Asian women: A pooled analysis of 13 cohort studies. *Breast Cancer Res.* **2024**, *26*, 158. [CrossRef]
165. Cubelos-Fernandez, N.; Davila-Batista, V.; Fernandez-Villa, T.; Castano-Vinyals, G.; Perez-Gomez, B.; Amiano, P.; Ardanaz, E.; Delgado Sillero, I.; Llorca, J.; Tardon, G.F.; et al. Burden of postmenopausal breast cancer attributable to excess body weight: Comparative study of body mass index and CUN-BAE in MCC-Spain study. *J. Epidemiol. Community Health* **2024**, *79*, 64–71. [CrossRef] [PubMed]
166. Holm, J.B.; Baggesen, E.; Cronin-Fenton, D.; Frystyk, J.; Bruun, J.M.; Christiansen, P.; Borgquist, S. Circulating C-reactive protein levels as a prognostic biomarker in breast cancer across body mass index groups. *Sci. Rep.* **2024**, *14*, 14486. [CrossRef] [PubMed]
167. Ogiya, A.; Kimura, K.; Ueno, T.; Iwase, T.; Ohno, S. Time trend of breast cancer-related lymphedema according to body mass index. *Eur. J. Surg. Oncol.* **2024**, *50*, 108350. [CrossRef]
168. Quartuccio, N.; Ialuna, S.; Pulizzi, S.; D'Oppido, D.; Antoni, M.; Moreci, A.M. The Correlation of Body Mass Index with Risk of Recurrence in Post-Menopausal Women with Breast Cancer Undergoing Fluorodeoxyglucose Positron Emission Tomography/Computed Tomography. *J. Clin. Med.* **2024**, *13*, 1575. [CrossRef]
169. Van Cauwenberge, J.; Van Baelen, K.; Maetens, M.; Geukens, T.; Nguyen, H.L.; Nevelsteen, I.; Smeets, A.; Deblander, A.; Neven, P.; Koolen, S.; et al. Reporting on patient's body mass index (BMI) in recent clinical trials for patients with breast cancer: A systematic review. *Breast Cancer Res.* **2024**, *26*, 81. [CrossRef] [PubMed]
170. Al-Muhtaseb, S.I. Serum and saliva protein levels in females with breast cancer. *Oncol. Lett.* **2014**, *8*, 2752–2756. [CrossRef] [PubMed]
171. Hamad, A.W.R.; Ibrahim, M.A.; Al-Mohtasib, S.I.; Al-Kobasi, K.N. Comparative Study on Saliva Proteins in Patients of Brain Tumors and Healthy Individuals. *Trends Med. Res.* **2009**, *4*, 16–23. [CrossRef]

172. Gao, W.M.; Kuick, R.; Orchekowski, R.P.; Misek, D.E.; Qiu, J.; Greenberg, A.K.; Rom, W.N.; Brenner, D.E.; Omenn, G.S.; Haab, B.B.; et al. Distinctive serum protein profiles involving abundant proteins in lung cancer patients based upon antibody microarray analysis. *BMC Cancer* **2005**, *5*, 110. [CrossRef] [PubMed]
173. Halliwell, B. Antioxidants: The basics–what they are and how to evaluate them. *Adv. Pharmacol.* **1997**, *38*, 3–20. [CrossRef] [PubMed]

life

MDPI

Article

Immediate Diagnosis of Breast Carcinoma on Core Needle Biopsy Using Ex Vivo Fluorescence Confocal Microscopy: Feasibility in a One-Stop Breast Clinic Workflow

Marie-Christine Mathieu [1,2], Voichita Suciu [1,2], Marie-Laure Tanguy [3,4], Neila Ines Ben Romdhane [1,2], Salma Moalla [5], Sana Harguem-Zayani [5], Remy Barbe [5], Corinne Balleyguier [5], Angelica Conversano [2,6] and Muriel Abbaci [2,7,*]

1 Department of Medical Biology and Pathology, Gustave Roussy, Université Paris-Saclay, 94805 Villejuif, France; marie-christine.mathieu@gustaveroussy.fr (M.-C.M.)
2 Surgery and Pathology Photonic Imaging Group, Gustave Roussy, 94805 Villejuif, France; angelica.conversano@gustaveroussy.fr
3 Department of Biostatistics and Epidemiology, Gustave Roussy, Université Paris-Saclay, 94805 Villejuif, France
4 Oncostat U1018, Inserm, Université Paris-Saclay, Labeled Ligue Contre le Cancer, 94805 Villejuif, France
5 Department of Radiology, Gustave Roussy, Université Paris-Saclay, 94805 Villejuif, France
6 Department of Breast and Plastic Surgery, Gustave Roussy, Université Paris-Saclay, 94805 Villejuif, France
7 UMS AMMICa 23/3655, Plateforme Imagerie et Cytométrie, Gustave Roussy, Université Paris-Saclay, 94805 Villejuif, France
* Correspondence: muriel.abbaci@gustaveroussy.fr

Abstract: Background: In the one-stop breast clinic setting, breast cytology traditionally provides immediate diagnosis of carcinoma. Fluorescence confocal microscopy (FCM) is an emerging optical technique enabling ex vivo analysis of breast biopsies in real-time. This study represents the first proof of concept for integrating FCM imaging into the routine workflow of breast core needle biopsies (CNB) at Gustave Roussy's one-stop breast clinic. Methods: Fifty women with breast masses underwent consecutive enrollment. Biopsies were stained with acridine orange and fast green, followed by imaging using the Vivascope 2500M-G4 (FCM). Interpretation was conducted by two pathologists in real time (PT1) or postoperatively (PT2). Concordance with definitive histology, the duration of the FCM protocol, and its impact on conventional histopathology, immunohistochemistry, and FISH analyses were evaluated. Results: In our study of 50 biopsies, a concordant diagnosis of malignancy was performed using FCM on the malignant cases at definitive histology in 93.5% (29/31 cases) and in 90.3% (28/31 cases) according to PT1 and PT2, respectively. When the FCM suspicious cases were added, FCM identified 100% (31/31 cases) and 96.7% (30/31 cases) of the malignant cases according to PT1 and PT2, respectively. A notable false positive case was identified as a complex sclerosing lesion. The median time for sample preparation (including tissue reception) was 5 min, while the median time for imaging acquisition with interpretation was 3 min for PT1, but 1 min required for interpretation alone by PT2. Histopathological alterations were not more prevalent in FCM-imaged biopsies compared to conventionally treated biopsies. The immunophenotyping and molecular assessment of tissue were preserved after FCM protocol. Conclusions: FCM shows promise as a new histological method for the immediate diagnosis of breast carcinoma on core needle biopsies in a one-stop clinic setting, while also preserving tissue specimens for final histology.

Keywords: pathology; diagnosis; breast cancer; confocal microscopy; fluorescence; one-stop breast clinic

1. Introduction

One-stop diagnostic facilities, providing same-day diagnosis and integrated multidisciplinary care, have been conducted and implemented since the 1990s to improve the quality and decrease the delays in achieving a diagnosis for suspected breast cancer [1]. When the clinical, radiological, and diagnostic findings consistently indicate malignancy, the patient is informed that the mass is highly likely to be malignant. This initiates a discussion about the potential treatment plan, including surgery, allowing the patient to begin adjusting to the likely diagnosis. The following week, the final histological results are presented at the multidisciplinary breast meeting, and the confirmed diagnosis is communicated to the patient. A consensus treatment plan is also finalized and can be promptly implemented. Gustave Roussy established a large-scale, multidisciplinary one-stop breast clinic in April 2004. Prompt and accurate diagnostic information is given directly from a pathologist specialized in cytology. This translates into improved management of breast cancer patients, characterized by earlier diagnosis and a reduced number of outpatient appointments. The implementation of immediate diagnosis has proven transformative, significantly reducing turnaround time and alleviating patient anxiety [1]. Touch imprint cytology (TIC) of core needle biopsy (CNB) allows for immediate cytological diagnosis and high quality histological assessment from a single diagnostic procedure [2–4]. This approach is used for the evaluation of suspicious breast masses categorized as BI-RADS 4b, 4c, and 5 according to the breast imaging reporting and data system, within the Gustave Roussy one-stop breast clinic [5].

Fifteen years ago, benchtop confocal microscopy revolutionized the imaging of breast cancer by providing high-resolution images of cells and tissue architecture [6]. Although capable of fluorescence imaging, the technology at the time was hindered by lengthy acquisition times and a limited field of view, which precluded its clinical application. However, recent advancements in ex vivo fluorescence confocal microscopy (FCM) have overcome these challenges. FCM now offers the rapid acquisition of high-resolution images from fresh, thick tissue specimens, facilitated by optimized ergonomics and user-friendly software [7]. In a recent study focused on breast cancer imaging, pathologists demonstrated high sensitivity and specificity in interpreting FCM images obtained from surgical specimens using dedicated large-sample imager [8]. This technology is particularly valuable for assessing tumor margins in breast lumpectomy procedures [9,10]. However, there is also significant potential in providing rapid histological interpretations of fresh samples shortly after patient examination, following a simple staining process. Elfgen et al. demonstrated high accuracy in interpreting breast carcinoma in core needle biopsies [11]. In oncology, the ability to obtain a reliable histopathologic diagnosis quickly could offer a substantial clinical advantage in the management of patients undergoing one-day breast diagnosis [12]. The recent literature on FCM for imaging breast surgical specimens further underscores the relevance of evaluating this technology for use in one-stop breast clinics [13].

We conducted the first proof-of-concept study to assess the feasibility of FCM imaging on breast CNB within the routine workflow of a one-stop breast clinic. Our study aimed to achieve several critical objectives. First, we sought to evaluate the concordance rate between FCM imaging diagnoses and definitive histology. This evaluation involved a blinded review of FCM images of breast biopsies by two pathologists. Second, we conducted a time analysis to assess how FCM could be integrated into the routine workflow of the one-stop breast clinic at Gustave Roussy. Lastly, we focused on evaluating the quality of samples post-imaging to determine their suitability for downstream histological examinations.

2. Materials and Methods

2.1. Study Population

This prospective ex vivo study was conducted at Gustave Roussy's one-stop breast clinic in Villejuif, France. Approval was obtained from the Gustave Roussy breast committee and the institutional review board for the present work (reference 2023-276). Between October 2023 and November 2023, fifty women with masses categorized as BI-RADS 3, 4a, 4b, 4c, or 5 were consecutively enrolled according to the Breast Imaging-Reporting and Data System of the American College of Radiology [14], as follows: BI-RADS 0—insufficient or incomplete study; BI-RADS 1—normal study; BI-RADS 2—benign features; BI-RADS 3—probably benign (<2% risk of malignancy); BI-RADS 4—suspicious features (divided into categories 4a, 4b, and 4c depending on the likelihood of malignancy); BI-RADS 5—

probably malignant (>95% chance of malignancy); and BI-RADS 6—malignant (proven malignant on tissue biopsy).

There was no selection based on BI-RADS classification or the size of the mass. In accordance with French Good Clinical Practices guidelines for non-interventional studies involving ex vivo human tissue, patients were provided with an information sheet during their medical visit, and this was subsequently added to their records.

2.2. Patient Evaluation

On the same day of their visit to the one-stop breast clinic, patients underwent ultrasound-guided biopsy following radiologic examination. Typically, three consecutive core biopsies (either 14G or 10G) were performed on the identified mass. Biopsy thickness varied between 1.6 mm and 2.5 mm based on needle gauge. All biopsies were 10–20 mm in length. Among these biopsies, two underwent touch imprint cytology for immediate diagnosis. Additionally, one of these two biopsies was analyzed using FCM.

Subsequently, all biopsies were formalin-fixed and paraffin-embedded (FFPE), and stained with hematoxylin/eosin/saffron (HES) for definitive histology according to established guidelines. To accurately assess the time required for FCM analysis of a single nodule within the clinic's workflow, patients with biopsies involving multiple breast nodules were excluded from this specific time measurement.

2.3. Confocal Microscopy

Biopsies were imaged using the Vivascope 2500M-G4 (Vivascope GmbH, Munich, Germany), a clinical-grade confocal scanning microscope. This device features two laser diodes operating at 488 nm and 638 nm for fluorophore excitation, combined with fluorescence signal collection. The microscope is equipped with a x38 objective water immersion with a numerical aperture of 0.85 (Caliber I.D. StableView). It achieves a maximum magnification of ×550, producing reconstructed images from a mosaic of individual scans, with a maximum total scan area of 34 × 25 mm. Each single field of view offers a resolution of 1024 × 1024 pixels, corresponding to a pixel size of 0.5 μm. The Vivascope 2500M-G4 system allows for z-plane imaging from 0 to 200 μm, utilizing fluorescence signal collection with an axial resolution of 4 μm. In addition to confocal imaging, the system features a camera that captures a white-light macro image of the sample. Integration of the confocal and macroscopic images is facilitated through precise correlation, aided by a pointer that enables seamless navigation between the two views. The final fluorescence images are presented in a format akin to H&E staining, where nuclei are depicted in purple, while cytoplasm and extracellular structures appear pink.

2.4. Tissue Processing and Staining

The fresh biopsy was carefully placed on a foam support and rinsed with 0.9% NaCl solution. Subsequently, the staining process involved topical application in the following sequence, repeated twice with brief intervals of few seconds between each step: rinsing with 70% ethanol, applying 5 drops of acridine orange at 0.025% (the solution was prepared by diluting a 1% stock solution (Morphisto, Offenbach, Germany) with 0.9% NaCl), followed by 5 drops of fast green at 0.067% (the solution was prepared by diluting a 0.1% stock solution (Morphisto, Offenbach, Germany) with 0.9% NaCl), and final rinsing with 0.9% NaCl solution.

Acridine orange is a cell-permeating nucleic acid-binding dye that emits green fluorescence when bound to DNA (λ excitation = 502 nm and λ emission = 525 nm). Previous studies have successfully described acridine orange dye for ex vivo microscopy for histologic diagnosis [10,15]. Fast green is highly selective for collagen (λ excitation = 619 nm and near-infrared fluorescence emission when bound to proteins) and was reported in previous fluorescence microscopy examination [16].

Next, the foam carrying the tissue sample was sandwiched between two glass slides, ensuring gentle and complete contact between the tissue, foam, and glass slides. Blu-tack was used to securely connect the glass slides in a "sandwich" configuration. This assembly, with the tissue in contact with the glass slide on the objective side, was then mounted onto the microscope sample holder for en face image acquisition.

Following image acquisition, the core biopsy was promptly fixed in 10% buffered formalin. Subsequently, each imaged biopsy was embedded individually in a paraffin block. Formalin-fixed paraffin-embedded tissue processing preserves tissue for long-term analysis. It involves several steps, fixation, dehydration, and inclusion in paraffin. Breast biopsies were typically fixed in formalin from 6 to 18 h. Sections were cut at two levels (3 μm thick, 60 μm apart) and stained with hematoxylin/eosin/saffron (HES) for definitive histology using standard methodologies. Other sections could be added for immunohistochemistry to refine the diagnosis and to evaluate biomarkers in invasive carcinomas.

The definitive histological diagnosis was performed on HES sections according to the WHO classification of breast tumors [17]. Immunohistochemistry was used to refine the diagnosis and to evaluate biomarkers in invasive carcinomas [18].

2.5. Image Review and Interpretation

Two pathologists conducted histomorphologic interpretation of the 50 FCM images in a blinded manner. Both pathologists were unaware of the touch imprint cytology diagnosis and the definitive histological diagnosis during their assessments.

PT1 (MCM): A senior pathologist, specializing in breast cancer and experienced in optical imaging [8,19], reviewed the FCM images immediately after acquisition within the clinical workflow of the one-stop breast clinic.

PT2 (VS): Another senior pathologist specializing in breast cancer, without prior experience in optical imaging, reviewed the images remotely, four weeks after the examination.

Neither pathologist had received specific training in interpreting FCM images of breast biopsies. They were provided with clinical and imaging data, including BI-RADS classification and patient records (age, medical history, mass localization, and size). The interpretation followed conventional histopathological criteria to establish diagnoses based on the FCM images.

Each pathologist completed the following interpretation sheet for the FCM images:

- The acquisition mode of the specimen was reported as either "Vivablock" for a single image at one depth, or "Vivacube" for a z-series comprising one image at three depths spaced 12 μm apart.
- The quality of the FCM image was assessed using the following categories: limited quality, moderate, good, and very good, facilitating confident sample interpretation.
- FCM image B-Classification according to the reporting system of Ellis for minimal invasive biopsies, as follows: B1: normal tissue; B2: benign; B3: benign but of uncertain malignant potential; B4: suspicious of malignancy; and B5: malignant [3].

Histopathological diagnosis on FCM images according to:

- **Normal**
- **Benign**: mastopathy, fibroadenoma, papilloma, inflammation, other benign conditions;
- **Atypical**: atypical ductal hyperplasia, atypical lobular hyperplasia, other atypical conditions;
- **Malignant**: ductal carcinoma in situ (DCIS), lobular carcinoma in situ (LCIS), invasive carcinoma of no special type (IC-NST), invasive lobular carcinoma (ILC), other malignant conditions.

The B classification and histopathological diagnosis on FCM images were compared directly with the definitive B classification and histological diagnosis performed on the corresponding HES slide, which served as the gold standard. Concordance between the diagnoses was based on whether the B-category (benign or malignant) assigned by FCM imaging matched that of the definitive histology. Minor discrepancies were noted when the B-category (benign or malignant) was the same in both FCM and histology, but the specific type of lesion differed. Major discrepancies were recorded when the B-category assigned by FCM imaging differed from the B-category in the definitive histological diagnosis (benign versus malignant).

2.6. Histology and Immunohistochemistry

The quality of the histopathological sections following fluorescence staining and "sandwich mounting" for FCM image acquisition was assessed.

Immunohistochemical (IHC) evaluation of biomarkers (estrogen receptor (ER), progesterone receptor (PR), HER2, and Ki67) was performed in invasive carcinomas and used for patient selection for chemotherapy, hormonal treatment, and anti-HER2. The staining was conducted on a Ventana platform (Ventana BenchMark Ultra system, Tucson, AZ, USA) following the manufacturer's instructions. The antibodies used included estrogen receptor (ER) (clone SP1, Ventana), progesterone receptor (PR) (clone 1E2, Ventana), HER2 (clone 4B5, Ventana), and Ki67 (clone MIB1, Dako), applied to all biopsies performed, including those analyzed via FCM.

Tumors were categorized as ER- or PR-positive based on French guidelines, which define positivity as at least 10% of invasive tumor cells displaying clear nuclear staining, regardless of staining intensity. HER2 expression was assessed using ASCO guidelines [20], with scores ranging from 0 to 3+. Ki67 positivity was determined by the presence of at least 20% invasive tumor cells displaying clear nuclear staining, independent of staining intensity.

For tumors scoring 2+ for HER2, fluorescence in situ hybridization (FISH) was performed using the Her2/Cep17-G probe kit (Vysis). A positive result was determined if HER2 gene amplification was detected by FISH.

The corresponding HES, IHC, and Ki67 slides were digitized using a digital pathology system, specifically scanned viaa NanoZoomer S210 Scanner (Hamamatsu Photonics, Massy, France).

3. Results

3.1. Clinical and Histopathological Data

Fifty patients were enrolled in the protocol. The BI-RADS classification of the 50 breast masses that underwent biopsies was as follows: BI-RADS 5—19 cases, BI-RADS 4—28 cases (6 BI-RADS 4a, 11 BI-RADS 4b, 11 BI-RADS 4c), and BI-RADS 3—3 cases. The final histological diagnosis revealed carcinoma in 35 patients (70%), including 2 cases of DCIS (ductal carcinoma in situ), 24 cases of IC-NST (invasive carcinoma of no special type), 6 cases of ILC (invasive lobular carcinoma), 1 case of micropapillary carcinoma, 1 case of metastasis from renal carcinoma, and 1 case of papillary carcinoma. Benign lesions were diagnosed in the remaining 15 patients (30%), comprising 2 cases of mastopathy, 4 cases of fibroadenomas, 2 cases of inflammation, 1 case of ductal hyperplasia without atypia, 1 case of cylindric metaplasia, 2 cases of adenosis, 1 case of a cyst, 1 case of gynecomastia, and 1 case of complex sclerosing lesion.

In four cases, carcinoma was not detected in the core biopsy imaged by FCM, but was found in another tissue fragment biopsy. Therefore, only 31 malignant cases were imaged by FCM in total.

3.2. FCM Image Interpretation

All 50 fresh biopsies underwent FCM imaging, resulting in a single image per sample presented to the pathologists. These 50 images were independently analyzed by PT1 and PT2. One image was deemed insufficiently informative by PT2 and therefore was not interpreted, primarily due to poor contact of the sample in the lower part (Figure 1A). The quality of the obtained images was considered good or very good in 41 cases (82%) for PT1 and 36 cases (72%) for PT2. In nine cases (18%) reviewed by both pathologists, the image quality was moderate. For PT2, the image quality was limited in five cases (10%), including the case that was not interpreted. Limited or moderate quality was attributed to fluorescence dye staining that was either too intense or too light, and horizontal lines observed on the FCM images due to ongoing construction work in the hospital, which introduced acquisition artifacts.

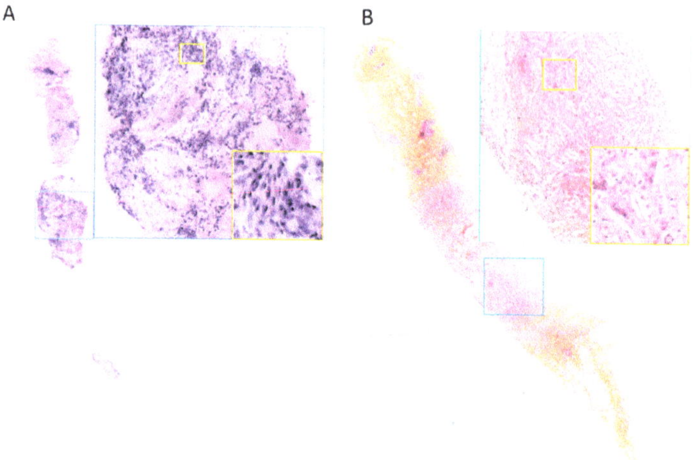

Figure 1. (**A**). Fresh tissue—FCM image not interpreted by PT2. (**B**). FFPE tissue—corresponding HES slide interpreted as inflammation by PT1, magnification ×15—blue inset magnification ×100 and yellow inset magnification ×400.

The correlation between the B-classification of FCM images and corresponding H&E-stained (HES) sections is summarized in Table 1.

Table 1. Correlation between B classification on FCM images and corresponding HES slides.

		Pathologist 1 B-Classification on FCM Images					Pathologist 2 B-Classification on FCM Images						Total
		B1	B2	B3	B4	B5	B1	B2	B3	B4	B5	NI	
B-classification on definitive histology	B1	3	-	-	-	-	2	1	-	-	-	-	3
	B2	-	10	-	4	1	-	11	3	-	-	1	15
	B3	-	-	-	-	1	-	-	-	-	1	-	1
	B4	-	-	-	-	-	-	-	-	-	-	-	-
	B5	-	-	-	2	29	-	1	-	2	28	-	31
	Total	3	10	0	6	31	2	13	3	2	29	1	50

With B1: normal, B2: benign lesion, B3: uncertain malignant potential, B4: suspicious lesions, and B5 malignant lesions. NI: biopsy not interpreted.

Among the 31 malignant cases (B5) at definitive histology, malignancy (B5) by FCM was diagnosed in 93.5% (29/31 cases) and in 90.3% (28/31 cases) according to PT1 and PT2, respectively (Table 1).

Initially, six biopsies were classified as suspicious (B4) by PT1 and two by PT2. Following confirmation that these suspicious B4 cases were indeed malignant (2/6 for PT1 and 2/2 for PT2), the overall diagnosis of malignancy improved to 100% (31/31 cases) for PT1 and 96.7% (30/31 cases) for PT2 (Figure 2A). PT2 did not identify one malignant biopsy out of thirty-one on FCM images.

For the remaining four B4 cases reviewed by PT1, a re-evaluation of FCM images with the assistance of corresponding HES sections and patient reports was conducted. Among these cases, PT1 noted foci of sclerosing adenosis that were challenging to differentiate from well-differentiated invasive carcinoma, and three cases of fibroadenoma were identified as

B4. However, in one benign case, PT1 observed granular material in the duct suggestive of tumor necrosis, while PT2 interpreted an intraductal proliferation as atypical ductal hyperplasia (Figure 2C).

The diagnostic agreement between the two pathologists regarding the FCM images for categories B1+B2+B3 versus B5 (benign versus malignant lesions) at both the one-stop and consolidated diagnoses was 97.5% (40/41 cases).

The correlation between the histological diagnosis on FCM images and the corresponding H&E-stained (HES) sections, considered the gold standard, is detailed in Table 2. Both pathologists (Figure 3) correctly diagnosed all 29 invasive carcinomas. Regarding histological subtypes, PT1 and PT2 correctly identified all 21 cases of invasive carcinoma of no special type (IC-NST). Among the lobular carcinomas, PT1 identified four out of six cases correctly, while PT2 identified three out of six cases correctly as lobular carcinoma. For special types, both pathologists classified a micropapillary carcinoma as IC-NST (Figure 2A). However, PT2 correctly identified a metastasis from renal carcinoma, whereas PT1 classified it as IC-NST or apocrine carcinoma.

Table 2. Correlation between histological diagnosis on FCM images and corresponding HES slides.

		Pathologist 1 FCM Image Interpretation					Pathologist 2 FCM Image Interpretation						Total
		Normal	Benign	Atypical/ Suspicious	DCIS	IC	Normal	Benign	Atypical/ Suspicious	DCIS	IC	NI	
Histological type on definitive histology *	normal	2	-	-	-	-	2	-	-	-	-	-	2
	benign	1	11	3	1	1		14	1	1	-	1	17
	atypical/ suspicious	-	-	-	-	-	-	-	-	-	-	-	-
	DCIS	-	-	1	-	-	-	1	-	-	-	-	1
	IC	-	-	-	-	29	-	-	-	-	29	-	29
	Total	3	11	4	1	30	2	15	1	1	29	1	49 *

* One case of papillary carcinoma could not be classified as in situ or invasive on biopsy at definitive histology. It was diagnosed as papillary carcinoma by PT1 and DCIS by PT2. NI: biopsy not interpreted.

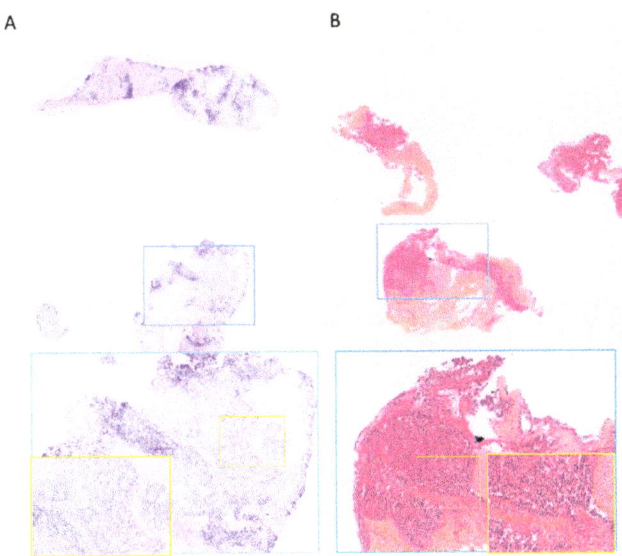

Figure 2. *Cont.*

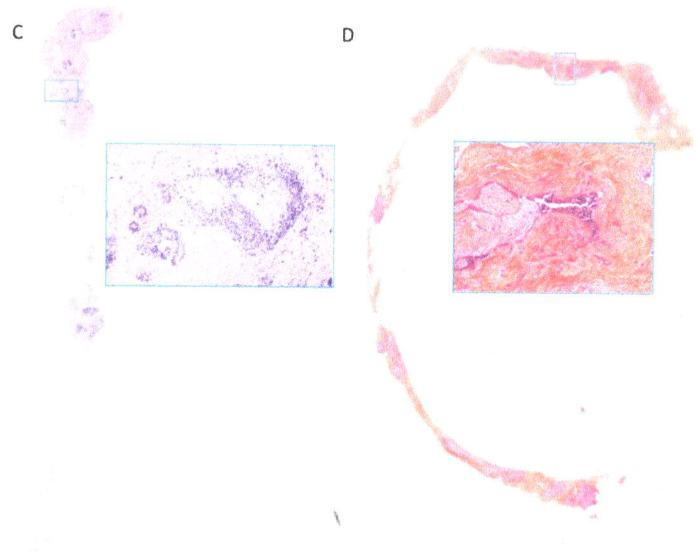

Figure 2. FCM images interpreted as suspicious (B4) by PT1 or PT2 (**A**). Fresh-tissue—papillary carcinoma. (**B**). FFPE tissue—corresponding histological section (**C**). Fresh tissue—Fibroadenoma (**D**). FFPE tissue—corresponding HES section. Magnification ×15—blue inset magnification ×100—yellow inset magnification ×200.

During a re-evaluation of the FCM images with the support of corresponding H&E-stained (HES) sections and patient reports, false negatives and false positives were analyzed (Table 3). PT2 encountered one false negative case on FCM, which was identified as DCIS (ductal carcinoma in situ) on definitive histology (Figure 4C), where only two ducts showed evidence of DCIS (Figure 4D). Upon re-examination of the FCM image, these ducts were not visible in the z-plan confocal image, leading to the false negative diagnosis for PT2. For PT1, aggregates of cells with large and irregular nuclei were observed surrounding the core biopsy, leading to a suspicious of malignancy (B4) classification (Figure 4C). The false positives included two benign lesions. PT1 diagnosed a fibroadenoma with significant lactating metaplasia as DCIS (Figure 4A,B), whereas PT2 recognized it correctly as lactating metaplasia. In another case (Figure 4E), a complex sclerosing lesion with florid epithelial hyperplasia was diagnosed as in situ carcinoma with invasion by PT1 and as DCIS by PT2 (Figure 4E,F).

Table 3. True negative, true positive, false positive, and false negative after suspicious cases exclusion (B1 + B2 + B3 versus B5).

	Pathologist 1 (n = 44)	Pathologist 2 (n = 47)
True negative	100% (13/13)	94.4% (17/18)
True positive	93.5% (29/31)	96.5% (28/29)
False positive	6.5% (2/31)	3.5% (1/29)
False negative	0 (0/13)	5.6% (1/18)

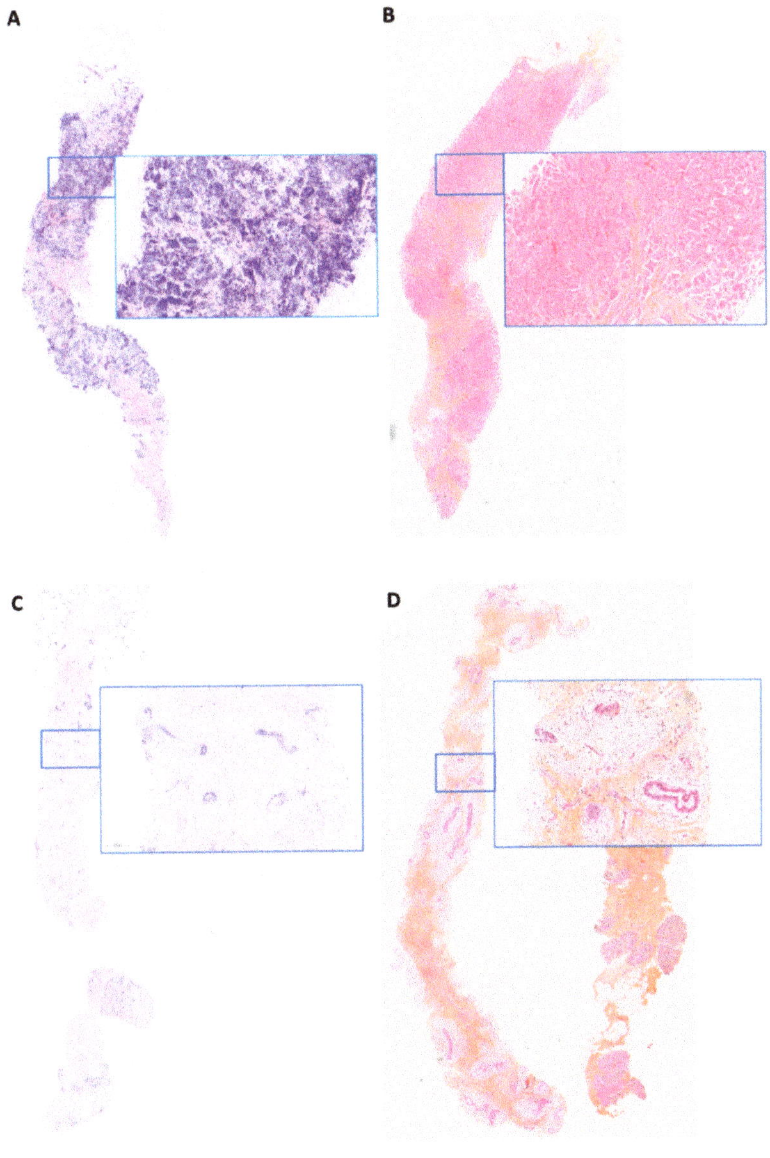

Figure 3. Examples of FCM images correctly interpreted by PT1 and PT2 (**A**). Fresh tissue—IC-NST. (**B**) FFPE tissue—corresponding HES slide diagnosed as IC—micro papillary subtype. (**C**). Fresh tissue—fibroadenoma (**D**). FFPE tissue—corresponding HES slide. Magnification ×15—blue inset magnification ×100.

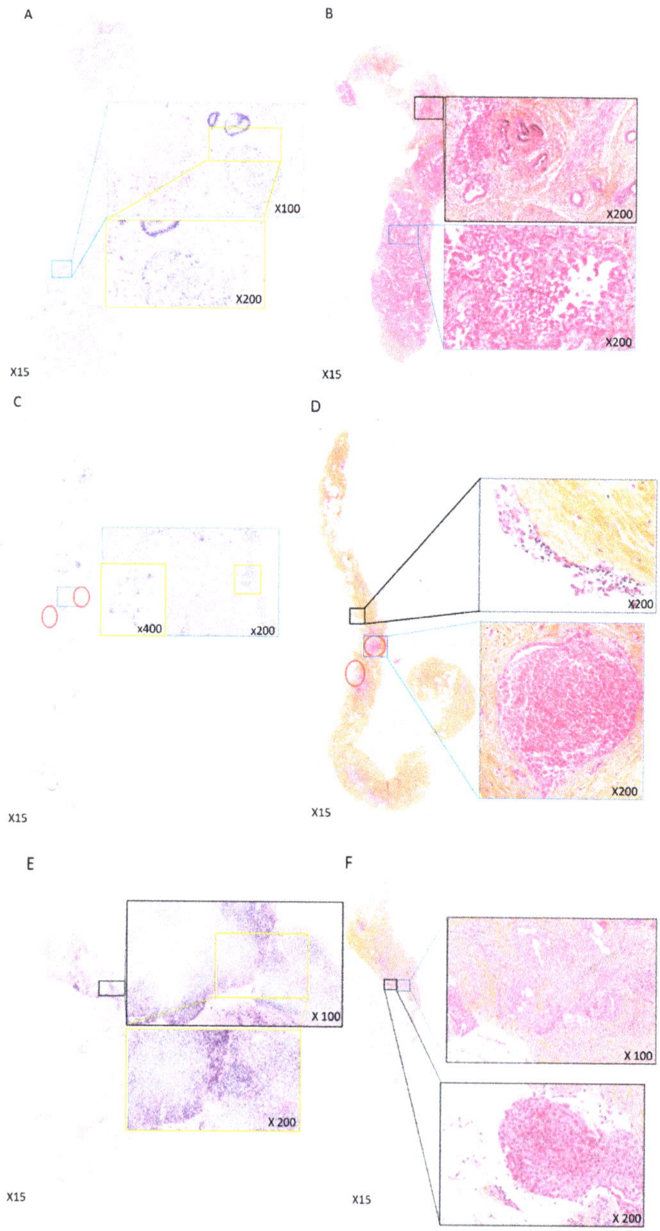

Figure 4. FCM images interpretation showed major discrepancies with the final diagnosis. (**A**). Fresh tissue—fibroadenoma with lactating metaplasia interpreted as DCIS by PT1. (**B**). FFPE tissue—corresponding HES section. (**C**). Fresh tissue—DCIS interpreted as suspicious by PT1 and as mastopathy by PT2. In red circles: clusters of suspicious cells observed by PT1. (**D**). FFPE tissue—corresponding HES section with the two ducts leading to the final diagnosis of DCIS circled in red. (**E**). Fresh tissue—complex sclerosing lesion interpreted as DCIS invasion by PT1 and as DCIS by PT2. (**F**). FFPE tissue—corresponding HES section.

3.3. FCM Protocol Duration

The duration of the FCM protocol from biopsy reception (including administrative tasks) to tissue preparation (fluorescence staining and "sandwich mounting") ranged from 2 to 17 min, with a median time of 5 min and a mean time of 5 min and 43 s (95% confidence interval (CI) [4 min and 44 s; 6 min and 43 s]). In two exceptional cases, a delay of 17 min occurred when biopsies from two patients arrived simultaneously for analysis. The second sample was processed only after the first one had been interpreted and fixed in formalin following image acquisition.

The process of image acquisition typically required one to two acquisitions, primarily to select the appropriate z-plane for obtaining a satisfactory image. The time for image acquisition ranged from 1 to 13 min, with a median of 3 min and a mean of 3 min and 43 s (95% CI [3 min and 7 s; 4 min and 30 s]) for PT1. For image interpretation by PT2, the time ranged from 1 to 6 min, with a median of 1 min and a mean of 1 min and 36 s (95% CI [1 min and 16 s;1 min and 55 s]). A delay of 6 min occurs for a FCM image case which was classified B4. Diagnosis based on FCM images was made following the first image acquisition in 38 cases (76%), after the second image acquisition in 9 cases (18%), after one image and a z-series in 1 case (2%), and after two images and one z-series analysis in 2 cases (4%), aimed at obtaining the most representative image of the sample.

3.4. Assessment of Biopsy Quality, Immunohistochemical Staining, and FISH Evaluation Following FCM Imaging

The histopathological analysis remained uncompromised by the FCM protocol (staining and imaging steps), as no significant changes were observed in the H&E-stained (HES) sections. Specifically, 37 FCM biopsies showed no alteration compared to 36 directly fixed biopsies. Light alterations, such as hyperchromatism of nuclei, cell crushing, and cell retraction, were reported in the remaining samples (13 with FCM protocol and 14 without) (Table 4). Importantly, no specific alterations attributable to the imaging procedure were identified.

Table 4. Alterations identified in the HES sections.

Alteration	FCM Biopsy	Directly Fixed Biopsy
No alteration	37	36
Nuclear hyperchromatism	8	10
Cells crushed	4	4
Cell retraction	1	-
Total	50	50

Concordance in the classification of ER and PR positivity or negativity was maintained between specimens imaged with FCM and those directly fixed in formalin.

For HER2, the scoring was consistent between specimens imaged with FCM and those directly fixed in formalin in nine out of twelve cases. One tumor was classified as score 0 on the imaged specimen and 1+ on the non-imaged specimen. Conversely, two specimens were classified as score 2+ on the imaged specimen and 1+ on the non-imaged specimen. In these two cases, membranous staining was complete and moderate in the imaged specimens, whereas it was incomplete and ranged from low to moderate in the non-imaged specimens.

Ki67 staining showed concordance in nine out of twelve cases. In the discordant cases, Ki67 expression percentages were 10%, 20%, and 22% in the imaged specimens compared to 20%, 15%, and 15% in the non-imaged specimens, respectively. The maximum difference in Ki67-positive cell percentages between imaged and non-imaged specimens was 15%, with a median difference of 0% (Table 5).

Table 5. Comparison of biomarkers expression between specimens imaged with FCM and directly fixed.

Biomarker	Specimen Imaged	Specimen Not Imaged
ER		
negative	3	3
positive	9	9
PR		
negative	2	2
positive	10	10
HER2		
score 0	1	2
score 1+	6	3
score 2+	3	5
score 3+	2	2
Ki67		
≤5	1	-
10–25	5	5
≥25	6	7
<20	3	2
≥20	9	10

FISH evaluation was conducted in three cases, with similar results observed between specimens imaged with FCM and those directly fixed in formalin. In all cases, HER2 gene amplification was not detected. The HER2/CTR17 ratio, used for determining FISH positivity, was concordant between the imaged and non-imaged specimens (Table 6).

Table 6. Comparison of HER2 ratio between specimens imaged with FCM and directly fixed.

Number of the Case	Ratio HER2/CTR17 in the FCM Sample	Ratio HER2/CTR17 in the Non-FCM Sample
18	0.89	0.62
42	1.29	1.35
43	1.00	1.18

4. Discussion

4.1. FCM Data Analysis Concordance Rate between FCM Imaging Diagnoses and Definitive Histology

This study represents the first demonstration of the feasibility of using FCM for the histological analysis of core biopsies from suspicious breast masses in a one-stop clinic setting. The interpretation process was straightforward and intuitive, requiring no prior training, as was the case for PT2. Nackenhorst's study indicated that the resolution of FCM images obtained with a Vivascope is comparable to that of frozen sections. However, FCM images typically provide less detailed visualization of cytoplasmic structures and the extracellular matrix compared to final histology [13]. Nevertheless, we achieved a high rate in detecting malignancy with FCM. When FCM suspicious and malignant cases are added, FCM identifies 100% (31/31 definitive malignant cases) and 96.7% (30/31 cases) according to PT1 and PT2, respectively. These results compare favorably to those reported by Elfgen, who achieved a malignancy detection concordance rate of 91% (63/69 cases) on CNB [11]. It is important to note that Elfgen's study included a smaller number of cases (n = 24) [11].

While 30.9% of FCM images exhibited limited to moderate quality, image interpretation was not hindered significantly. Only one case could not be interpreted by PT2. Independently of the Vivascope system, ongoing construction near Gustave Roussy hospital caused significant microvibrations during FCM image acquisition. These vibrations affected FCM acquisitions by introducing repeated horizontal lines into the images, which were identified as artifacts in the final FCM images. Additionally, in some cases, fluorescent staining appeared either too intense, heterogeneous, or inadequate due to incomplete contact of the biopsy with the imaging window. A learning curve was observed for tissue preparation, particularly for the sandwich mounting technique and image acquisition, aimed at optimizing FCM image quality. Importantly, an antivibrational table has been integrated into the Vivascope system to mitigate horizontal line artifacts in future studies.

PT1 and PT2 failed to identify the sole case of DCIS on the FCM image, as the two invaded ducts identified on the corresponding HES section were absent in the confocal image. It is noteworthy that aggregates of tumor cells around the core biopsy were diagnosed as suspicious of malignancy by PT1. Although the promising global results from PT2 did not highlight the importance of a learning curve, the experienced PT1 suspected malignancy in the DCIS case due to the presence of detached, suspicious cells, which are frequent in biopsy tissues containing DCIS with necrosis. The two false positive cases were not attributable to FCM imaging but rather represented challenging diagnostic cases well known by pathologists. A common FCM false positive was a complex sclerosing lesion, where the final histopathological diagnosis had required the immunohistochemical identification of myoepithelial cells to confirm the benignity. Furthermore, PT1's second false positive was a fibroadenoma with significant lactating metaplasia, a rare lesion characterized by large nuclei, prominent nucleoli, and anisonucleosis, contributing to the misinterpretation. In this case, the young age of the patient provided a crucial clue for guiding PT2 towards a diagnosis of benignity. Additionally, for false positive cases, reviewing medical files and experience of FCM images particularly in challenging cases will be helpful for improving the diagnosis.

4.2. Time Analysis and Samples Processing

Our study demonstrated rapid breast biopsy analysis, with processing, acquisition, and interpretation completed in under 8 min by a pathologist. This swift turnaround time aligned with the goals of a one-stop breast clinic, enabling immediate diagnosis. The FCM installation in the one-stop breast clinic and its evaluation in "real conditions" for data acquisition and image interpretation streamline the intense workflow. Moreover, our feasibility study shows that FCM could fit into the existing clinical workflow providing a low occurrence of unsatisfactory samples (1/50 for PT2) or false negative cases leading adaption capabilities. Research on other cancers has explored the potential of FCM for rapid, intraday diagnosis; Razi et al. emphasized the high diagnostic quality of FCM in skin cancer imaging [21]. Amandoeira et al. reported on FCM's application in imaging solid pancreatic lesions via endoscopic ultrasound-guided fine-needle biopsy, achieving 83.8% adequacy compared to final histology [22]. Titze et al. demonstrated that combining FCM with MRI-guided targeted prostate biopsy enables rapid preliminary diagnosis and enhances clinical management, with intraday diagnosis of clinically significant prostate carcinomas achieved in 68% of cases [23]. The short learning curve for FCM was evident, as pathologists became proficient in image acquisition and data interpretation after just half a day of training. This suggests that integrating FCM into routine practice would not require extensive staff training.

Compared to touch imprint cytology, FCM could offer most of the time a complete histological diagnosis, with final histology serving as a confirmatory step. FCM provides additional architectural information, potentially enabling more precise identification of carcinoma type (in situ vs. invasive). This technology, offering "multiple virtual sections" through depth imaging in a sample, provides a distinct advantage over imprint cytology. FCM analysis was predominantly conducted on a single optical plane, but multiple optical planes were utilized when necessary. This capability allowed for image acquisition at different depths within the tissue, providing pathologists with the option to interpret images with the highest available information, sometimes from deeper layers of the tissue. In this study, multiple optical planes were employed 12 times to interpret lesions. Consequently, the representativeness of lesions in biopsy images could be enhanced by FCM without requiring additional tissue processing.

Finally, FCM did not modify the quality of the specimen for definitive histology, ER, PR, HER2, Ki-67 immunohistochemical staining, and HER2 FISH evaluation in our study. FCM does not impose any restrictions on subsequent conventional histology, immunohistological staining, or FISH examinations. Importantly, the entire sample analyzed by FCM remained available for subsequent analyses, which is crucial for managing small tissue samples. These findings align with previous studies by Verri et al. and Titze et al., which also assessed the quality of tissue specimens following FCM imaging [24,25]

4.3. FCM for One-Stop Breast Clinic

Recent studies have highlighted the utility of FCM for ex vivo analysis, demonstrating high sensitivity and specificity in the pathology management of diverse conditions. Histopathological diagnosis is based on the identification of morphological criteria validated within the pathologists' community. Therefore, the recognition of these same criteria in fluorescence imaging by FCM is a guarantee of rapid adoption of the technology. We first describe the application of FCM in one-stop breast clinic with an integrated multidisciplinary care. FCM offers a powerful tool for the diagnosis of breast cancer due to its ability to provide high-resolution and at multiple depths images of biopsy. The high-resolution images enable detailed visualization of cellular structures and features that may be challenging to discern with imprint cytology. Currently, breast mass diagnosis via touch imprint cytology (TIC) of core needle biopsy (CNB) demonstrates high sensitivity (92–97%) and specificity (89–96%) rates [2–4]. Inadequate cellularity is observed in 5 to 11.4% of cases with TIC [2–4]. Based on our first results, FCM technology has the potential to serve as an alternative to cytological diagnosis, particularly in pathology settings lacking cytopathologists, by providing optical histological sections from breast biopsies within minutes. Failed staining was not observed in our study and was not reported in the literature [8,13,26]. However, in the event of a failed fluorescence staining, biopsies would not be taken again from the patient. The protocol would then be the same as for touch imprint cytology currently, the biopsy analysis would be considered inadequate, and the report would state 'awaiting definitive histology'.

FCM enables the classification of carcinomas into in situ and invasive types, as well as the identification of various subtypes of invasive carcinomas. This information is valuable for the patient management. The rapid diagnosis allows a first orientation of the patient to a standard care pathway and decrease the delay of patient care. Usually, if two pathologists disagree on a diagnosis of an FCM image, a third pathologist should be consulted. However, in a one-stop breast clinic, time constraints may need to wait definitive histology. Nonetheless, the rapid FCM diagnosis will always be confirmed by definitive histology and complemented by immunohistochemical or FISH evaluation. With FCM system featuring two excitation wavelengths, future developments may incorporate fluorescence morphological imaging combined with immunofluorescence imaging in each emission channel. This integration could offer complementary information.

A very promising aspect of FCM is its potential role in extemporaneous telepathology, which is becoming increasingly relevant as pathology departments worldwide consolidate into fewer, more specialized diagnostic centers [13]. Given the anticipated shortage of pathologists in the future, this approach could be crucial. Tissues could be processed and imaged at radiology centers without on-site pathology departments and then analyzed in real-time by specialized pathology centers remotely. Our study suggests that FCM represents a significant advancement in the imaging arsenal of one-stop breast clinics, enabling patients to receive histological results on the same day.

A limitation of our study is that we did not assess the cost-effectiveness of FCM. We plan to address this in future research. While FCM represents a substantial investment, a single device can be shared for various one-day diagnostic applications, such as breast, thyroid, prostate, and pancreatic cancers [22,25,27]. A biopsy sample imaged by FCM is preserved in its entirety for subsequent conventional histological and biological analysis. By providing real-time information, FCM could eliminate the need for multiple biopsies to acquire cancer tissue, sparing patients from the discomfort and risks associated with multiple biopsies. In a therapeutic context, real-time imaging of surgical margins during breast-conserving surgery using FCM is being investigated in numerous trials to identify and remove any residual cancer cells [9,10,28,29]. This would reduce the need for multiple surgeries and increase the likelihood of complete resection. Despite the cost, the return on investment could be significant, particularly if FCM is integrated into a diagnostic and therapeutic approach to enhance the perspective of healthcare provider.

5. Conclusions

Based on the study results, our work demonstrates the feasibility of immediate histological diagnosis of core biopsies of suspect breast masses using FCM for the first time in a one-stop clinic. FCM provides rapid histological interpretation without the need for tissue section preparation, delivered within minutesafter patient examination. Our findings underscore the potential to integrate FCM procedures for patients requiring CNB, offering immediate histological interpretations within the routine workflow of a one-stop breast clinic.

To further assess whether FCM analysis can facilitate clinicians in delivering definitive diagnoses to patients within a single day, our next study will focus on measuring the diagnostic quality parameters of FCM in a larger patient cohort within a one-stop breast clinic setting, including considerations of financial sustainability.

Author Contributions: Conceptualization, M.-C.M., M.A., V.S. and A.C.; methodology, M.-C.M., M.A., V.S., S.M., S.H.-Z., R.B. and C.B.; software, M.A. and N.I.B.R.; validation, M.-C.M. and M.A.; formal analysis, M.-C.M., M.A., V.S. and M.-L.T.; investigation, M.-C.M.; resources, C.B.; data curation, M.A. and A.C.; writing—original draft preparation, M.-C.M., M.A. and V.S.; writing—review and editing, N.I.B.R., S.M., S.H.-Z., R.B., C.B. and A.C.; visualization, M.-C.M., V.S. and M.-L.T.; supervision, M.-C.M. and M.A.; project administration, M.A. and A.C. All authors have read and agreed to the published version of the manuscript.

Funding: This research received no external funding.

Institutional Review Board Statement: This study was conducted in accordance with the Declaration of Helsinki, and approved by the Institutional Review Board of Gustave Roussy (2023-276/29/09/2023).

Informed Consent Statement: In accordance with French Good Clinical Practices guidelines for non-interventional studies involving ex vivo human tissues, patients were provided with an information sheet during their medical visit, and this was subsequently added to their records.

Data Availability Statement: The datasets generated and analyzed during the current study are available from the corresponding author upon reasonable request.

Acknowledgments: The authors gratefully acknowledge the medical and paramedical staff of the one-stop breast clinic at Gustave Roussy for their assistance in patient inclusion, the technicians of the Department of Biopathology for processing the LYRIC patient samples, and Vivascope GmbH (Munich, Germany) for providing the Vivascope confocal microscope 2500M-G4 on loan.

Conflicts of Interest: The authors declare no conflicts of interest.

References

1. Delaloge, S.; Bonastre, J.; Borget, I.; Garbay, J.R.; Fontenay, R.; Boinon, D.; Saghatchian, M.; Mathieu, M.C.; Mazouni, C.; Rivera, S.; et al. The Challenge of Rapid Diagnosis in Oncology: Diagnostic Accuracy and Cost Analysis of a Large-Scale One-Stop Breast Clinic. *Eur. J. Cancer* **2016**, *66*, 131–137. [CrossRef]
2. Kulkarni, D.; Irvine, T.; Reyes, R.J. The Use of Core Biopsy Imprint Cytology in the "one-Stop" Breast Clinic. *Eur. J. Surg. Oncol.* **2009**, *35*, 1037–1040. [CrossRef] [PubMed]
3. Masood, S.; Feng, D.; Tutuncuoglu, O.; Fischer, G.; Bakhshandeh, M.; Bertholf, R.L.; Wolfson, D. Diagnostic Value of Imprint Cytology during Image-Guided Core Biopsy in Improving Breast Health Care. *Ann. Clin. Lab. Sci.* **2011**, *41*, 8–13.
4. Kehl, S.; Mechler, C.; Menton, S.; Weiss, C.; Wasgindt, S.; Sütterlin, M.; Trunk, M. Touch Imprint Cytology of Core Needle Biopsy Specimens for the Breast and Quick Stain Procedure for Immediate Diagnosis. *Anticancer Res.* **2014**, *34*, 153–157. [PubMed]
5. D'Orsi, C.J.; Sickles, E.A.; Mendelson, E.B. *ACR BI-RADS® Atlas, Breast Imaging Reporting and Data System*; American College of Radiology: Reston, VA, USA, 2013.

6. Schiffhauer, L.M.; Boger, J.N.; Bonfiglio, T.A.; Zavislan, J.M.; Zuley, M.; Fox, C.A. Confocal Microscopy of Unfixed Breast Needle Core Biopsies: A Comparison to Fixed and Stained Sections. *BMC Cancer* **2009**, *9*, 265. [CrossRef] [PubMed]
7. Ogrzewalla, H.; Möhrle, M.; Metzler, G.; Eigentler, T.; Münch, A.K.; Forchhammer, S. A Feasibility Study for Immediate Histological Assessment of Various Skin Biopsies Using Ex Vivo Confocal Laser Scanning Microscopy. *Diagnostics* **2022**, *12*, 3030. [CrossRef]
8. Conversano, A.; Abbaci, M.; Van Diest, P.; Roulot, A.A.; Falco, G.; Ferchiou, M.; Coiro, S.; Richir, M.; Genolet, P.-M.M.; Clement, C.; et al. Breast Carcinoma Detection in Ex-Vivo Fresh Human Breast Surgical Specimens Using Fast Slide-Free Confocal Microscopy Scanner: The HIBISCUSS Project. *Br. J. Surg. Open* **2023**, *7*, zrad046. [CrossRef]
9. Wernly, D.; Beniere, C.; Besse, V.; Seidler, S.; Lachat, R.; Letovanec, I.; Huber, D.; Simonson, C. SENOSI Confocal Microscopy: A New and Innovating Way to Detect Positive Margins in Non-Palpable Breast Cancer? *Life* **2024**, *14*, 204. [CrossRef]
10. Sandor, M.-F.F.; Schwalbach, B.; Hofmann, V.; Istrate, S.-E.E.; Schuller, Z.; Ionescu, E.; Heimann, S.; Ragazzi, M.; Lux, M.P. Imaging of Lumpectomy Surface with Large Field-of-View Confocal Laser Scanning Microscope for Intraoperative Margin Assessment—POLARHIS Study. *Breast* **2022**, *66*, 118–125. [CrossRef]
11. Elfgen, C.; Papassotiropoulos, B.; Varga, Z.; Moskovszky, L.; Nap, M.; Güth, U.; Baege, A.; Amann, E.; Chiesa, F.; Tausch, C. Comparative Analysis of Confocal Microscopy on Fresh Breast Core Needle Biopsies and Conventional Histology. *Diagn. Pathol.* **2019**, *14*, 58. [CrossRef]
12. Ragazzi, M.; Longo, C.; Piana, S. Ex Vivo (Fluorescence) Confocal Microscopy in Surgical Pathology: State of the Art. *Adv. Anat. Pathol.* **2016**, *23*, 159–169. [CrossRef] [PubMed]
13. Nackenhorst, M.C.; Kasiri, M.; Gollackner, B.; Regele, H. Ex Vivo Fluorescence Confocal Microscopy: Chances and Changes in the Analysis of Breast Tissue. *Diagn. Pathol.* **2022**, *17*, 55. [CrossRef] [PubMed]
14. Daly, C.; Puckett, Y. *New Breast Mass*; StatPearls: Treasure Island, FL, USA, 2022.
15. Abbaci, M.; Villard, A.; Auperin, A.; Asmandar, S.; Moya-Plana, A.; Casiraghi, O.; Breuskin, I. Ultra-Fast Confocal Fluorescence Microscopy for Neck Lymph Node Imaging in Head and Neck Cancer. *Oral Oncol.* **2024**, *154*, 106862. [CrossRef] [PubMed]
16. Hartmann, D.; Buttgereit, L.; Stärr, L.; Sattler, E.C.; French, L.E.; Deußing, M. Intraoperative PRO Score Assessment of Actinic Keratosis with FCF Fast Green-Enhanced Ex Vivo Confocal Microscopy. *Appl. Sci.* **2024**, *14*, 1150. [CrossRef]
17. Tan, P.H.; Ellis, I.; Allison, K.; Brogi, E.; Fox, S.B.; Lakhani, S.; Lazar, A.J.; Morris, E.A.; Sahin, A.; Salgado, R.; et al. The 2019 World Health Organization Classification of Tumours of the Breast. *Histopathology* **2020**, *77*, 181–185. [CrossRef] [PubMed]
18. Zhou, Y.; Tao, L.; Qiu, J.; Xu, J.; Yang, X.; Zhang, Y.; Tian, X.; Guan, X.; Cen, X.; Zhao, Y. Tumor Biomarkers for Diagnosis, Prognosis and Targeted Therapy. *Signal Transduct. Target. Ther.* **2024**, *9*, 132. [CrossRef]
19. Scholler, J.; Mandache, D.; Mathieu, M.C.; Ben Lakhdar, A.; Darche, M.; Monfort, T.; Boccara, C.; Olivo-Marin, J.-C.; Grieve, K.; Meas-Yedid, V.; et al. Automatic Diagnosis and Classification of Breast Surgical Samples with Dynamic Full-Field OCT and Machine Learning. *J. Med. Imaging* **2023**, *10*, 034504. [CrossRef]
20. Wolff, A.C.; Elizabeth Hale Hammond, M.; Allison, K.H.; Harvey, B.E.; Mangu, P.B.; Bartlett, J.M.S.; Bilous, M.; Ellis, I.O.; Fitzgibbons, P.; Hanna, W.; et al. Human Epidermal Growth Factor Receptor 2 Testing in Breast Cancer: American Society of Clinical Oncology/College of American Pathologists Clinical Practice Guideline Focused Update. *J. Clin. Oncol.* **2018**, *36*, 2105–2122. [CrossRef]
21. Razi, S.; Ouellette, S.; Khan, S.; Oh, K.S.; Truong, T.M.; Rao, B.K. Role of VivaScope 2500 Ex Vivo Confocal Microscopy in Skin Pathology: Advantages, Limitations, and Future Prospects. *Ski. Res. Technol.* **2023**, *29*, e13388. [CrossRef]
22. Amendoeira, I.; Arcidiacono, P.G.; Barizzi, J.; Capitanio, A.; Cuatrecasas, M.; Di Matteo, F.M.; Doglioni, C.; Fukushima, N.; Fulciniti, F.; Ginès, A.; et al. New Digital Confocal Laser Microscopy May Boost Real-Time Evaluation of Endoscopic Ultrasound-Guided Fine-Needle Biopsy (EUS-FNB) from Solid Pancreatic Lesions: Data from an International Multicenter Study. *eBioMedicine* **2022**, *86*, 104377. [CrossRef]
23. Titze, U.; Titze, B.; Hansen, T.; Barth, P.J.; Ali, F.A.; Schneider, F.; Benndorf, M.; Sievert, K.D. Ex Vivo Fluorescence Confocal Microscopy of MRI-Guided Targeted Prostate Biopsies for Rapid Detection of Clinically Significant Carcinomas—A Feasibility Study. *Cancers* **2024**, *16*, 873. [CrossRef] [PubMed]
24. Titze, U.; Sievert, K.-D.D.; Titze, B.; Schulz, B.; Schlieker, H.; Madarasz, Z.; Weise, C.; Hansen, T. Ex Vivo Fluorescence Confocal Microscopy in Specimens of the Liver: A Proof-of-Concept Study. *Cancers* **2022**, *14*, 590. [CrossRef] [PubMed]
25. Verri, M.; Scarpino, S.; Naciu, A.M.; Lopez, G.; Tabacco, G.; Taffon, C.; Pilozzi, E.; Palermo, A.; Crescenzi, A. Real-Time Evaluation of Thyroid Cytology Using New Digital Microscopy Allows for Sample Adequacy Assessment, Morphological Classification, and Supports Molecular Analysis. *Cancers* **2023**, *15*, 4215. [CrossRef] [PubMed]
26. Eissa, A.; Puliatti, S.; Peñaranda, N.R.; Resca, S.; Di Bari, S.; Vella, J.; Maggiorelli, S.; Bertoni, L.; Azzoni, P.; Bonetti, L.R.; et al. Current Applications of Ex-Vivo Fluorescent Confocal Microscope in Urological Practice: A Systematic Review of Literature. *Chin. Clin. Oncol.* **2024**, *13*, 52. [CrossRef]
27. Falagario, U.G.; Selvaggio, O.; Sanguedolce, F.; Milillo, P.; Sighinolfi, M.C.; Bruno, S.M.; Recchia, M.; Bettocchi, C.; Busetto, G.M.; Macarini, L.; et al. One-Day Prostate Cancer Diagnosis: Biparametric Magnetic Resonance Imaging and Digital Pathology by Fluorescence Confocal Microscopy. *Diagnostics* **2022**, *12*, 277. [CrossRef]

28. Togawa, R.; Hederer, J.; Ragazzi, M.; Bruckner, T.; Fastner, S.; Gomez, C.; Hennigs, A.; Nees, J.; Pfob, A.; Riedel, F.; et al. Imaging of Lumpectomy Surface with Large Field-of-View Confocal Laser Scanning Microscopy "Histolog® Scanner" for Breast Margin Assessment in Comparison with Conventional Specimen Radiography. *Breast* **2023**, *68*, 194–200. [CrossRef]
29. Leff, D.R. Technologies and Techniques to Improve Precision in Breast Conserving Surgery. *J. Surg. Oncol.* **2024**. [CrossRef]

life

MDPI

Case Report

Cyclin-Dependent Kinase 4/6 Inhibitors Combined with Radiotherapy in Curative Breast Cancer Patients Induced Pneumonitis: A Case Report

Pei-Yu Hou [1,2]

[1] Department of Radiation Oncology, Far Eastern Memorial Hospital, New Taipei City 220216, Taiwan; jcgv03@mail.femh.org.tw

[2] Department of Computer Science and Engineering, Yuan Ze University, Taoyuan 320315, Taiwan

Abstract: Background: The role of CDK4/6 inhibitors (CDK4/6i) has expanded from the treatment of advanced breast cancer to early-stage disease, as recent studies have demonstrated their therapeutic benefits. However, evidence regarding the safety of combining CDK4/6i with adjuvant radiation therapy (RT) in a curative setting remains limited. This study aims to present clinical experiences of pulmonary toxicity following the combined use of adjuvant RT and CDK4/6i. Case presentation: We report a case of an Asian female with left breast cancer who underwent a modified radical mastectomy followed by adjuvant chemotherapy, RT, endocrine therapy, and CDK4/6i (abemaciclib) treatment. Cancer therapy-induced grade 2 pneumonitis was impressed by clinical signs and image findings. A 57-year-old postmenopausal woman was diagnosed with left breast invasive lobular carcinoma, hormone receptor–positive, human epidermal growth factor receptor 2–negative (HR+/HER2−), K67 index of 5–10%, and classified as pT3N3aM0 (stage IIIC). She received adjuvant chemotherapy with FEC followed by docetaxel, endocrine therapy with letrozole, and adjuvant RT of 50.4 Gy in 28 fractions to the left chest wall and regional nodal irradiation. Abemaciclib was initiated after completing RT. Treatment-related pneumonitis developed five months after RT and abemaciclib use. Conclusions: In breast cancer patients receiving a combination of RT and CDK4/6i as curative adjuvant treatment, pulmonary toxicity is a concern and requires careful monitoring, particularly in Asian populations.

Keywords: breast cancer; radiation therapy; cyclin-dependent kinase 4/6 inhibitors; pneumonitis

1. Introduction

In recent years, significant advancements have been made in novel systemic treatments for breast cancer. One of these new approaches is the use of cyclin-dependent kinase 4/6 inhibitors (CDK4/6i), which are small-molecule inhibitors targeting CDK4 and CDK6. These inhibitors help overcome resistance to hormonal blockades in advanced breast cancer. The most commonly used CDK4/6i include ribociclib, palbociclib, and abemaciclib. The combination of CDK4/6i with endocrine therapy has significantly improved progression-free survival (PFS) and overall survival (OS) in patients with hormone receptor (HR)-positive, human epidermal growth factor receptor 2 (HER2)-negative advanced breast cancer. Several phase III randomized controlled trials (RCTs) have demonstrated their efficacy as a first-line treatment for patients with advanced or metastatic breast cancer [1–5].

In early-stage breast cancer, the monarchE trial established the invasive disease-free survival (iDFS) benefit of the addition of two years for adjuvant abemaciclib to endocrine therapy in HR-positive, HER2-negative patients with node-positive disease and a high risk of recurrence, defined as ≥four pathological involved lymph nodes (LNs) or one to three LNs and either tumor size ≥ 5 cm or grade 3 [6,7]. The additional two-year abemaciclib to endocrine therapy had reduced the absolute risk of disease recurrence at four years by 6.4% (85.8% in the abemaciclib plus endocrine therapy group vs. 79.4% in the endocrine therapy alone group, hazard ratio (HR) 0.664, $p < 0.0001$) [8]. Additionally, the NATALEE trial demonstrated that the addition of three years of ribociclib combined with an aromatase inhibitor (AI) significantly prolonged iDFS in patients with HR-positive, HER2-negative stage II or III early breast cancer [9]. The iDFS benefit at three years was 3.3% (90.4% with ribociclib plus AI and 87.1% with AI alone, HR 0.75, $p = 0.003$). As a result, the role of CDK4/6i has expanded from the treatment of advanced breast cancer to early-stage disease and is increasingly being used in curative treatment settings. According to National Comprehensive Cancer Network clinical practice guidelines in 2025, two years of adjuvant abemaciclib in combination with endocrine therapy (Tamoxifen or AI), or three years of adjuvant ribociclib combined with AI is recommended for eligible HR-positive, HER2-negative high-risk patients.

Radiation therapy (RT) is a key adjuvant treatment for breast cancer, particularly following breast-conserving surgery in patients with early-stage disease. Radiation-induced pneumonitis is a recognized toxicity associated with chest irradiation. Symptoms such as shortness of breath or a non-productive cough typically manifest within one to six months after treatment. The diagnosis of therapy-related pneumonitis requires the exclusion of other potential etiologies and should be based on a comprehensive evaluation of clinical symptoms, temporal association with treatment, and radiological findings. Corticosteroids remain the mainstay of treatment for symptomatic patients.

Previous studies evaluating the safety of combining CDK4/6i with RT have primarily focused on patients with advanced or metastatic breast cancer, where RT is typically administered at relatively low doses for palliative purposes [10–14]. These studies have reported acceptable toxicity profiles when CDK4/6i are combined with palliative RT, with the most adverse effects being hematologic, gastrointestinal (GI), or skin related. However, one case series documented severe grade 5 pneumonitis [14]. Although recent literature and clinical guidelines increasingly support the use of CDK4/6i in the treatment of early-stage breast cancer, there is a lack of evidence regarding the safety of combining these agents with adjuvant RT, which typically involves moderate to high radiation doses. It remains unclear whether this combination increases the severity or alters the pattern of toxicity. Here, we report a case of an Asian female who developed pneumonitis after receiving adjuvant chest RT and CDK4/6i in a curative setting. Through this case, we aim to contribute clinical insights into the safety of combining RT with CDK4/6i in early-stage breast cancer.

2. Case Presentation

2.1. Patient Population

We retrospectively reviewed the patient with breast cancer, who was receiving adjuvant RT and CDK4/6i at Far Eastern Memorial Hospital (FEMH) in 2024 and reported the associated cancer treatment-related toxicity. The severity of toxicity was classified according to Common Terminology Criteria for Adverse Events (CTCAE) v5. This study was approved by the Human Experimentation Committee of Far Eastern Memorial Hospital (FEMH No.: 114014-C).

2.2. RT Treatment Plan

Computed tomography (CT) simulation with a 2.5 mm slice thickness (Discovery CT590 RT, GE Healthcare, Chicago, IL, USA) was performed for RT treatment planning to delineate the target volume and adjacent OARs. The patients were placed in the supine position and allowed to breathe freely during CT simulation. The clinical tumor volume (CTV) included the chest wall (CW), supraclavicular and infraclavicular regions, any part of the axillary bed at risk, and internal mammary region for regional nodal irradiation (RNI). The planning target volume (PTV) was defined as the CTV plus a 5–6 mm margin to allow for setup errors. Daily image guidance with CT was used to define the anatomy and internal soft tissue position. The RT prescription was a conventional dose of 45–50.4 Gy in 25–28 fractions with a daily fraction to the chest wall and RNI.

2.3. RT Techniques and Dosimetry Evaluation

Helical tomotherapy (HT) was planned with the aim of accomplishing better homogeneity and conformity of target coverage while sparing adjacent normal organs to minimize exposure to the heart, lungs, and other organs at risk (OARs). A dose–volume histogram (DVH) was used to evaluate the radiation dose constraints of the target volumes and OARs. The following criteria were required: at least 100% of the CTV volume was to receive 100% of the prescription dose, at least 95% of the PTV volume was to receive 95–100% of the prescription dose, and the maximal dose to the PTV region should be less than 110% of the prescription dose. The constraints of OARs were a mean lung dose $\leq$ 20 Gy, V20 $\leq$ 35%, a mean heart dose $\leq$ 20 Gy, V25 < 10%, a maximal spinal cord dose < 45 Gy, and as low a dose as possible to the contralateral breast, with the usual requirements for other normal organs.

A tomotherapy Hi Art Planning system (version 5.1.3, Tomotherapy, Inc., Madison, WI, USA) was used to plan HT, which was delivered by using a Tomotherapy® Hi-Art or HD system (Tomotherapy®; Accuray Inc., Madison, WI, USA).

3. Case

A 57-year-old post-menopausal woman had an underlying disease of dyslipidemia under medical control. She did not have a history of smoking or any pre-existing pulmonary disease. She had been diagnosed with left breast invasive lobular carcinoma, Grade 2, ER-positive (100%), PR-positive (100%), HER-2-negative (Score 0), K67 index of 5–10%. She underwent left modified radical mastectomy with tissue expander placement, and the pathological staging was pT3N3aM0 (stage IIIC). She completed adjuvant chemotherapy, consisting of four cycles of fluorouracil/epirubicin/cyclophosphamide (FEC), followed by four cycles of docetaxel, and then initiated endocrine therapy with letrozole at a daily dose of 2.5 mg. After completing chemotherapy, she received adjuvant RT with 50.4 Gy in 28 fractions to the left CW and 45 Gy in 25 fractions to the RNI. RT was delivered using HT. The radiation dose to OARs for the lungs was as follows: the ipsilateral lung received a mean dose of 11.9 Gy, with V20 of 21.9%, V10 of 38%, and V5 of 62%. For the whole lung, the mean dose was 7.5 Gy, with V20 of 10.3%, V10 of 18.6%, and V5 of 36.2%.

The dose distribution of the initial 45 Gy covering the CW and RNI is shown in Figure 1. The tissue expander had ruptured before referring for RT preparation. The left CW CTV had covered the whole expander, as shown in Figure 1.

CDK4/6i abemaciclib (150 mg twice daily) was initiated after the completion of RT. Five months later, the patient presented with exertional dyspnea. A chest CT scan with and without intravenous contrast was performed. The image of the lung window revealed subsegmental atelectasis, consolidation, and fibrotic changes in the left upper lung and lower lung, with suspected radiation- and CDK4/6i-induced pneumonitis (Figure 2). The radiological changes observed on the chest CT were characteristic of radiation pneumonitis and corresponded with the irradiated area of the left chest wall. The timing of onset was consistent with the typical presentation of pneumonitis following cancer therapy. As the patient showed no signs of infection, common differential diagnoses such as pneumonia or respiratory tract infection were considered unlikely. The patient had no history of chronic obstructive pulmonary disease or heart failure, and cardiac function tests performed before and after cancer treatment demonstrated preserved ventricular function, thereby excluding the exacerbation of these comorbidities. During the follow-up period, there was no evidence of cancer recurrence, effectively ruling out lymphangitic carcinomatosis. Consequently, a diagnosis of therapy-related pneumonitis was established. The severity was graded as Grade 2 according to version 5 of the CTCAE. The spirometry test revealed a mild restrictive lung defect with a forced expiratory volume in one second (FEV_1) of 1.68 L (88% of predicted) and a forced vital capacity (FVC) of 1.93 L (84% of predicted). Lung diffusion testing indicated a moderate impairment in the diffusion capacity for carbon monoxide (DLCO: 47% of predicted). Oral prednisolone therapy was initiated. At the three-month follow-up, a chest CT demonstrated prominent fibrosis, consolidation, and traction bronchiectasis in the left lung (Figure 3). The patient's dyspnea improved after one month of oral betamethasone titration therapy. The positive response to corticosteroid therapy further supported the diagnosis of therapy-induced pneumonitis. The timeline of breast cancer treatment and the subsequent development of pneumonitis is listed in Figure 4.

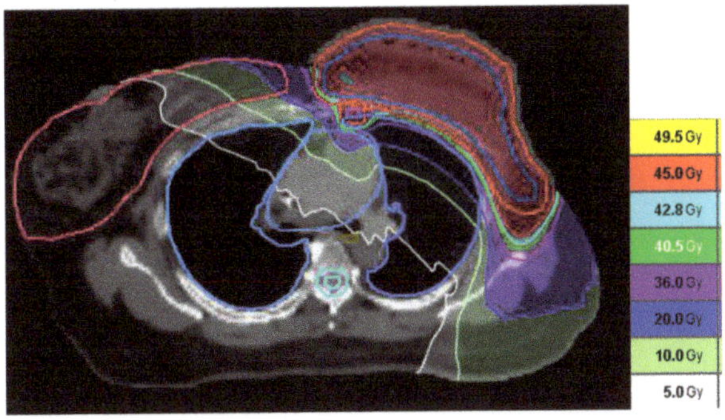

Figure 1. The dose distribution of RT planning.

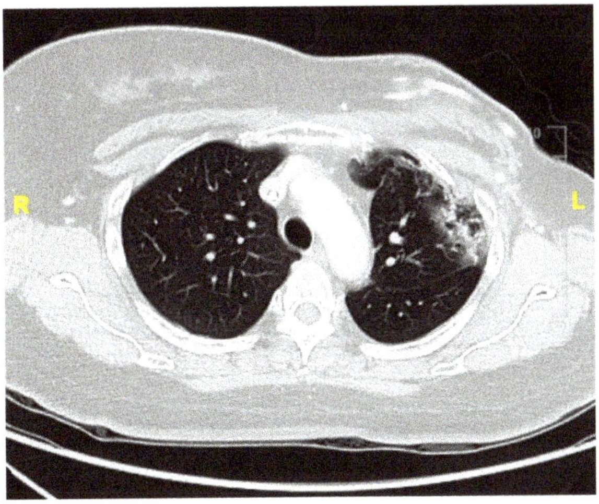

Figure 2. Chest computed tomography (CT) scan of lung window showed pneumonitis with subsegmental atelectasis, consolidation, and fibrotic changes in the left lung. (R: the right side. L: the left side.)

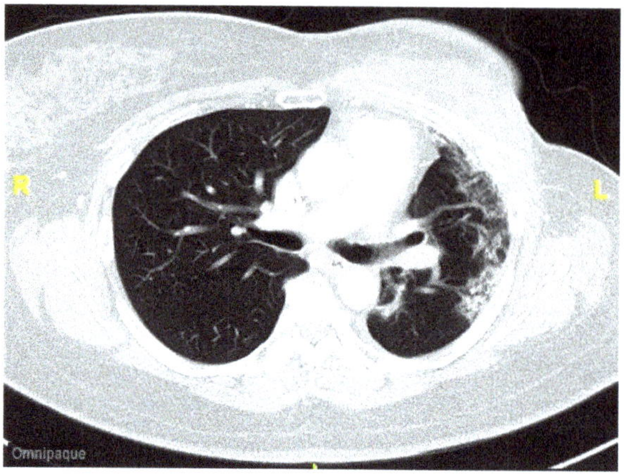

Figure 3. Chest CT of lung window showed prominent fibrosis, consolidation, and traction bronchiectasis in the left lung approximately 8 months after the completion of RT and initiation of abemaciclib. (R: the right side. L: the left side.)

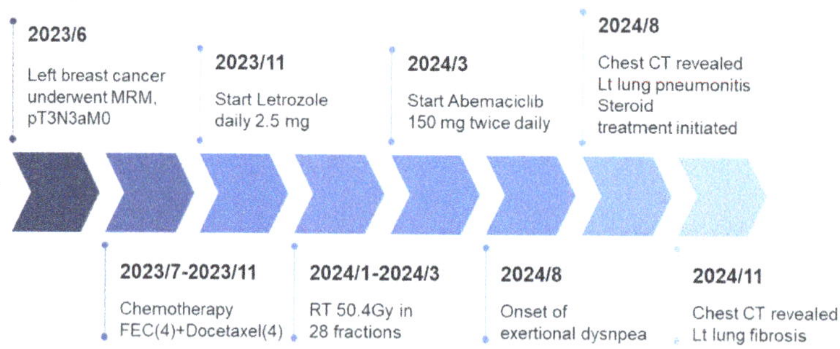

Figure 4. The timeline of breast cancer treatment and subsequent development of pneumonitis.

4. Discussion

For patients with advanced breast cancer, treatment strategies include loco-regional therapies such as surgery and RT, as well as systemic therapies such as chemotherapy, targeted therapy, endocrine therapy, and immunotherapy. These novel cancer treatments have contributed to improved disease control and reduced mortality. This study reports a case of an Asian female with breast cancer who received a comprehensive post-operative treatment regimen, including chemotherapy, RT, endocrine therapy, and the recently approved CDK4/6i for curative adjuvant therapy. The patient developed treatment-related pneumonitis five months after receiving RT and abemaciclib.

For patients with advanced-stage disease or LN involvement, post-operative RT with RNI significantly reduces mortality and recurrence [15,16]. CDK4/6i combined with endocrine therapy has demonstrated PFS and OS benefit for HR-positive and HER2-negative advanced breast cancer, as confirmed by the PALOMA, MONALEESA, and MONARCH trials [1–5]. However, these treatments are associated with certain unavoidable side effects. As multimodal treatments become increasingly common in aggressive cancer management, clinicians must be cautious of unpredictable toxicity patterns and severity.

Radiation pneumonitis is a well-known concern following chest irradiation, with its incidence correlating with the lung radiation dose [17,18]. The other risk factors associated with the development of radiation pneumonitis include poor performance status, older age at diagnosis, smoking history, baseline pulmonary function, and the use of concurrent medications such as chemotherapy, immunotherapy, and target therapy. The dosimetric constraints recommended by the QUANTEC and ESTRO guidelines aim to minimize pulmonary toxicity, with suggested limits of a mean lung dose < 23 Gy and V20 < 35%. With the evolution of radiation techniques from three-dimensional conformal radiation therapy (3D-CRT) to modern modalities such as intensity-modulated radiotherapy (IMRT), volumetric modulated arc therapy (VMAT), and particle therapy, the ability to meet these constraints has improved, resulting in a reduced incidence of radiation pneumonitis. In the reported case, the radiation plan adhered to these constraints, with an ipsilateral lung mean dose of 11.9 Gy and V20 of 21.9%, and a whole lung mean dose of 7.5 Gy with V20 of 10.3%. The primary risk factor identified in this case was the use of concurrent medication. CT remains the primary imaging modality for the diagnosis of radiation pneumonitis. In the acute phase (within 1–6 months post-RT), typical radiological findings include ground–glass opacities (GGO) and consolidation localized to the irradiated lung regions. In the late phase (6–12 months post-RT), characteristic features include lung fibrosis, volume loss, consolidation, and traction bronchiectasis. In this case, CT imaging demonstrated findings consistent with the typical radiological features of both the acute phase (at 5 months after completion of RT) and late phase (at 8 months after RT). Furthermore, the patient's good response to corticosteroid therapy supported the diagnosis of radiation pneumonitis.

While lung toxicity is relatively rare when CDK4/6i is used alone, ILD or pneumonitis is an uncommon but potentially severe adverse effect. The mechanism of CDK4/6i-induced ILD remains unclear. A preclinical study in mice suggested that palbociclib may augment inflammatory cell recruitment and enhance pulmonary inflammation [19]. Whether combining chest RT with CDK4/6i increases the risk of lung toxicity remains controversial [10,12–14,20,21]. Small case series have reported severe pneumonitis in patients receiving combination of CDK4/6i and chest RT, primarily in the context of metastatic breast cancer treated with low-dose palliative RT. Given the increasing use of CDK4/6i as a first-line treatment and their application in curative adjuvant settings, the potential for lung toxicity warrants further investigation.

The most common side effects of CDK4/6i are hematologic and GI toxicities [22]. The U.S. Food and Drug Administration (FDA) has issued warnings about the potential for drug-induced lung toxicity associated with CDK4/6i, as reported in the FDA Adverse Event Reporting System (FAERS). ILD appears to be more frequently associated with abemaciclib (2.1%) compared to palbociclib or ribociclib (0.3%) [23], although the overall incidence remains low [21,24–26]. Notably, FAERS data suggest that CDK4/6i-related ILD is more common in Asian populations, particularly among Japanese patients, accounting for nearly half of reported cases. In the monarchE trial, which evaluated the efficacy of abemaciclib as an adjuvant treatment, 95.4% of patients in both the abemaciclib arm and control arms had received prior RT. ILD occurred in 2.7% of patients in the abemaciclib arm and 1.2% in the control arm. Among different ethnic groups, ILD incidence was higher in Asian patients (6.6%) compared to the overall cohort (2.3%) [6,20,27]. A Japanese study reported an even higher incidence of CDK4/6i-induced ILD than previously documented in the PALOMA and MONARCH trials, with ILD occurring in 8.0% of patients (13% of abemaciclib-treated cases and 6.5% of palbociclib-treated cases), and a median onset time of approximately six months [20]. Given that our patient was an Asian female treated with abemaciclib, the risk of ILD may have been elevated.

Regarding other safety concerns associated with combining RT and CDK4/6i, case series and systematic reviews have reported GI and skin toxicities when these treatments are used either sequentially or concurrently. Notably, these toxic effects may persist for several months following treatment [10,14,28]. Currently, strong evidence or consensus on the toxicity of combining RT with CDK4/6i in the adjuvant setting remains lacking.

It remains unclear whether temporary suspension of CDK4/6i during RT is necessary or whether concurrent or sequential administration is feasible. These issues warrant further investigation, as an increasing number of breast cancer patients will undergo multimodality treatments in the modern era.

5. Conclusions

A multimodality therapy approach is crucial in breast cancer treatment to maximize disease control and minimize therapy-related side effects. In breast cancer patients receiving a combination of RT and CDK4/6 inhibitors as curative adjuvant treatment, pulmonary toxicity is a concern and requires careful monitoring, particularly in Asian populations, even using modern RT techniques to limit the lung exposure dose.

Funding: This research received no external funding.

Institutional Review Board Statement: This study was approved by the Human Experimentation Committee of Far Eastern Memorial Hospital (FEMH No.: 114014-C, approval date: 27 January 2025).

Informed Consent Statement: Patient consent was waived due to the retrospective nature of the study, which involved a review of medical records and did not impact the patient's medical rights or treatment strategy. A waiver of informed consent was approved by the Human Experimentation Committee of Far Eastern Memorial Hospital.

Data Availability Statement: The data presented in this study are available on request from the corresponding author. The data are not publicly available due to patients' privacy and medical ethics.

Acknowledgments: This work was supported by Far Eastern Memorial Hospital.

Conflicts of Interest: The author declares no conflicts of interest.

References

1. Finn, R.S.; Martin, M.; Rugo, H.S.; Jones, S.; Im, S.-A.; Gelmon, K.; Harbeck, N.; Lipatov, O.N.; Walshe, J.M.; Moulder, S. Palbociclib and letrozole in advanced breast cancer. *N. Engl. J. Med.* **2016**, *375*, 1925–1936. [CrossRef] [PubMed]
2. Goetz, M.P.; Toi, M.; Campone, M.; Sohn, J.; Paluch-Shimon, S.; Huober, J.; Park, I.H.; Trédan, O.; Chen, S.-C.; Manso, L. MONARCH 3: Abemaciclib as initial therapy for advanced breast cancer. *J. Clin. Oncol.* **2017**, *35*, 3638–3646. [CrossRef] [PubMed]
3. Hortobagyi, G.N.; Stemmer, S.M.; Burris, H.A.; Yap, Y.-S.; Sonke, G.S.; Paluch-Shimon, S.; Campone, M.; Blackwell, K.L.; André, F.; Winer, E.P. Ribociclib as first-line therapy for HR-positive, advanced breast cancer. *N. Engl. J. Med.* **2016**, *375*, 1738–1748. [CrossRef]
4. Slamon, D.J.; Neven, P.; Chia, S.; Fasching, P.A.; De Laurentiis, M.; Im, S.-A.; Petrakova, K.; Bianchi, G.V.; Esteva, F.J.; Martín, M. Overall survival with ribociclib plus fulvestrant in advanced breast cancer. *N. Engl. J. Med.* **2020**, *382*, 514–524. [CrossRef] [PubMed]
5. Im, S.-A.; Lu, Y.-S.; Bardia, A.; Harbeck, N.; Colleoni, M.; Franke, F.; Chow, L.; Sohn, J.; Lee, K.-S.; Campos-Gomez, S. Overall survival with ribociclib plus endocrine therapy in breast cancer. *N. Engl. J. Med.* **2019**, *381*, 307–316. [CrossRef]
6. Johnston, S.R.; Harbeck, N.; Hegg, R.; Toi, M.; Martin, M.; Shao, Z.M.; Zhang, Q.Y.; Martinez Rodriguez, J.L.; Campone, M.; Hamilton, E. Abemaciclib combined with endocrine therapy for the adjuvant treatment of HR+, HER2−, node-positive, high-risk, early breast cancer (monarchE). *J. Clin. Oncol.* **2020**, *38*, 3987–3998. [CrossRef]
7. Johnston, S.R.; Toi, M.; O'Shaughnessy, J.; Rastogi, P.; Campone, M.; Neven, P.; Huang, C.-S.; Huober, J.; Jaliffe, G.G.; Cicin, I. Abemaciclib plus endocrine therapy for hormone receptor-positive, HER2-negative, node-positive, high-risk early breast cancer (monarchE): Results from a preplanned interim analysis of a randomised, open-label, phase 3 trial. *Lancet Oncol.* **2023**, *24*, 77–90. [CrossRef]
8. Rugo, H.; O'Shaughnessy, J.; Boyle, F.; Toi, M.; Broom, R.; Blancas, I.; Gumus, M.; Yamashita, T.; Im, Y.-H.; Rastogi, P. Adjuvant abemaciclib combined with endocrine therapy for high-risk early breast cancer: Safety and patient-reported outcomes from the monarchE study. *Ann. Oncol.* **2022**, *33*, 616–627. [CrossRef]
9. Slamon, D.; Lipatov, O.; Nowecki, Z.; McAndrew, N.; Kukielka-Budny, B.; Stroyakovskiy, D.; Yardley, D.A.; Huang, C.-S.; Fasching, P.A.; Crown, J. Ribociclib plus endocrine therapy in early breast cancer. *N. Engl. J. Med.* **2024**, *390*, 1080–1091. [CrossRef]

10. Becherini, C.; Visani, L.; Caini, S.; Bhattacharya, I.S.; Kirby, A.M.; Marta, G.N.; Morgan, G.; Salvestrini, V.; Coles, C.E.; Cortes, J. Safety profile of cyclin-dependent kinase (CDK) 4/6 inhibitors with concurrent radiation therapy: A systematic review and meta-analysis. *Cancer Treat. Rev.* **2023**, *119*, 102586. [CrossRef]
11. Kim, K.N.; Shah, P.; Clark, A.; Freedman, G.M.; Dastgheyb, S.; Barsky, A.R.; Dreyfuss, A.D.; Taunk, N.K. Safety of cyclin-dependent kinase4/6 inhibitor combined with palliative radiotherapy in patients with metastatic breast cancer. *Breast* **2021**, *60*, 163–167. [CrossRef] [PubMed]
12. Al-Rashdan, A.; Quirk, S.; Roumeliotis, M.; Abedin, T.; Amaro, C.P.; Barbera, L.; Lupichuk, S.; Cao, J.Q. Radiation therapy with cyclin-dependent kinase 4/6 inhibitors: A multi-institutional safety and toxicity study. *Int. J. Radiat. Oncol. Biol. Phys.* **2022**, *114*, 399–408. [CrossRef]
13. Kubeczko, M.; Gabryś, D.; Gawkowska, M.; Polakiewicz-Gilowska, A.; Cortez, A.J.; Krzywon, A.; Woźniak, G.; Latusek, T.; Leśniak, A.; Świderska, K. Safety and feasibility of radiation therapy combined with CDK 4/6 inhibitors in the management of advanced breast cancer. *Cancers* **2023**, *15*, 690. [CrossRef] [PubMed]
14. David, S.; Ho, G.; Day, D.; Harris, M.; Tan, J.; Goel, S.; Hanna, G.G.; Srivastava, R.; Kruss, G.; McDowell, L. Enhanced toxicity with CDK 4/6 inhibitors and palliative radiotherapy: Non-consecutive case series and review of the literature. *Transl. Oncol.* **2021**, *14*, 100939. [CrossRef] [PubMed]
15. Poortmans, P.M.; Weltens, C.; Fortpied, C.; Kirkove, C.; Peignaux-Casasnovas, K.; Budach, V.; van der Leij, F.; Vonk, E.; Weidner, N.; Rivera, S. Internal mammary and medial supraclavicular lymph node chain irradiation in stage I–III breast cancer (EORTC 22922/10925): 15-year results of a randomised, phase 3 trial. *Lancet Oncol.* **2020**, *21*, 1602–1610. [CrossRef]
16. Whelan, T.J.; Olivotto, I.A.; Parulekar, W.R.; Ackerman, I.; Chua, B.H.; Nabid, A.; Vallis, K.A.; White, J.R.; Rousseau, P.; Fortin, A. Regional nodal irradiation in early-stage breast cancer. *N. Engl. J. Med.* **2015**, *373*, 307–316. [CrossRef]
17. Marks, L.B.; Bentzen, S.M.; Deasy, J.O.; Bradley, J.D.; Vogelius, I.S.; El Naqa, I.; Hubbs, J.L.; Lebesque, J.V.; Timmerman, R.D.; Martel, M.K. Radiation dose–volume effects in the lung. *Int. J. Radiat. Oncol. Biol. Phys.* **2010**, *76*, S70–S76. [CrossRef]
18. Bledsoe, T.J.; Nath, S.K.; Decker, R.H. Radiation pneumonitis. *Clin. Chest Med.* **2017**, *38*, 201–208. [CrossRef]
19. Birnhuber, A.; Egemnazarov, B.; Biasin, V.; Bonyadi Rad, E.; Wygrecka, M.; Olschewski, H.; Kwapiszewska, G.; Marsh, L.M. CDK4/6 inhibition enhances pulmonary inflammatory infiltration in bleomycin-induced lung fibrosis. *Respir. Res.* **2020**, *21*, 167. [CrossRef]
20. Funasaka, C.; Naito, Y.; Kusuhara, S.; Nakao, T.; Nakajima, H.; Kawamoto, M.; Baba, K.; Mamishin, K.; Kondoh, C.; Harano, K. Clinical features of CDK4/6 inhibitor-related interstitial lung disease in patients with breast cancer: A case series study. *Jpn. J. Clin. Oncol.* **2023**, *53*, 105–114. [CrossRef]
21. Algwaiz, G.; Badran, A.A.; Elshenawy, M.A.; Al-Tweigeri, T. Ribociclib-induced pneumonitis: A case report. *Breast Care* **2021**, *16*, 307–311. [CrossRef]
22. Onesti, C.E.; Jerusalem, G. CDK4/6 inhibitors in breast cancer: Differences in toxicity profiles and impact on agent choice. A systematic review and meta-analysis. *Expert Rev. Anticancer Ther.* **2021**, *21*, 283–298. [CrossRef] [PubMed]
23. Raschi, E.; Fusaroli, M.; Ardizzoni, A.; Poluzzi, E.; De Ponti, F. Cyclin-dependent kinase 4/6 inhibitors and interstitial lung disease in the FDA adverse event reporting system: A pharmacovigilance assessment. *Breast Cancer Res. Treat.* **2021**, *186*, 219–227. [CrossRef] [PubMed]
24. Ahsan, I.; Malik, F.; Jafri, S.I. Palbociclib related pnemotoxicity: A rare side effect. *Am. Thorac. Soc.* **2017**, *195*, A5546.
25. Gong, J.; Cho, M.; Yu, K.W.; Waisman, J.; Yuan, Y.; Mortimer, J. A single institution experience with palbociclib toxicity requiring dose modifications. *Breast Cancer Res. Treat.* **2018**, *168*, 381–387. [CrossRef]
26. Jazieh, K.A.; Budd, G.T.; Dalpiaz, N.; Abraham, J. Can CDK4/6 inhibitors cause fatal lung injury? *Expert Rev. Anticancer Ther.* **2019**, *19*, 917–919. [CrossRef] [PubMed]
27. Toi, M.; Harbeck, N.; Puig, J.; Cruz, J.; Seo, J.; Takahashi, M.; Hulstijn, M.; Twum, E.; Regev, A.; San Antonio, B. 44O characterization of venous thromboembolic events (VTE), elevated aminotransferases (EAT) and interstitial lung disease (ILD) in monarchE. *Ann. Oncol.* **2021**, *32*, S39–S40. [CrossRef]
28. Franco, R.; Cao, J.Q.; Yassa, M.; Hijal, T. Safety of CDK4/6 inhibitors combined with radiotherapy in patients with metastatic breast cancer: A review of the literature. *Curr. Oncol.* **2023**, *30*, 5485–5496. [CrossRef]

MDPI AG
Grosspeteranlage 5
4052 Basel
Switzerland
Tel.: +41 61 683 77 34

Life Editorial Office
E-mail: life@mdpi.com
www.mdpi.com/journal/life

www.ingramcontent.com/pod-product-compliance
Lightning Source LLC
Chambersburg PA
CBHW061449210926
51850CB00008B/1654
9783725874125